Salt and the Colombian State

Pitt Latin American Series
John Charles Chasteen and Catherine M. Conaghan, Editors

SALT and the COLOMBIAN STATE

Local Society and Regional Monopoly in Boyacá, 1821–1900

JOSHUA M. ROSENTHAL

University of Pittsburgh Press

Published by the University of Pittsburgh Press, Pittsburgh, Pa., 15260

10 9 8 7 6 5 4 3 2 1

Library of Congress Cataloging-in-Publication Data

Rosenthal, Joshua M.
Salt and the Colombian state : local society and regional monopoly in Boyaca, 1821–1900 / Joshua M. Rosenthal.
p. cm. — (Pitt Latin American series)
Includes bibliographical references and index.
ISBN 978-0-8229-6180-2 (pbk. : alk. paper)
1. Salt industry and trade—Colombia—History—19th century. 2. Salt mines and mining—Government policy—Colombia—History—19th century. 3. Government monopolies—Colombia—History—19th century. 4. Boyacá (Colombia : Dept.)—History—19th century. 5. Colombia—Politics and government—19th century. 6. Colombia—Economic conditions—19th century. I. Title.
HD9213.C72R67 2012
338.2'76320986109034—dc23 2011039630

Contents

Acknowledgments *vii*

Note on Sources *xi*

Chapter 1. The Salt Monopoly, the State, and Boyacá *3*

Chapter 2. Change and Community in La Salina *17*

Chapter 3. Making Salt in a Ministry Works *41*

Chapter 4. The Ministry Monopoly and the Market Monopoly *62*

Chapter 5. La Salina and Colombian History to 1857 *94*

Chapter 6. La Salina, Boyacá, and Colombia after 1857 *113*

Notes *145*

Bibliography *201*

Index *219*

Acknowledgments

So many years have passed since this book's beginning that attempting to thank those who helped me along the way is daunting.

My parents are first. Though Naomi and Joel T. Rosenthal are guilty of making an academic life look almost easy, they provide recompense with a number of other fine qualities. In this context, the most charming has been their willingness to dispense with decades of experience and declare this study worthy of publication from the very beginning. To my siblings, Jessica Benson (author of the most readable books produced by the family to date) and Matthew Rosenthal, I'll say that one of the best parts of graduate school was getting to spend all that time together. Things are a little more global now, but I appreciate your support and company throughout the venture.

I am proud to claim both David Garrett and Nora Jaffary as friends from that era of my life, David from the first week of classes and Nora from a few years later. They continue to be dear friends and, for me, role models. Through Nora I met Ed Osowski, whom I am proud to call a friend as well. Linda Green also provided much-needed words of encouragement for this project when I was a graduate student, at a time when such offerings were rare. New York did have other diversions, which may have slowed my research but were beneficial in other ways. I would like to thank Grupo Capoeira Brazil, led by Mestre Caxias, and through this group the capoeiristas I met and spent time with across the United States and Brazil.

From the moment I arrived in Bogotá in 1994 I have been treated better than I have deserved by more people than I can name here. The place where the most important of these interactions took place was the Archivo General de la Nación, which remains an amazing place to work. In particular I would like to thank Adelida Sourdis, Yaneth Sandoval, Nazly González, Carlos Puentes, Enrique Ballesteros, and Leonardo Quintero. I am also grateful to the staff of the Biblioteca Nacional and the Academia Colombiana de Historia. Finally, the friendship and support of Alberto Flores was crucial in helping me navigate this year in Bogotá. In Tunja Celina Trimiño opened her home, for which I am still thankful. I would also like to thank the staffs of the Archivo Regional de Boyacá, the Archivo Departamental de Boyacá, the library at the Universidad Pedagógica Tecnológica de Colombia, and everyone associated with the masters program in history there.

I have benefited from the opportunity to present parts of this work at various conferences. Here I am grateful to the Conference on Latin American History, especially the Gran Colombia committee; LASA, particularly the Colombia section; the Asociación de Colombianistas; the New England Council on Latin American Studies; and the Society of Latin American Studies in the United Kingdom. Less institutionally, I have benefited from the company, ideas, and friendship of the community of historians and scholars who study Colombia. First I would like to acknowledge the extraordinary generosity of Jane Rausch, who has supported this project since we met at the AHA in Atlanta in 1996. She has read many drafts and is generally too kind to mention it when my work falls short of the standards she maintains. I would like to thank David Sowell, who has read at least two drafts of this work over the years, for reasons I have never been able to fathom, though self-interest prevented me from asking why he was willing to do so. Nancy Appelbaum was my first Colombianist friend, and I owe her for many things, including my introduction to Riosucio (whose people I would like to acknowledge for their generosity, hospitality, and the good times I have had there). I would also like to thank Michael Francis, Richard Goulet, David Sanders, Richard Stoller, Victor Uribe-Uran, Hayley Froysland, Brett Troyan, Robert Karl, Mary Roldán (a walking postdoc in Colombian history, and a generous one who has yet to collect on the many favors she has rendered), Claudia Leal, Shawn Van Ausdal, Javier Guerrero Baron, Hermes Tovar, Marcela Echeverri, Lina del Castillo, Meri Clark, Jamie Sanders, Kris Lane, Hillel Soifer, Robert Karl, Rebecca Tally, and Aims McGuiness. In one way or another I have been aided by their work and company. I am sorry for any names I may have left off this list. I would also like to Marc Chernick and Tulia Camacho, who is missed, for sharing so much with me over the years.

Finally, I have been lucky enough to hold jobs that have allowed me to continue writing. Thanks first to the history department of Montclair State University and Amy Srebnick for some career-saving employment; Trinity College, for my first full-time job; and SUNY Oneonta, where I began my tenure-track career. There I owe special thanks to Thomas Beal and K. O. O'Mara, whom I wish I could see more often than I do. Finally, Western Connecticut State University is a great place to work. Perhaps it is the institution as a whole, but I cannot but help think that the example of my colleagues Burt, Michael, Kevin, Wynn, Kate, Leslie, Abubaker, Marcy (who has gone above and beyond collegiality with her copyediting), and Jennifer has been a crucial part of finishing this book. Thanks also to Tai Lucas.

In writing this book I have benefited from access to the library at SUNY Albany and the University of Massachusetts at Amherst. I also owe a general debt to the Five College Libraries of Western Massachusetts and the many cafes and wifi spots of the area, particularly Amherst Coffee, where a great deal of this manuscript has been rewritten more than once.

The shortcomings or errors in this book are not a reflection on any of these people, who did their best to correct them. In this line, Bruce Bethell did a wonderful job of copyediting. Bruce and all of the people mentioned above are innocent of the any imperfections that follow. Those I claim as my own. I would also like to thank Joshua Shanholtzer and the rest of the staff at the University of Pittsburgh Press.

The greatest debt of course is to my family, Amber, Zeke, and Samara. If they have given little directly to this project, they have done everything that made working on it possible. Zeke and Samara, with their love of books and hilarity, did their share, and then there is Amber. It is easy to switch jobs and play the supporting, trailing spouse when your wife's career lands you where we are now; any sacrifices on my part have been a mirage. Thanks for building the life we live and sharing it with me. I dedicate this book to you.

Note on Sources

In writing this work I sought to use citations and bibliographic references that meet the standards of historical scholarship and the customary practices for the relevant Colombian materials, but careful readers may find passages where I seem to fall short of this goal. These and other matters bear some explanation. First, citations of texts in the Archivo General de la Nacíon de Colombia (AGN) do not include a letter *v* to indicate the reverse (verso) side of a numbered sheet. This approach in part reflects the binding of the *legajos* themselves. These enormous bound volumes often contain more sheets than the sheet numbers imply. It is fairly common to find a numbered sheet followed by several additional sheets, sometimes written on and sometimes blank, of varying size. Anyone seeking to trace individual documents or correspondence would have to leaf through a few pages no matter what notation is used. Second, after I conducted my initial research, the AGN staff began transferring Republican holdings onto microfilm; researchers must now consult spools of microfilm rather than original documents. The two systems do not always match with respect to systems of organization. For example, legajos of Gobernaciones specific to an individual province, such as Tunja, or Tundama, or Boyacá, were organized in collections running in the single or low double digits. Currently all these provincial series are microfilmed and organized as a single collection, Gobernaciones (location SR046), comprising 516 volumes. The staff of the AGN will aid any researcher seeking to navigate between the older organization (documented in older indexes that were still available for consultation in 2007) and the current system. I suspect that similar consolidations will occur as other collections are transferred to microfilm. The citations in this study will serve as a guide, but readers should be alert to these changes.

Finally, it was often difficult or impossible to ascertain authorship when dealing with the voluminous correspondence from La Salina on which my research relied, for much of it had been copied and forwarded. As a result, my reconstruction of events sometimes refers to official posts rather than to an individual person. I am confident that this failing has not produced any historical inaccuracies, but it is one that must be acknowledged and, I hope, tolerated by readers.

Salt and the Colombian State

FIGURE 1.1. "La Salina de Chita, 1806" by Ignacio Caicedo.
AGN, Mapas y Planos, Mapoteca no. 4, ref. 94 A.

CHAPTER ONE

The Salt Monopoly, the State, and Boyacá

In 1806, when Ignacio Caicedo drew a sketch of La Salina de Chita, in eastern Colombia, it wasn't much of a town, though that hardly mattered to him. An administrator who oversaw the sale of salt, all of which was controlled in theory by a Spanish royal monopoly, Caicedo was more concerned with documenting how salt was made than with explaining any other aspect of local life. Now in the Mapoteca of Colombia's Archivo General de la Nación (AGN), his sketch conveys something of the topography of the surrounding valley, which had long before been cut by the rushing waters of the Casanare River as it carved a passage from the highlands of Boyacá to the plains of Casanare. Caicedo set the town's "ruined Church" and other buildings in the background, emphasizing the elements of salt making. Here and there in the sketch he depicted mules, either fully loaded and led by muleteers or grazing freely. There were stacks of timber taller than a person, as well as men cutting or carrying wood. In the shelter of an open shed a figure tended to an oven where brine was being slowly cooked into the salt cakes characteristic of the region. In another spot, people stooped over a low-lying well fed by a thermal spring with a high saline content, a *vertiente*—the reason people had settled in this particular spot. Similar springs dot the Colombian highlands north and east of Bogotá, as well as the terrain where the mountains descend to the Llanos, Colombia's plains (running from the Southwest to the Northeast). Many springs had been used for making salt even before the Spanish came to the Americas, but few were as naturally rich as this one or those in La Salina's outlying barrio of Córdoba.

For Caicedo and his administrative successors in republican Colombia, this natural bounty was provocative and even frustrating. When he looked at salt making in La Salina, Caicedo saw what he thought should be rather than what was.[1] In 1801 the noted naturalist Alexander von Humboldt had visited the great saltworks in Zipaquirá, a city just north of Bogotá, and subsequently published a description of the works complete with recommendations for improvements. This prompted a concerted effort to increase the efficiency of salt production in New Granada and to earn more revenue for the Spanish royal treasury. In the report Caicedo sent to the viceroy of New Granada, in Bogotá, he called for more guards in La Salina. His sketch included an ambitious proposal: a dotted line indicated a wall, taller than a person and wide enough to walk on, to be built on the banks of the Casanare River. The proposed addition would protect the pool of salt water from the river during the four-month rainy season, when the swollen Casanare rose to engulf the spring, rendering salt making difficult, if not impossible. Obviously, making salt all year would be one of the most effective ways to produce more money for the royal treasury. With this in mind, Caicedo proposed an ambitious construction project through which the state would elaborate its material presence in this out-of-the-way spot.

Through the tumult of independence and the fitful birth of a republic in the 1810s and early 1820s, Caicedo's successors, particularly those who worked for the Ministerio de Hacienda, Colombia's finance ministry in the 1830s, held similar ideas and made similar proposals. After some flirtation with privatizing government-owned saltworks and a decade of arrangements where individuals ran saltworks as autonomous renters, the Finance Ministry fulfilled Caicedo's vision. In 1836 the ministry placed the works at La Salina under what it termed "direct administration." Ministry employees oversaw the construction of a wall like one the Caicedo had proposed. They also supervised improvements to local roads, the construction of an *almacén* (a combination office, store, and warehouse), and the building of new ovens for cooking salt. The goal was to create a profitable industry that would function all year. In 1837 an enthusiastic official wrote, "In the parish of La Salina . . . the state has a salt mine [*sic*] that without a doubt is worth one million pesos."[2] In the strictest sense his estimate was correct. In time La Salina did deliver more than a million pesos to the ministry, but it took far longer than anyone at that time anticipated. Moreover, many ministry officials felt that far too much of the million pesos had been used to finance salt production or support related administrative duties.

Despite these frustrations, administrators built up the ministerial bureaucracy in La Salina and extended its institutional presence deeper into the salt industry in a number of ways—among other things, by arranging for land purchases and supervising further construction projects. The ministry's control over the salt industry bound La Salina to the variegated rhythms of Colombia's economic and political history through the end of the century. A report issued after the devastating Thousand Days War (1899–1902) described the town in

1904: "The only things that remain standing are things that were too difficult to destroy or those things that would yield absolutely nothing of worth."[3] This civil war, the last to occur anywhere during the nineteenth century, appears to have undone the work imagined by Caicedo almost one hundred years earlier.

A Salt Monopoly

Considering La Salina's history renders questions about the state concrete rather than theoretical. The value of theoretical accounts of state processes notwithstanding, pursuing the concrete allows us to ground various points in material life rather than abstract concepts. A saltworks of the preindustrial world is particularly well suited to a practical inquiry into the nature and operational methods of a state. Salt-production centers, such as La Salina, have been loci of state power throughout human history. In contrast to its contemporary status as a cheap and abundant product, salt was a valuable article in the preindustrial world, making it an inviting object for taxation. Salt, or more precisely, the monopolies exercised over the production and sale of salt, was an important element in the global history of states and state building. Salt trades and salt monopolies played important roles in pre-Columbian Mesoamerica, in imperial China, and in early modern France, to cite only a few examples. In considering salt's global history, S. A. M. Adshead observes that "salt administrations belong to the adolescence or senescence of central power."[4] In this taxonomy, the survival of the Colombian salt monopoly for decades after independence implies that the state blended the senescence of imperial Spain and the immaturity of its replacement for decades. In other words, Adshead's argument predicts continuities in Colombia, which is just what occurred.

Where other Spanish American countries moved away from salt taxes, in Colombia the government maintained monopoly control over the production of salt after the Wars of Independence, in parts of the country continuing through the economic reforms of the mid-nineteenth century and the doctrinaire liberalism of the 1860s and 1870s. While many factors were involved in its survival in each case, the monopoly endured because it generated desperately needed revenue for the national government. Decade after decade, the pointed criticism of national elites and the widespread resentment of the tax were not enough to overcome fiscal reality. From the 1820s until 1900 profits from ministry saltworks managed under the monopoly accounted for about 10 percent of Colombia's total federal revenue. The impact of this 10 percent was enhanced because the monopoly was a consistent revenue producer. Receipts from taxes on imports and exports provided much more than the salt monopoly but tended to fluctuate. The steadiness inherent in salt sales was part of the monopoly's worth. For example, the government might issue promissory notes redeemable for salt at government sites as a guarantee for loans made during moments of civil war.[5]

This steady stream of revenue was produced by a handful of saltworks scattered throughout Colombia's eastern highlands, the most important being the operation at Zipaquirá, a city just under fifty kilometers of easy travel north of Bogotá. There the Finance Ministry's administration oversaw sales of, on average, more than eight million kilograms of salt a year. Zipaquirá was the site where the great rebel army of the Comuneros, who rose in part to protest the introduction of new monopolies and taxes, camped as its leaders debated seizing a defenseless Bogotá in 1781. Not only did it attract Alexander von Humboldt when he toured New Granada, but today Zipaquirá boasts an internationally known cathedral (the Catedral de Sal) dug out of its vast underground salt banks. In fact, the works at Zipaquirá casts a shadow that at times makes it hard to see the historical importance of La Salina. Salt from Zipaquirá was traded across central Colombia and consumed by a huge portion of the national population. Tellingly, however, while there are studies on Zipaquirá in the colonial era and on republican saltworks, little scholarship exists on the great works after independence.

There is, however, a meticulous study by Anuar Herna Peña Díaz on smaller saltworks, *Sal, sudor y fisco: el proceso de institucionalización del monopolio de la sal en las salinas de Chámeza, Recetor y Pajarito, 1588–1950,* which serves as an introduction to this issue after independence. Peña Díaz argues that the ministry-run saltworks of Chámeza, Recetor, and Pajarito provide a vantage point for examining the institutional links between Bogotá and Casanare. He also stresses the degree to which the history of the salt monopoly over several centuries documents the process of institutional centralization based in Bogotá. Placing the history of the salt monopoly in a broader context, as I do here, extends these arguments into a larger discussion. Peña Díaz focuses on institutional growth as a narrative structure, but the institutional nature of the salt monopoly can serve as evidence that the project of state building was an important factor in nineteenth-century Colombia. In fact, the fitful process of state building, both as a part of daily life in La Salina and as an element of politics in Bogotá, embodied much of Colombia's national history.

In Colombia the nineteenth century was marked by disjuncture and contradiction. The radical promises of change and equality implicit in independence and the subsequent waves of change clashed with the efforts of those who felt themselves responsible for governing this unruly and economically underdeveloped society. The salt monopoly provided a template for this conflict. A powerful or wealthy state would not have maintained the salt monopoly. Even other Spanish American governments, which were generally neither wealthy nor powerful, ended their salt monopolies relatively quickly. The Colombian Finance Ministry's failure to free itself from this fiscal anachronism was an ongoing reminder of its own limitations, its inability to implement the fundamental reforms so many elites deemed necessary for progress. Moreover, the salt monopoly did not simply survive; it grew. This growth demanded the creation of an institutional bureaucracy to manage the monopoly, a bureaucracy that embodied the contradictory role of the state in the nineteenth century. The bureaucracy's raison d'être was to provide the state with necessary revenue, yet its inefficiencies and proclivity for corruption hindered the execution of these tasks. At the same time, the growth of this bureaucracy was a crucial element of state growth, particularly in the realm of information gathering. Reports on salt making and related policies filled the pages of official newspapers and other government publications, forming part of a public discussion about the proper role for the state.[6]

The history of salt making, of ministerial efforts to control salt making, and of local responses in La Salina runs through three intertwined stories. The first concerns fiscal policy and the way its implementation shaped life in La Salina, the second centers on regional responses to the state's monopoly control, and the third is a chronological account of the course of national politics in La Salina and Boyacá. Each story draws on the same cast of characters, including the Finance Ministry as an institution, finance ministers, ministry employees across Boyacá, entrepreneurs, local contractors, residents of La Salina, migrant workers, and communities in Boyacá's eastern cantons. Taken together, these stories provide a portrait of Colombia as a place where state action and state weakness dramatically shaped history, though, as was the case throughout Latin America, not in the manner envisioned by these would-be architects of the state.

The State

Questions concerning the state and state formation loom large in examinations of nineteenth-century Latin America published during the two decades spanning the turn of the twentieth century. Although scholars disagree about the exact nature of the state in the republican era, a number of them argue that measuring individual states as successes or failures by some Weberian rubric is a fruitless pastime. Attempts to conceptualize states by this dichotomy lead almost inevitably to the apparent contradiction where, in the words of Fernando López-Alvez, Latin America appears as a region of both "weak states with ill-trained bureaucracies" and "centralist, corporatist institution[s]."[7] But the problem lies in the attempt to cast these conditions as contradictory forces rather than as two facets of the republican state. As Stacey Hunt has argued, "few states, if any, meet Weber's ideal in absolute terms. When held up to this ahistorical myth, the 'failure' of states is the rule rather than the exception."[8] The history of La Salina demonstrates that Hunt's urge to reconcile seeming contradictions applies equally to the quandary described by López-Alvez. The centralist institution of the salt monopoly was an ill-trained bureaucracy, and the institutional state manifest in La Salina was neither a success nor a failure. It was, however, a significant historical actor.

The means for examining the state in this fashion were laid out by Oscar Oszlak in "The Historical Formation of the State in Latin America: Some Theoretical and Methodological Guidelines for Its Study." In this essay Oszlak argues for a methodology that distinguishes between the abstract state, which is the entity that seeks a monopoly on the legitimate use of force and the corresponding authority, and the material, institutional state, which is an interlocking collection of buildings, bureaucracies, mail routes, and other pedestrian items. "The state apparatus manifests itself as a multifaceted and complex social actor, in the sense that its various units and arenas of decision and actions express a diffuse and, at times, ambiguous presence in the network of social relations."[9] I am particularly concerned with the third of Oszlak's four steps for research, documenting the degree to which a state developed public institutions with "(a) a recognized capacity to extract, on a regular basis[,] resources from society; (b) a certain degree of professionalization of their functionaries; and (c) a certain measure of centralized control over their multifarious activities."[10]

Remarkably, given the rich social science literature available, historians seldom focus on the state, particularly the material or institutional state. Historians often consider the state as one element that serves as an analytical pivot in a broader topic, something to touch on before turning to other issues, usually ones relating to political culture.[11] This tendency resulted from the paradigm shift in Latin American historiography during the 1990s, when approaches that privileged labor, the particularities of class, and various versions of dependency theory ceded intellectual terrain to a focus on discourse and identities as a means of charting popular contributions to politics, national formation, and political culture. While the general turn toward examining political culture has produced a rich understanding of Latin America as a place with a dynamic political history, it has not brought a commensurate appreciation for the state as a historical actor or for state building as a historical force.[12] A focus on the state, even an inchoate state, therefore permits a fresh look at Colombia's early republic. More specifically, fiscal policy, often examined in terms of economic history, offers a rich source for examining contact between the institutional state and society.[13]

Thus the material reality of salt and the institution that developed around the salt monopoly provide a concrete base for a discussion that can otherwise be overly theoretical.[14] For example, there were marked continuities in the Colombian state despite political fluctuations and various new constitutions. This continuity was manifest most clearly in the material presence of the institutional state dedicated to managing the salt monopoly. Conceptualizing state building as an institutional process while charting its material presence provides a methodology for recasting the state as an important historical force in the nineteenth century.[15] This approach has value beyond Colombia, for state presence was weaker there than in other parts of Spanish America.[16] If the state can be recast as a relatively important historical actor in Colombia, then the same approach will likely shed new light on the rest of the region.

The competition between state agents and other actors was a quotidian reality focused on economic results. More specifically, these competitions took place in carefully demarcated spaces defined by the institutional state as vital elements of state infrastructure. Just as elites found that subordinate actors took advantage of new political spaces and discourses to promote their own interests, the Finance Ministry found that the institutional spaces of the salt monopoly—the sites of production, sale, and transport—were transformed into arenas for popular defiance. As in the political arena, it was easier to define a space than to control what occurred in it. But these moments of contestation most clearly show the utility of eschewing success or failure when considering state building, for during them the state's roles as an entity with power and a concrete institutional presence ran together. Thus, while Oszlak's bifurcation of an abstract and material state provides a useful analytical tool, this distinction did not play out in real life. In the experience of people who lived in La Salina, the institutional state was not an abstraction but rather a daily practice. The question of state legitimacy, so prone to theoretical parsing in scholarship, was a matter of profits and the prices buyers paid for salt.

Boyacá and Colombia

Within Colombian historiography we need to address a number of questions beyond the role of the state. The dominant themes of scholarship on the nineteenth century include the development of political parties; the pattern of civil wars; regionalism; and more recently, the role of ethnicity in all these issues.[17] Colombian history contains no single significant event (such as the Mexican Revolution) or figure (such as Juan Manuel de Rosas in Argentina) that throws stark light on the meaning of the state.[18] Equally important, the broken topography of Colombia and the intensity of its regionalism and regional histories make it a challenge to even conceptualize a national state.[19] An emphasis on the monopolies of the Finance Ministry, however, offers one method for charting the reach of the institutional state. The ministry's bureaucracy was in a sense an archipelago. Scattered across Colombia's daunting terrain were regional boards, offices, contractors, subcontractors, and collection centers, linked precariously by mail routes and mule paths.[20] Though the network was designed to pass instructions from core to periphery and carry information and receipts in the other direction, these islands of the bureaucratic state were also supposed to be points of authority and control. In practice, however, these islands often developed into focal points of defiance to the state and its authority.

La Salina was one such place. Moreover, the steady correspondence written by Finance Ministry employees and forwarded to Bogotá makes it possible to document two important dynamics. First, profits from salt revenue were intended to stimulate market activity, but the monopoly demanded that the state control as many local resources as possible in an aggressively antiliberal fashion, thereby limiting the markets.[21] Second, elites, whether critics or ministry officials, sought to promote both moral and economic progress, which they often conflated. In La Salina, these dynamics often intertwined, heightening expectations and frustrations.

This history would have been notable in any region, but the setting in Boy-

TABLE 1.1. *Population in selected Colombian provinces and Colombia, 1810–1905*

	Province					
Year	*Antioquia*	*Boyacá*	*Cauca*	*Cundinamarca*	*Santander*	*Colombia*
1810	111,000	231,000	200,000	189,000	237,000	1,309,000
1835	159,000	288,000	210,000	255,000	280,000	1,571,000
1843	190,000	332,000	269,000	288,000	324,000	1,814,000
1851	243,000	381,000	312,000	319,000	382,000	2,094,000
1870	366,000	499,000	435,000	414,000	433,000	2,713,000
1887	520,000	615,000	635,000	550,000	565,000	3,666,000
1898	620,000	685,000	800,000	630,000	640,000	4,262,000
1905	897,000	503,000	765,000	631,000	550,000	4,144,000

Source: Jorge Orlando Melo, "La evolución economica de Colombia, 1830-19300," in *Manual de historia de Colombia*, 4th ed., ed. Jaime Jaramillo Uribe, 3 vols. (Bogotá: Tercer Mundo, 1992), 2:138.
Note: This table does not include Bolívar, Tolima, Magdalena, or Panama, which broke away in 1905.

acá, one of Colombia's most populous regions, makes it particularly important, for the historiography of this region is underdeveloped by almost any standard, adding weight to any attempt to further the understanding of Boyacá's internal heterogeneity and its role in national politics. While the Universidad Pedagógica y Tecnológica de Colombia at Tunja has published well-researched regional studies, often based on sources in the Archivo Regional de Boyacá, and although a body of work that celebrates Boyacánse accomplishments does exist, this literature does not equal the other regional historiographies of Colombia.[22] Boyacá is too often portrayed as a region where history was received rather than generated, in stark contrast to Antioquia, Cauca, Santander, or the Caribbean, but its size alone means that it could never be entirely absent from nation affairs (see table 1.1). An understanding of Boyacá is thus necessary for understanding the history of La Salina and the history of Colombia. As with all Colombian history, a review of the relevant geography is a necessary prelude to the history that follows.

As the Andes run north along the western coast of South America to the city of Pasto in southern Colombia, they split into three separate ranges. The eastern range is the largest of the three, sprawling hundreds of kilometers east to west; it contains the bulk of Colombia's densely settled highland plains and valleys. Of the three ranges, this one comes closest to reaching Colombia's Caribbean coast, though it splits again near Cucuta. From here one line of waning peaks continues north only to end with the dramatic coda of the towering Sierra Nevada de Santa Marta, while the other arm runs into Venezuela. The central and eastern ranges are not as high or as broad as their eastern sibling, but they too boast snow-capped peaks and rich highland valleys. The three ranges are separated by the Cauca and Magdalena Valleys, making east-west travel difficult and time consuming. Colombia's other major geographic features include deserts, swamps, plains whose constant flooding makes them ill-suited for settle-

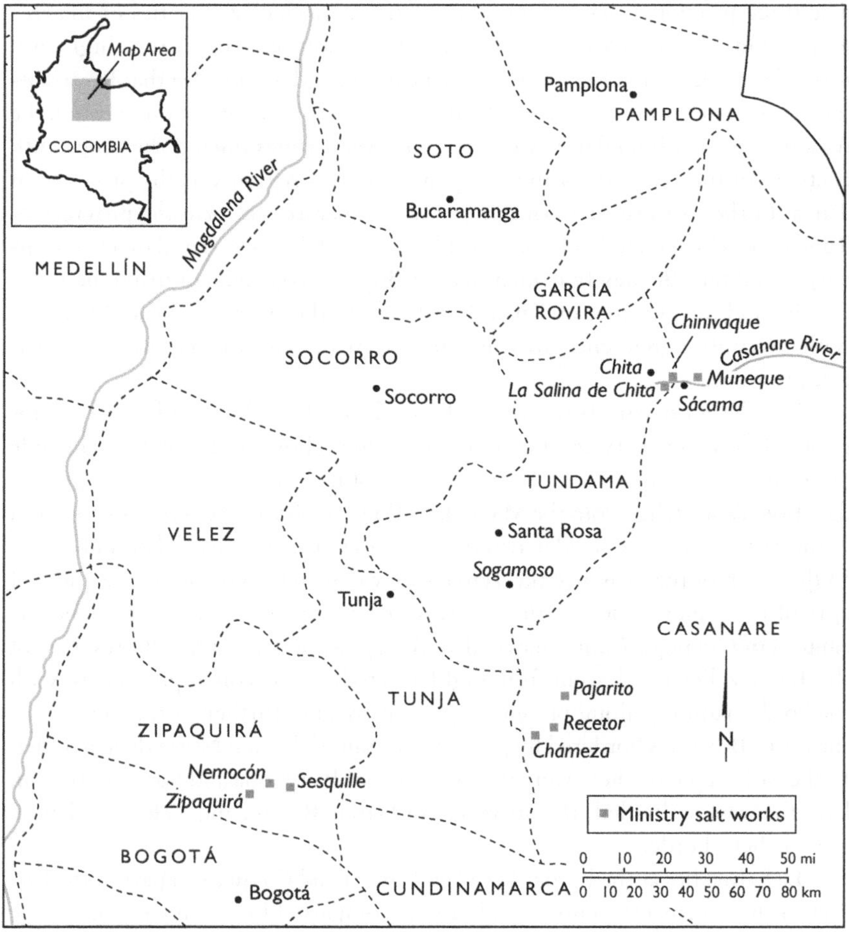

FIGURE 1.2. "The Eastern Highlands until 1856."

ment, a Caribbean coast, a Pacific coast, and areas with some of the highest annual precipitation in the world. This variation carves the country into regions with distinct demographic, cultural, political, and economic histories. Accordingly, regionalism and regional differences are important topics when considering the country's political and historical development, though scholars disagree over the precise number of regions.[23]

One important element documented in regional literatures but rarely detailed for Boyacá is that even the most recognized region was a composite of microregions and that regional identities were in part social constructions.[24] Seen from a national perspective, Boyacá is often grouped with Cundinamarca and Santander as the territories making up the eastern highlands.[25] During the nineteenth century, Boyacá was sometimes linked to Santander and the combination dubbed "the North." At times this region was taken to include Cundinamarca.

The three provinces shared a number of traits and had strong ties of interregional trade, but on closer examination the notion that the eastern highlands formed a single region loses most of its meaning, as does the idea that Boyacá itself was internally homogeneous.[26] Adding to the complexity of these issues, the region's political boundaries were altered several times during the nineteenth century. In the years following independence, this region was the province of Tunja. In the 1850s the eastern cantons were separated to form the province of Tundama, which included La Salina. Then, in the late 1850s, Tundama was reincorporated into the newly named state of Boyacá. To avoid confusion between the city and the province of Tunja, I will refer to the entire region as "Boyacá," with intermittent references to Tundama even in decades when it was not a separate province.

For the most part Boyacá is and has always been highlands, with major population centers between 1,900 to 3,000 meters above sea level. At an altitude of not quite 1,500 meters, La Salina was one of its lowest towns. In the present day, Boyacá stretches from the Magdalena River valley in the west to Venezuela in the east, though the slender arm that reaches to the west was a late addition.[27] In the south of the province the Tensa Valley descends to the Llanos, an area of agricultural colonization in the nineteenth century. The valley contained the cantons of Guateque (sometimes called Tensa), Garagoa, and Miraflores. East of the Tensa Valley the descent from highlands to llanos becomes precarious, with the border running along present-day Casanare and, further east, Arauca.[28] To the north Boyacá is bordered by present-day Santander and Norte de Santander. In the early republic these were the provinces of Pamplona and Socorro, though later these were divided; the provinces of Garcia Rovira and Velez also lay on the northern border.

La Salina lies at the border between Boyacá and Casanare, a part of the former in the nineteenth century and now encompassed by a slender spike of the latter that wedges between Boyacá and the department of Arauca. The town lies in a valley that climbs west to the highland cantons that formed Tundama: Santa Rosa, Sogamoso, El Cocuy, and Soatá.[29] Trade linked Tundama to the cities of San Gil, Socorro, Málaga, and Pamplona in present-day Santander and Norte de Santander. In the 1850s this area comprised the provinces of Socorro, Soto, Pamplona, and García Rovira.

A number of countervailing forces were woven into this geography. One pair lay in the contrast between La Salina's institutional links to Tunja and Bogotá and its regional market orientation with the cantons of Tundama and what is now Santander. Another arose from its location on one of Colombia's internal frontiers. The highlands to the north and west had long been settled, and some cantons were marked by increasing smallholding and subsistence agriculture. In contrast, the llanos to the south and east remained an open frontier, a place of people and nature equally untamed. To would-be state architects, Casanare married the threat of disease and a wilderness not worth taming to the endless promise of bountiful nature.[30]

Focusing on La Salina and Tundama corrects the misleading tendency to treat the area around Tunja as representative of Boyacá as a whole.[31] In scholarship, literature, and popular culture, this monolithic Boyacá is often portrayed as the home of a peasantry that, though generally passive, was fierce when it was aroused—a feat that only the region's conservative clergy could accomplish.[32] However accurate this stereotype might be, conflating Tunja and its environs with all of Boyacá obscures a reality of internal variation and intraprovincial rivalry. The simplified version of a homogeneous Boyacá dovetails with the view that the whole region reached its zenith in the seventeenth century and then entered a four-hundred-year decline.[33] This idea itself has a long pedigree. Writing in the mid-nineteenth century, Manuel Ancízar observed that "Tunja is for *Granadinos* an object of respect" because of its past, but he did so only after he had asked, "What is left for Tunja, city without arts, without agriculture, without commerce?"[34] In the late twentieth century, Germán Colmenares began his study of colonial Boyacá by explaining, "The city of Tunja does not provoke sympathy from visitors. . . . It is not easy to convey what Tunja signified in the early years of New Granada."[35]

Had it been located in a moderately prosperous region, the saltworks at La Salina might have been only a single element in a larger economic picture, but in Boyacá it stood alone.[36] Writing in 1823, Gaspard-Théodore Mollien, a French traveler, reported, "Today [Tunja] is no more than a dead city."[37] The intendant of Boyacá, Bernardino Soban, penned a similar report in 1830, but he also proposed a plan to better the situation:

> As peace has returned to this department, I have tried in many ways to promote and secure its prosperity. Since taking charge I have been struck by how [in] the whole [province] and especially in this [canton] and this city there is no industry, there is no commerce, there is no agriculture, nor are there mines or arts, or any other elements of prosperity that exist in the other parts of the Republic. The subsistence of this city is very precarious as it depends solely on the fact that it is the Capital of the department, and is affected by any change in circumstances. The only money that circulates here is the salary of the departmental employees. . . . Seeing such a state of misery, and the exposure to absolute destruction, I have used my limited understanding to find a source of wealth and abundant public prosperity. My solution is to establish in this city a textile factory for wool that would increase the value of the province's most abundant resource, encourage raising sheep, attract commerce as the base of other ventures, [and] employ a thousand workers, increasing the number dependent on agricultural products, which in turn will stimulate farming.[38]

But the prospects for reviving eastern Colombia's textile industry were not good. In his annual report for 1831, Finance Minister José Ignacio de Márquez noted the influx of cheap imported textiles and the damage they had done to local industry: "If there is any doubt that this is the truth, one has to do no more than tour the industrious provinces of Socorro, Tunja, Bogotá, and Pamplona, previously quite productive and today abandoned and poor."[39] In 1833 the newspaper *El Constitucional de Boyacá* complained that imported textiles had "damaged [the area's] few factories," where cotton and wool had formerly "flourished."[40] There is no evidence that Soban's plan went anywhere.

The dominant activity in the region was agriculture, dedicated mostly to crops consumed locally. Animal husbandry and ranching were also present to an extent. One of Boyacá's industries was fattening cattle from Casanare, thinned by the arduous journey from the plains, which were then sold in Santander or Cundinamarca. Salt was a key element of that industry. In a survey of industry for Boyacá, Agustín Codazzi mentioned the manufacture of shoes; hats; honey; wax; sugar products (mainly *aguardiente*) in regions with lower altitudes; some ceramic items, including tiles; prepared wood; and metal objects, which were produced by a few dozen blacksmiths.[41] His optimism for the prospects of the regional economy did not preclude frank descriptions, as when he conceded that "properly speaking," Miraflores, in the Tensa Valley, had no true manufacturing: "Everything in this canton is to be created. There are hardly 100 pieces of cotton cloth woven in a year, of which half, worth about 300 pesos, are exported."[42]

La Salina, in the canton of El Cocuy, stood out. A special tax on individual wealth, established in 1821 but not collected in Boyacá until in 1823, offers a benchmark for comparing the canton of El Cocuy's economy to the rest of Boyacá's. By canton, Sogamoso had the highest average payment, followed by El Cocuy, Santa Rosa, and Garagoa, with Soatá and Tensa falling below the provincial rate (see table 1.2).[43]

TABLE 1.2. *Special contribution of 1823*

Canton	*Population*[a]	*Tax*[b]	*Individuals taxed*	*Percentage taxed*	*Average payment*[c]
Santa Rosa	23,221	1,341	497	2.0	2 p 6 r
Soatá	27,683	1,230	503	1.8	2 p 3 r
Sogamoso	30,592	1,552	419	1.4	3 p 6 r
Cocuy	20,220	792	293	1.5	2 p 6 r
Tensa	28,506	1,449	612	2.1	2 p 3 r
Garagoa	12,893	655	252	2.0	2 p 5 r
Total	143,115	7,019	2,576	1.8	2 p 6 r

Source: Population figures in this and the following tables come from the census of 1835.

Note: The legislation was passed in 1821, but collections occurred only in 1823.

[a] Population figures include only the districts reporting in the tax levy, not total population of Boyacá.

[b] In pesos.

[c] Per individual, in pesos and reales.

TABLE 1.3. *Occupation by canton in Boyacá in 1845*

WESTERN BOYACÁ				
Canton	*Landowners*	*Laborers*	*Traders*	*Artisans*
Centro	2,940	6,039	630	1,443
Leyva	1,700	2,310	118	59
Ramiriqui	1,923	6,095	145	163
Total	6,563	14,444	893	1,665
TUNDAMA				
Canton	*Landowners*	*Laborers*	*Traders*	*Artisans*
El Cocuy	1,620	2,065	60	103
Santa Rosa	835	3,130	600	471
Soatá	334	3,430	124	179
Sogamoso	4,375	8,085	1,343	1,520
Total	7,064	16,710	2,127	2,273
VALLE DE TENSA				
Canton	*Landowners*	*Laborers*	*Traders*	*Artisans*
Garagoa	379	3,668	52	40
Tensa	403	6,345	360	996
Total	782	10,013	412	1,036

Sources: AGN, SR, Gobernación de Tunja, leg. 30, fols. 469–568, Oct.–Nov., 1845, eighty-eight responses to a questionnaire entitled "A Table showing what the district of________ produces in agriculture, livestock, industry, and mining."

A district survey from 1845 adds information on the varying social composition of each canton. The Canton del Centro, along with those of Tunja, Santa Rosa, Sogamoso, and Tensa, had occupational breakdowns suitable for centers of trade and artisan production, while the high ratio of landowners to laborers in Leyva and El Cocuy reflected advanced smallholding (see table 1.3).

El Cocuy was the smallest of Boyacá's eastern cantons, those that would form Tundama in the 1850s, and its habitable highlands were densely settled. Most of its towns were located more than 2,000 meters above sea level, except for Espino and La Salina, which stood at 1,994 and 1,439 meters, respectively.[44] Codazzi described the climate in the canton of El Cocuy as "cool and healthy," except for La Salina, which because of its lower altitude was "warm and healthy." El Cocuy's economic mainstays were typical highland crops: potatoes, maize, and beans. In the municipalities of El Cocuy, Chita, Guicán, and Chiscas, sheep and goats were notable. Local industry included small-scale textile production and two forges. The canton traded with its neighbors, the Tunebo Indians (a relatively unacculturated group of eastern Boyacá and Santander) and distant Bogotá. Manufactured goods from Santander or Bogotá and agricultural products from warmer climates were brought in, as were cattle from Casanare. Wheat, handmade textiles, and salt were also exported. None of this suggests a robust economy, but Codazzi noted that the canton's "healthy climate" and rich agri-

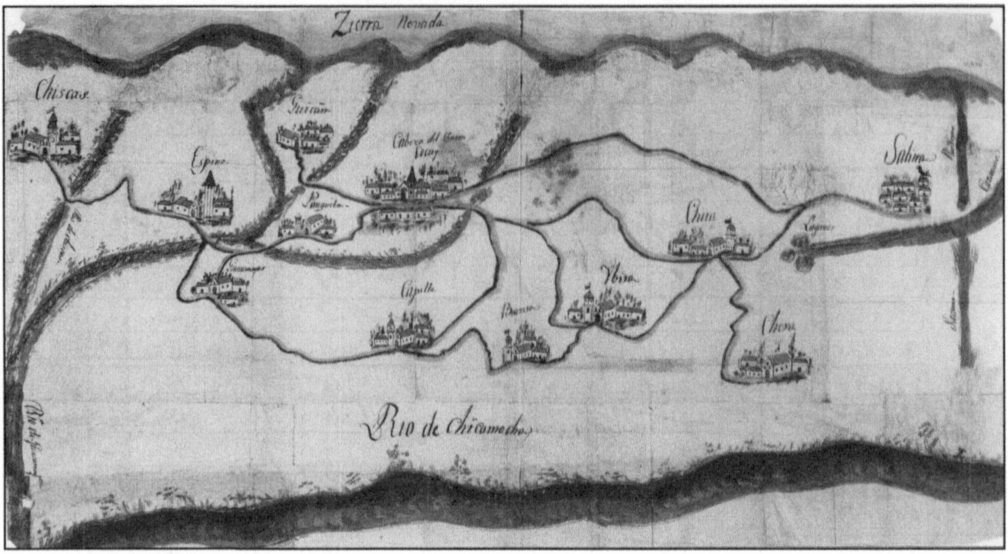

FIGURE 1.3. "Canton del Cocuy, 1825."
AGN, Mapas y Planos, Mapoteca No. 4, ref. 130 A.

culture would allow for a rapid population increase that would be "white and robust."[45]

Manuel Ancízar, who was a part of Codazzi's Comisión Coregráfica, was similarly optimistic about El Cocuy's potential, in part because its population comprised "well-formed whites and Indians, patient and vigorous, in whom routine guides the duties of the soul, and humility is the least of all their virtues."[46] Ancízar even had optimistic words for Chita, La Salina's neighbor and the source of its official name, which was generally viewed as a troublesome town. He offered his praise, however, in the racialized terms of the era: "The entire district includes a little more than 7,900 white *vecinos,* robust with glowing cheeks. . . . The town, like all those old Indian towns, resents its origin and shows it in the disorder of its streets and the terrible state of its houses. In contrast, all the people have a healthy disposition and simple manners, virtues that, united with honest labor, would produce a population morally unsurpassed yet removed from the superficial nature of the cultures of other nations."[47] For these men, El Cocuy resembled the cantons of Santander, a region possessing enormous potential. Such developments, however, lay in an ever-receding future. In fact, the saltworks of La Salina was the canton's most important industry. In this impoverished world, already full of promise and disappointment, the story of salt and the state played out.

CHAPTER 2

Change and Community in La Salina

In the final analysis, the republican government is, by its essential nature, not a distinct entity from society, nor does it destroy, as occurs at times with other political forms, the property and liberty of the citizens. It is simply an organization that has as its purpose a method for arranging for all . . . a means of satisfying certain needs inherent in human nature.

—Salvador Camacho Roldán, *Memoria de Hacienda y Fomento de 1871*

People made salt in the spot that would become La Salina de Chita long before the Spanish arrived and gave the town its name. Its pre-Columbian residents were the Lache, an indigenous group that formed part of the polity called the confederación del Cocuy. Relatively little is known about these people, who were linked to the Chibcha-speaking Muisca of the highlands to their west, but they did make an impression on the sixteenth-century Spaniards, who deemed the Lache to have been less civilized than the other indigenous groups they had encountered in the region.[1] Their description of the Lache focused on the group's acceptance of homosexuality, apparent worship of shadows and rocks, and participation in the ritual called *muma,* which purportedly consisted of striking one another.[2] Among the few goods that the Lache traded was salt.[3]

The pre-Columbian salt trade played a part even in the Spanish conquest of the Colombian highlands, though the starring role fell to salt from Zipaquirá, the grand and ancient production center lying just north of Bogotá. In 1536–1537 the Spanish expedition led by Jiménez de Quesada, the conquistador of New Granada, was traveling on the Magdalena River when it met highland traders who had two cakes of salt from Zipaquirá.[4] The Spaniards recognized this finished salt as a product of a culture more sophisticated than those they had encountered in the lowland valley of the Magdalena, which meant that this was a society worth conquering. Quesada left the valley and followed the salt trade route back to the highlands that would later form the heartland of the Spanish viceroyalty of New Granada.

In time Spanish dominion spread over the highlands until it included present-day La Salina. Initially the Spanish called the town "el pueblo de la Sal." Later, when it became part of the *encomienda* of Chita, the name was amended to La Salina de Chita.[5] As subjects of the Spanish Crown, the indigenous residents of La Salina paid part of their royal tribute in salt.[6] Apparently this obligation was not an excessive burden. Tribute lists from the seventeenth century document a small but notable increase in the local population, from fifty-one households in 1602 to fifty-eight in 1636 and seventy-seven in 1690.[7] This moderate increase shows how salt making set La Salina's history apart from broader trends in Colombia and across the Americas, where the demographic collapse of the indigenous population was almost universal. Salt making allowed La Salina to retain its population because legal residence as an *indio* included access to salt making.[8]

There is little indication the Spanish Crown sought greater involvement than that during the seventeenth century, though one official did see fit to mention "a great spring of salt water that thickens in the sun and makes a very white salt . . . that is normally taken to Pamplona."[9] There are two archival references to La Salina from this century. Both refer to problems between the encomendero, a position held by a Spaniard that conferred authority over the local indigenous population, and the indigenous residents.[10] The royal government was a distant force. It was not until the eighteenth century that the Bourbon reforms, which remade the relationship between ruler and ruled across Spanish America, brought La Salina into a closer contact with the colonial state.

The first Bourbon reform in this story was the 1723 attempt to separate New Granada from the viceroyalty of Peru, though this newly designated viceroyalty did not function as an autonomous entity until 1739. Several decades then passed before a series of laws concerning trade, taxation, and corporate landholding began to reshape life in New Granada.[11] In the eastern highlands the greatest impact resulted from the implementation of monopolies, including that on salt, and a reorganization of the *alcabala,* a sales tax.[12] In the three decades of Spanish rule following the introduction of these taxes, the tobacco monopoly was the royal treasury's most lucrative revenue source, followed by the alcabala, customs, and the monopoly on liquor.[13]

Eighteenth-century records show increasing contact between the residents of La Salina and representatives of the colonial state, though the evidence often comes from accusations that *vecinos,* or nonindigenous legal residents, had abused the local *indios,* or indigenous ones.[14] In 1774 the indios of La Salina brought suit against one Salvador García for cattle theft. Only four years later José María Campuzano y Lanza, a *corregidor* from Tunja, reported that the Crown official in La Salina, Joaquín Ascarza, had "extorted the indians and mistreated them with blows and lashes."[15] Campuzano removed Ascarza from La Salina and ordered him to pay damages. Campuzano, however, was there not to protect the indigenous community of La Salina but to get rid of them. He was one of the officials charged with removing indigenous communities from their communal lands (*resguardos*) on the grounds that others would put them to more productive use. Earlier attempts at such removals had been tentative, involving fewer than a dozen communities, but in the 1770s, under the direction of Viceroy Francisco Antonio Moreno y Escandón, the campaign grew teeth; sixty communities were removed and resettled onto only twenty-seven resguardos.[16]

La Salina's indigenous population had dwindled, making it an obvious target for resettlement, particularly as the town had a nonindigenous population of 302.[17] Once removed from their ancestral lands, the 153 legally recognized indigenous residents of La Salina were moved to the neighboring resguardo of Chita. Campuzano described this occasion, in the process revealing that two centuries after the conquest, some elements of Lache culture endured, as had the ironies of Spanish humanism: "As the Judge drew up the list, a group of Indians presented themselves, . . . young men . . . with ruptured abdomens from the terrible blows they give one another in the brutal, national game of their festivals that is called Muma. Outlawed, as it causes such terrible abuse, those who practice it are punished with 100 lashes and fifteen days in prison."[18]

The various accounts of the resettlement contain no references to salt making or the thermal springs. The value of the resguardo was set at 1,200 pesos, the price paid by the vecino community to obtain control over the land and the right to exploit the springs that came with it. This rate was lower than that set for the other resguardos in the region, but Campuzano claimed that this was "the most it could be worth." In the end the vecino community paid only 1,000 pesos for the resguardo and attendant privileges. Perhaps this was a spectacularly low price for the basis of the industry that would produce more wealth than any other in eastern Boyacá in the coming decades, but it must be noted that the heirs of these purchasers did not claim much of that wealth. Their collective control of the salt making proved ephemeral.

Campuzano explained the benefits of the resettlement for everyone: "Despite the natural repugnance the [Indians] have shown against leaving their town, it is indispensable to move them [to the resguardo of Chita,] where there are better resources, both spiritual and temporal. The greater good does not lie in satisfying their capricious desires but in obliging them [to accept] what they rationally deserve, with the least burden to the Royal Treasury."[19] Amaya Roldán, a parish priest and local historian from Chita, provides a different perspective: "The Indians were given shares in the resguardo [of Chita], but these were so small and insufficient that after a few months, compelled by need and nostalgia, they returned to La Salina, but the whites were now in possession and would not let them enter. Resigned to their luck, the natives retired to the outskirts of La Salina, on the edges of Rodrigoque, where, in order not to starve, they turned to making ceramics that they sold to whites for salt making."[20] It remains unclear why La Salina's indigenous population declined in the eighteenth century after its slow growth in the seventeenth century, though the drop may have been linked to the growth of the vecino population and the broader process of miscegenation underway in the eastern highlands. In an article on the resguardo of Chita during the late eighteenth century, Alba Luz Bonilla de Pico has documented the multifaceted nature of vecinos settling in a legally recognized indigenous town.[21] While this dynamic would have worked differently in La Salina, the same chaotic forces were in play. La Salina had ceased to be a "pueblo de indios" well before the removal of the indigenous community in 1778, but with that resettlement the transformation was formalized legally, and a community of vecinos gained control over the lands of the former resguardo. The vecinos would maintain control over salt making for two generations before they too were swept aside.

Other resettlements were only one part of a larger program of reform that not only affected society across New Granada but also established the structures that would underpin republican fiscal policy. At the moment of their implementation, it was the reaction against new and increased market taxes and monopolies on tobacco, salt, and playing cards that proved most explosive. In 1781, in the province of Socorro, protests against the alcohol and new tobacco monopoly grew into a series of local uprisings that by the middle of the year was a single movement, the Comuneros Rebellion.[22] Viewed from the continental perspective, this rebellion was overshadowed by the nearly simultaneous events of the Tupac Amaru and Tupac Katari rebellions in the viceroyalty of Peru, which were larger, more destructive, and more deeply threatening both to Spanish imperial control and to the fundamental societal order of Spanish America. But the Comuneros Rebellion was no small affair, and its forces were, in the interior of Colombia, unstoppable. As the revolt spread from Socorro to include residents of other regions, the rebels elaborated on their protests against the tobacco monopoly. Among the expanded list of conditions circulated by the rebels was a demand that the resguardos be returned to the deposed indigenous communities. Another demand was the abolition of the new monopolies.[23]

Eventually the rebels fielded an army of 20,000: 10,000 from Socorro and the other provinces of what is now Santander, 6,000 allies from Boyacá, and 4,000 indios. Initially the government could muster only a small force to oppose the rebels, and the rebels defeated it when the two armies met in battle in the province of Vélez. The Comuneros troops then marched for Bogotá unopposed. As would almost any army approaching Bogotá from the north (i.e., any campaign against the capital emerging from what is now Santander or Boyacá), they followed a route running through Zipaquirá. The rebels camped there, enjoying the resources of the saltworks and the prospect of a short march over flat terrain to the defenseless capital. Then, with Bogotá in their grasp, the leaders ended the campaign. One reason was that the archbishop of Bogotá, Antonio Caballero y Góngora, had capitulated to a list of rebel demands. Another was that the leaders of the Comuneros feared the social unrest that had been unleashed, as well as the inevitable settling of scores to come. The moment of crises passed, though not without some residual fighting. In 1772 the archbishop was appointed viceroy of New Granada. Though he granted a general pardon, his promises to meet rebel demands ultimately proved empty. The resguardos were never returned to the original inhabitants, and he met little opposition when he reinstated the monopolies and taxes that had originally stirred the protest.[24]

It remains unclear how the much the rebels' demands reflected the interest of either La Salina's former indigenous community or its vecinos.[25] Lists of officers in the Comunero army mention one "captain" from La Salina and another from neighboring Chita, which shows that some of the rebellion's participants came from the area. Still, although the interests of people around and in La Salina may have been bound up with the demands of the Comuneros, the fighting took place on the other side of Boyacá.[26] It was the aftermath of the rebellion, the implementation of objectionable Bourbon taxes, that directly affected life in La Salina.

Taking a more active role in overseeing the saltworks than it had played prior to 1781, the royal treasury levied a one peso tax on every load (a *carga,* which was ten *arrobas,* or 125 kilograms) of salt sold in the town.[27] Around this practice the royal treasury implemented what was later called the *antiguo sistema,* the "old system." In this arrangement, Laso de la Vega "rented" the springs, presumably collecting the tax and overseeing other matters, and the Crown maintained its legal control over the source of salt as a subsoil mineral resource.[28] The local *cosecheros* (literally, "harvesters") made salt independently and, crucially, retained free access to all the resources necessary for production. The production system was essentially the one sketched by Caicedo in 1806, though the first steps had been taken down the path to direct state management.

As is often the case with state building, these changes yielded an extensive impact. In the first decade of the nineteenth century, the ongoing reforms and the less dramatic but notable maturing of late colonial New Granada made for an accelerated pace of change in La Salina. In one telling example, the vecinos of La Salina hired Nicolas Llanos to represent their interests. Among his efforts on their behalf was a petition to have La Salina elevated to an independent parish.[29] But as would often prove to be the case over the following decades, the impulse toward communal action was overshadowed, at least in retrospect, by the actions of the state.

Ignacio Caicedo took up his post as the treasury's administrator in 1804. Caicedo had served in Zipaquirá and was ordered to report on conditions in La Salina. He was also aware of Humboldt's proposals for improving conditions there. Caicedo was undoubtedly thinking of the complex organization of the saltworks and the need for structural improvements as he considered La Salina's rustic nature. His correspondence documented the workings of the industry and cataloged a series of measures he believed were necessary for improving management of the saltworks.[30] In addition to proposing the previously mentioned wall, Caicedo called for more guards. He explained that the administrator could not both manage sales and ensure control over all the salt made. The solution to these problems was, typically, the extension of the state's bureaucratic presence. The cost incurred by employing more guards, he argued, would be recouped by the increased profits for the treasury.

Caicedo was also the first to report on local poverty in relation to salt making. He explained that most residents of La Salina were poor and suffered in the winter, when the industry flagged. Caicedo also mentioned problems stemming from newly arrived outsiders who attempted to claim a share of salt-making resources. Finally, he mentioned that corruption was a problem in the administration, though he did not provide details. Each of these issues would figure prominently in correspondence from La Salina during the coming century. Virtually every national official who served in the town contributed to an almost endless litany of complaints about the state of the wall, local poverty, the need for more guards, fraud, and problems stemming from outsiders. Caicedo's comments were unique only in being new and original, whereas those of his republican successors were repetitive and increasingly dire.

The history of La Salina and the salt monopoly fades from view during the independence era. Antonio Nariño, the herald of Colombian independence, included calls for fiscal reform that in theory might have affected the salt monopoly prior to the Wars of Independence, but such points are swept aside by heroic narratives of patriotism and struggle. Shortly after Caicedo made his sketch of La Salina, the Spanish empire entered a political crisis that brought more change to the Spanish world than had any other event since its creation in the decades following Columbus's arrival in the Americas. The crisis of legitimacy sparked when Napoleon forced the abdications of Carlos IV and his son Ferdinand VII led to a series of constitutional promulgations in Colombia. During the heady days of Colombia's "Patria Boba," incipient and competing governments sought to establish themselves. Spain's weakness allowed for some triumphs on the part of American patriots. After Ferdinand VII returned to the throne, Spain sought to reestablish its rule over its American colonies. Beginning in 1815 Spanish forces succeeded in retaking control of New Granada, though the stakes of this war and the execution of American patriots have earned this campaign the title of a "reconquest."[31]

There is little record of La Salina in this era. The town's only possible notoriety stems from Simon Bolívar, the ultimate leader of the independence effort in northern South America, who considered using it for his route when planning a campaign from the llanos of Venezuela and Colombia against the Spanish-held highlands of Colombia in 1819.[32] However, because the town was occupied by Royalist forces, Bolívar opted to pass through the Pisba and Tasco Valleys, two other corridors that linked the llanos to the highlands.[33] This campaign culminated in Bolívar's victory at the Battle of Boyacá. This triumph gave the Americans control over territory that would become Venezuela and Colombia and led to a constitutional congress held in Angostura in 1819. As the light of the republican era finally broke over the land, fiscal policy settled into a routine of vacillation between abolishing colonial revenue sources, such as Indian tribute, and retreating from such ambitious reforms. In the first year of the new republic, Alejandro Osorio wrote the first ministerial report on finance, the *Memoria del Ministro de Guerra y Hacienda de 1819*. Osorio described the precarious state of government finances with frequent references to what "his Excellency" (i.e., Bolívar) had already done. The chief characteristic of this inchoate fiscal system was a willingness to exploit any source of revenue. Osorio justified this approach by explaining that "the Supreme chief of the Republic" had ordered that existing taxes be maintained because "the days when the people fight for their liberty are not the days of reform, nor is the moment when they acquire this gift the moment to enjoy all of its benefits."[34]

The report included a few points about the salt monopoly. One was a flattering description of the way Bolívar had organized affairs in Zipaquirá. Another explained that "[his] Excellency, wishing to foster the growth at [La Salina] de Chita, and considering how this town suffered during the era of [Spanish rule], . . . exempted them of all special contributions."[35] This was a small gesture toward recognizing the needs of the community over the immediate balance sheet of the national government, one never repeated in later years. The only official reference to the effect of these measures was a note that the reorganization of Zipaquirá under the administration of Cristóbal Vergara had increased monthly profits from around 5,000 pesos to as much as 16,000 pesos.[36]

The next year delegates met at the Congreso de Cúcuta and passed the "Fundamental law of the Republic of Colombia," formally establishing the nation of Colombia; this united polity, which included present-day Ecuador, Venezuela, Panama, and Colombia, is now dubbed "Gran Colombia" to distinguish it from its smaller successor, Colombia. Naturally, organizing the Finance Ministry was a part of this process.[37] The notable patriot and former vice president José María del Castillo y Rada led the republic's Finance Ministry after overseeing its establishment.[38] In his first annual report, written in 1823, Castillo y Rada offered a grim assessment of the salt rent: "As for the salt administration, there is general confusion. The secretary of my department still does not know how it is run or managed in all the provinces of the Republic. . . . On this subject . . . nobody can provide specific details."[39]

This confusion did not preclude a sense of optimism about the salt monopoly as a source of revenue. In 1823 Congress sought, unsuccessfully, to rent out the works at Zipaquirá to a private contractor as a means of paying off the new nation's foreign debt.[40] That same year Congress passed legislation solidifying the government's control over salt making, thereby claiming the control established by its colonial predecessor. As always, this claim took material form in La Salina. For the first time there were references to an office where administrators oversaw the sale of salt. Nonetheless, residents, who were paid two pesos for each carga of salt they delivered to ministry officials, retained control over daily production.[41] The compromise at the heart of the "old system" had survived into the independence era, though there were proposals that the ministry take direct control over production as well.[42]

The first republican administrator of record was Pedro Ignacio Valderrama, a local landowner and the town's wealthiest resident. In correspondence sent in 1825, Valderrama addressed the possibility that the Finance Ministry might take direct control over the area's salt making. He explained that locals would object to allowing a contract signed in Bogotá set the prices for the timber used for fuel or the ceramics crucial to salt production.[43] If he was prescient in his predictions, it was because he was anticipating what he and his clients would do to defend their control over these supply industries if the ministry exerted direct control over production. His alternative proposal, which was ignored, was that the ministry give the administrator, meaning Valderrama himself, more cash to buy salt and sell it at market prices.

Valderrama was dismissed from his post in 1826, part of a ministerial effort to find entrepreneurs who would rent saltworks for an entire decade, though his clear determination to maintain his position of unchallenged primacy probably played a part in the change. He remained an important figure in La Salina, dominating ceramics and timber production for years to come. The switch to ten-year rentals was intended as a decisive step that, by putting private actors in a position to profit from production and reducing the ministry's involvement, would push salt making to a more market-oriented mode. The rental was also seen as a first step toward abolishing the salt monopoly as a whole.

It took two years for a new contractor to arrive. In 1828 Antonio Malo, a powerful figure from Tunja with national political ties, took a rental contract with the Finance Ministry.[44] Malo's actions and plans echoed Caicedo's and anticipated the changes of the next two decades. Whereas Valderrama had been intent on keeping La Salina isolated and under his personal control, Malo advocated an increased institutional infrastructure and sought to raise the town's profile within the national fiscal network. He had some success in these areas, but neither these advances nor his contract from Bogotá helped him alter who controlled salt making or how salt was made. The ministry refused permission for some of his improvements on the grounds that higher production levels would allow salt from La Salina to intrude on Zipaquirá's legally protected market.[45] During this same period, the ministry sought greater control over the salt market by reintroducing a system whereby its officials issued *guias,* "officially stamped invoices that listed their merchandise."[46] Guias were intended to reduce contraband and to help officials control regional distribution. In La Salina, supplying salt for the ranches of Casanare was always a primary concern.[47]

Malo's rental ended in 1836. During his eight-year term, the contract had been honored despite the breakup of Gran Colombia, a period of dictatorial rule by Bolívar, and the formation of another new nation, the Republic of New Granada, which comprised, as it would for the rest of the century, the present nations of Colombia and Panama. It is a testament to the continuities inherent in this salt administration that the political upheaval had little discernible impact on the monopoly, though the rental system insulated saltworks from political vicissitudes to a degree. On April 22, 1836, the last year of President Santander's term in office, Congress passed a law that tightened the Finance Ministry's power over salt production. Under its direct administration, an *elaborador,* a contractor, controlled production and sold salt to the ministry's salt administration, which then sold the salt to the public for a price that had been set in Bogotá.[48] Most contractors agreed to terms that called for minimum production levels, with fines assessed for underproduction and a higher rate paid for each arroba delivered over contractually set levels; in La Salina 125,000 kilograms a month was a fairly standard rate set in contracts.[49]

The distinction between a contractor and a renter depended in part on technical issues. Legally, rental contracts gave the Finance Ministry a less significant administrative presence, and guard units were less directly involved in preventing contraband.[50] A rental agreement required the renter to make a fixed monthly payment to the government, ensuring a steady stream of revenue. In theory, this would reduce the ministry's worries about drops in sales or other problems. Rental agreements often included a production ceiling, so even wildly successful renters were theoretically limited in what they might earn. Generally, the ministry worked with an elaborador in eras when it had the capacity or inclination to be directly involved with day-to-day affairs in La Salina, when the institutional state felt energetic and expansive. The ministry searched for a renter when it could not manage the works or the politics of the day demanded that the national government reduce its presence. Direct administration was a clear message that the new Colombian state deemed close contact with La Salina to be necessary, appropriate, and feasible.

A Local Institution

Direct administration would significantly affect life in La Salina, though the entire institutional apparatus envisioned by the Finance Ministry would not be in place for some time. The essential elements, however, were installed in 1836. The rental contract in force in La Salina that year expired at midnight on June 9.[51] Ignacio Morales was already subcontracting for the renter Antonio Malo; when he formally became La Salina's first contractor, however, the occasion merited a ceremony that was held at six o'clock in the morning on June 10. Several notable figures participated in this public celebration of the ministry's assumption of control, including Vicente de Azuenaga, representing the "Supreme Government"; Antonio Pompeyo, who acted as an "agent" for Morales; a "rich Bogotano" involved in national politics; and the alcalde Justo Olguin.[52] The administrative staff, which consisted of an administrator who ran the office and a bookkeeper or *almacenista* who oversaw public sales and received salt, began work on June 14.[53] Finally, a guard unit was charged with various matters.

Administrators and the officers in the guard unit received salaries and a percentage of the profits from salt sales. The men who held these positions had to pay a deposit as a guarantee for their conduct and a guard against corruption when they took their posts, with the administrator paying the largest deposit and standing to earn the most.[54] Given this situation, the goals of Finance Ministry employees differed little from those of the private contractors with whom they dealt. This similarity in their goals was rarely noted in official correspondence, though it influenced how relations between the two groups functioned. Positions in the administration were taken by people who could provide a deposit, often individuals who had worked in Zipaquirá or served in the military. Having connections with influential people who could vouch for one's patriotism and moral character was also important in obtaining such a position.[55] This system contributed to local conflicts, for ministry appointments bypassed the networks of influence and opportunity that ran across the eastern highlands. Apparently few men from La Salina or El Cocuy had the wealth or connections to hold these positions after 1836.

The novelty of direct administration and its specific applications in La Salina were evident in one of the first tasks the administration undertook: finding suitable storage space for the salt it planned to warehouse. Although problems with storing salt had certainly presented themselves before direct administration, after June 1836 reports mention problems that caught the administration by surprise. Employees sent notes explaining that as many as six buildings were needed to store salt but that the owners of local structures were reluctant to rent them to the administration, for they believed it would not repair corrosion caused by salt.[56] The first contractor even complained that administration employees were harassing residents in their attempt to find buildings. He claimed that employees had forced "citizens" from their homes and melodramatically speculated that the administration would next turn to churches for salt storage, taking "God from his temple."[57]

When renting a building for the almacén proved tricky, the Finance Ministry turned to construction. With this move the ministry quite literally built the state. A plan from 1839 contains the design for a large building with various rooms arranged around a patio: three storerooms (two measuring eleven by six meters, and the other, sixteen by six and one-half meters), a room for sales, another for the guards, and a forge. The plan also called for paved streets around the building and a lodging house for buyers from out of town.[58] Nothing of the kind was ever built. The almacén, like the rest of the infrastructure in La Salina, was often in bad shape.[59] At times it was in such disrepair that it was in danger of collapsing, and on one occasion a significant portion did fall down.[60] On another occasion the building was so decrepit that, according to employees, people were stealing salt directly from storerooms.[61] Another official, observing that the storeroom was letting in rain, commented that it was "ridiculous to call [the buildings] almacenes."[62]

The other important manifestation of the state that came with direct administration was the guard unit. Various units were charged with protecting the integrity of Finance Ministry rents in the early republic, a reminder that the quest to claim a monopoly on the use of force coincided with immediate economic goals. In the years directly after independence, saltworks were monitored by a single unit operating out of Zipaquirá. These men were charged with patrolling the salt trade and salt production in Cundimarca and Boyacá, an almost impossible task.[63] In 1845, in response to La Salina's growing importance, the Finance Ministry established the "piquete de resguardo de las Salinas de Boyacá," which operated under the authority of the Junta de Hacienda de Boyacá. These guards were charged with maintaining security on local roads and bridges, preventing smugglers or troublemakers from entering town, and acting as an all-purpose security force. On occasion guards also sealed small springs to prevent their illegal exploitation.[64]

Over the next two decades, administrators echoed Ignacio Caicedo's petition of 1806 in requesting more guards. By the late 1840s more than a dozen guards were assigned to the Boyacá unit.[65] By the 1860s, after two decades of unrest during which saltworks constituted prominent military targets, the government began to station regular troops in La Salina. Simultaneously the Finance Ministry expanded the guard unit until it numbered over twenty. A decree organizing the unit in 1866 explained that guards were to "lend a strong hand to the orders of the government."[66] From that point on, though the number of guards varied, it rarely fell below twenty.[67]

As with every other element of the Finance Ministry's presence, the guards brought particular problems with them. For example, the unit operated under an autonomous command structure. Lieutenants stationed at La Salina reported directly to their commander, who was stationed sometimes there and sometimes in Sogamoso or Santa Rosa, rather than to the administrator.[68] This did little to establish the administrator's authority. Further problems stemmed from the unit's composition and the perennial question of whether to hire regional residents as guards. Officials opined that locals who served as guards were more likely to participate in contraband and profiteering than were outsiders.[69] When Castillo y Rada laid out his vision for the ministry in 1826, he specified that guards in such units should be respectable, zealous, "robust, educated, of noble sentiments, and generous."[70] Few records show guards living up to these standards, though in fairness, few accounts show anyone in La Salina living up to Castillo y Rada's expectations.

Of course, the reports tended to mention the guards' failings, not their successes in doing their assigned jobs. The collective record describes poorly armed, poorly paid men charged with fulfilling a difficult and unpopular task.[71] There were regular complaints about guards drinking, thieving, and trying their hands at dealing in contraband.[72] Sometimes these offenses resulted in dismissals. Less frequent, but still notable, were reports of crimes that had nothing to do with the salt industry. These ranged from harassing or attacking locals to sexual assault and murder.[73] As far as I can determine from the record, residents generally disliked guards and at times acted on these sentiments. In one case a guard was murdered.[74]

Mediating between the two groups was difficult. One administrator reported that he had ordered the guards to sleep in the fábrica partly to secure the facility but also to protect them from residents, who "hated them like death." One responded that he was not "such an idiot" that he would obey this order, and he left.[75] Other reports describe fighting among the guards.[76] Overall the guard unit proved a double-edged sword for the Finance Ministry. It contributed to the general atmosphere of instability around the town, but like every other element of the industry, whether bureaucratic or material, it was seen, symbolically and practically, as indispensable to the business of controlling salt production and marketing. The springs were to be walled in, the sale of salt was to be controlled by ministry employees, and as much of the area as possible should be overseen by armed agents of the state.

Along with the people who took formal positions with the Finance Ministry came another, informal group, the entrepreneurs who succeeded Antonio Malo. The most notable, though not the most typical, were the Espinosa family. Both Bonifacio Espinosa and his son Honorato Espinosa served as contractors in La Salina. Bonifacio's other sons also worked in La Salina, and one, Bernardo Espinosa, held positions in the Finance Ministry and published a newspaper in Bogotá.[77] In a marked contrast to the other important contractors who had success in La Salina, the Espinosas seem never to have controlled, either as renters or owners, the forested lands around town, typically a crucial element in wringing profits from a contract. Working without obvious control over these forests, the Espinosas used contacts, expertise, and ruthlessness to achieve a level of success that only a few others equaled in La Salina.

Bonifacio Espinosa came to La Salina with a contract to run the mail between La Salina and the town of El Cocuy just as direct administration began, in 1836.[78] He also served in the almacén in 1836 and 1837, until he was replaced by Juan Nepomuceno Solano, whose own career highlights include credible accusations of corruption and charges that he absconded with funds during a rebellion.[79] Following that, Espinosa largely disappears from the record in La Salina until 1845, when he took a production contract. In 1847 the Finance Ministry granted an extension on this contract that was to run until 1852, but complaints about his tenure led to its revocation in 1850.

Part of the problem seems to have been the way Espinosa treated people. In 1846 the administrator wrote that though Espinosa managed his workers well, his attitude toward officials was disquieting: "One of the things that is most troubling . . . is his custom of not answering questions no matter how important or urgent they may be."[80] The next two years brought a stream of similar notes: Espinosa did not fulfill his contractual obligation to plug unused springs or to carry out minor repairs; he instructed his workers to bring salt to the scale in the almacén but not to help move it into the warehouse after weighing; he had not repaid a loan of 6,000 pesos.[81] In 1847 the town *cabildo* (i.e., the municipal council) wrote with a number of complaints, stressing that his low payments for wood might prompt workers to leave the town.[82] Two years later the administrator reported that Espinosa had still not repaid the 6,000 peso loan, that the salt he produced was low in quality, and that workers had in fact fled because of the poor treatment they received at his hands.[83]

Nonetheless, although the complaints about Espinosa's actions were a factor, the Finance Ministry revoked his contract primarily because it was seeking better terms. Following a public invitation for potential contractors to submit proposals, the contract was awarded to Mariano Sáenz de Santamaría, whose bid was chosen over those of four other applicants, including one from Honorato Espinosa and his brothers.[84] Whether Bonifacio Espinosa's behavior in La Salina worked against the bid submitted by his children is not clear, because Santamaría promised to produce more salt for less money than any of his competitors. He also promised to extend the industrial infrastructure in La Salina in an unrealistic fashion and to move the almacén to the town of Chita. The latter project included overseeing improvements on the road linking La Salina and Chita, for 100,000 kilograms of salt would need to be moved from one to the other at the expense of the administration. Santamaría defaulted on this contract in less than a year. By March 1852 the *Gaceta Oficial* again published a notice that the ministry was accepting new production proposals.[85] Given the terms of his contract, Santamaría would probably have failed sooner or later, but he encountered more problems than just the ones he created with his unrealistic proposals. Soon after the new contract was announced, a suspicious fire broke out in the woods near the fábrica, a vital resource for cooking salt. There is no proof, but it is no stretch to imagine that the Espinosas or one of their clients committed arson. Bearing out Valderrama's earlier warnings about local responses to greater ministry involvement, any new actor without local support encountered opposition that was hard to see from Bogotá.

Honorato submitted a contract proposal in 1856 and, together with his brother Bernardo, another in 1858, but neither was accepted.[86] For almost a decade the Espinosas did not hold a significant position in La Salina. In this decade of administrative chaos, the Reyes y Niño Company came to prominence. In the late 1860s, when the Finance Ministry was able to bring some order to the administrative side of the industry in La Salina, Honorato Espinosa took up the position of contractor.[87] His tenure replicated his father's. A steady stream of complaints ensued, particularly from locals who felt he mistreated them as a matter of policy. I discuss the specifics of these practices later, but the significant point in this context is that he succeeded after a number of other contactors had failed. In a fashion unmatched by any other individual, family, or company, the Espinosas were able to set the terms for the industry. Other contractors came and went, sometimes succeeding and sometimes failing at the tricky business of salt making, but the Espinosas, who arrived at the establishment of direct administration, usually succeeded. In this they were, in a sense, their own institution.

La Salina's Declining Economy

Finance Ministry employees commonly reported that La Salina was virtually an empty town, one filled only with migrant workers and dealers in contraband. This profile was trotted out as an explanation whenever troubles involving production and sales arose. Even regional officials made these claims. In 1843 an official from El Cocuy wrote claiming that La Salina had only twenty permanent residents, the remaining population consisting of a "continual concurrence of people" there to work or trade.[88] Belying such comments are estimates that put the population of the town at around one thousand (see table 2.1).

Though hardly vacant, then, La Salina was a small town. More significantly, its population did not increase over the course of the century, even by the standards of Boyacá, which was one of Colombia's slowest-growing regions.[89] In his statistical survey of 1875, Anibal Galindo reported that between 1851 and 1875 Boyacá's population increased by 29 percent, while La Salina's population, moving from 1,150 to 1,252, increased by less than 9 percent, the smallest growth in the region.[90] This stagnation reflected a transient population; residents were as likely to leave the town as they were to settle there. References to out-migration appeared frequently, though generally in comments about the threat that depopulation posed to maintaining a viable workforce. For example, when Bonifacio Espinosa, who was not sympathetic to the daily concerns of his workers or residents, warned that limiting common access to timber would ensure the town's depopulation, he was thinking about the laborers he needed to fulfill his contract.[91]

TABLE 2.1. *Population estimates for La Salina*

Year	*Population*	*Year*	*Population*
1816[a]	900	1855[e]	1,150
1835[b]	637	1869[f]	600–1,000
1843[b]	1,032	1871[b]	1,250
1850[c]	1,230	1881[g]	1,552
1851[d]	1,150	1905[b]	1,252

[a] Tovar Pinzón and Tovar Mora, *Convocatoria al poder del número,* 383.
[b] Contraloría General de la República de Colombia, *Geografía económica de Colombia,* 3:125, 130, 145, 150 (the figure for 1835 and 1843 is repeated for 1848 in AGN, Archivo Anexo, caja 23, carpeta 5, 1848).
[c] Agustín Codazzi, *Geografía física y política,* 3:306.
[d] Oficina de Estadística Nacional de Colombia, *Anuario estadística de Colombia 1875,* 33.
[e] AGN, SR, Gobernación de Tundama, leg. 5, fol. 849, March 1855.
[f] AGN, SR, Salinas de Boyacá, leg. 10, fol. 375, November 1869.
[g] *El Boyacense,* June 17, 1881, 6845-47.

Low wages and, in the second half of the century, fear of military impressments were cited as the chief spurs for out-migration. In 1874 one official explained that the current shortage of workers resulted from "the high cost of coffee being grown in Casanare," which had "increased the cost of wages to an exorbitant level."[92] Casanare also figures in an 1891 report where an employee wrote, "Without a doubt many of the laborers needed in production have immigrated to Casanare or any other place where they can be free of impressments."[93] At the same time, the wages offered in La Salina and more generally the prospect of gaining something from the salt market were clear attractions for workers from throughout the region. Many probably worked there for short periods of time or for a season, but some settled in La Salina. Writing in 1881, the administrator Santiago Izquierdo expressed hope that a new wave of migrants might provide the foundation of a permanent settlement.[94] These comments suggest that La Salina simultaneously lost and gained residents. The town may also have served as a conduit for migration from the highlands of Boyacá to the llanos of Casanare, a way station for earning money before moving on. Such a profile shows La Salina's population to have been not stagnant but rather stable in a kind of dynamic equilibrium.

More concrete than these reports about overall quantities were complaints that La Salina was losing, or would soon loose, its "better sort." This group, somewhere above salt makers and day laborers but below the elites with political connections who held contracts, dwindled after 1836. The cabildo's 1847 complaint about Espinosa was quite specific on this point:

> Since Bonifacio Espinosa set foot here we have seen a multitude of afflictions in this poor town. In the first place there has been a complete emigration: Jesús Urrea, Rafael Estaba, José María Alvarez, Vicente Salinas, Nepomuceno Valderrama, Joaquín Murillo, Sixto Olguin, Juan de la Cruz Heras, Eustaquio Rincón, Espiritu Santo Vargas, Justa Franco and Juana Sandoval. All of these people, the leaders of this town, [have left] with their families, and now there remain only four employees, who wish to leave as they cannot bear the continue harassment of the contractor. The same people of the town are leaving daily.[95]

An alcalde's report two decades later portrays a town without this population segment and enumerates the attendant problems:

> As they are so poor, residents only barely manage to miserably support themselves. They have no property, and the only commerce known in the district is that established by the government in the saltworks. . . . [T]hose who have [some wealth] . . . carry the burden of sustaining a primary school, three secretaries, the cabildo, the judiciary, and the alcaldia. [They must also] maintain in good condition the jail, the school building, [as well as] the mail, and other costs.[96]

In this version, the "better sorts" disappeared in La Salina, but accounts of public disputes sometimes referred to "influential people" who either calmed or inflamed the gathering crowd. It remains unclear, however, whether this newer, unseen middle sector came from established families or comprised newer residents making a living from the salt industry.

Less murky than the status of this middle sector was the economic decline of La Salina in absolute terms. At the time of independence it had been, by the standards of Boyacá, a town of some means. The special contribution of 1823 (see table 1.2) provides a benchmark for measuring the ensuing eight-decade decline. Under the 1823 tax, fifty-nine residents of La Salina paid a total of 204 pesos in taxes. The average payment of 3.5 pesos was higher than the 2.7 peso average payment for both the province of Boyacá and the canton of El Cocuy.[97] In part this high average resulted from Pedro Ignacio Valderrama's personal fortune. His property in La Salina alone was assessed at 12,000 pesos.[98] But there were fifty-eight other residents wealthy enough to be taxed. The ratio between the number of residents taxed and total population shows that the percentage of people with wealth worth taxing was greater in La Salina than in any other town in Boyacá (see table 2.2; the population figure used here, however, comes from 1835, which makes the claim a bit speculative).

According to these figures, almost one-tenth of La Salina's residents had personal wealth worth taxing. While there are no details, individuals in this sizable middle sector undoubtedly earned their livings from work related to salt making or the salt market in one fashion or another. Beginning early in the century, compelling evidence suggests that the town possessed little in the way of economic activity apart from salt making and its related endeavors. An 1816 survey of local industry reported that La Salina, with a population of 900, had 190 houses, 20 horses, and 60 cows, but no goats or sheep, and no maize or wheat was cultivated.[99] Anyone who had money had gained it through salt: making it, trading it, or working in one of the industries that supported its production.

Various reports and petitions from La Salina after 1836 document the increasing competition for resources relating to these industries and the resulting impact on local life. In one instance, just as direct administration was being implemented, the vecinos of La Salina—children and grandchildren of the group that had purchased the rights to the common lands in 1778—entered a dispute with Pedro Ignacio Valderrama over the town's common lands. In 1836 Valderrama, who controlled land in the *vereda* (rural barrio) of Rodrigoque, purchased most of what had been the resguardo from Trinidad Alarcón.[100] The petitioners, who claimed to be the rightful owners of the land, turned to the government for help. They sought an alliance that would benefit both the state and the community, though not without setting certain terms:

> With the following conditions, [the government] will give the community the complete area of population . . . (being less important than the most insignifi-

TABLE 2.2. *The contribution of 1823 and population in 1835*

Town	*Population*	*Residents taxed*	*Percentage taxed*	*Total town payments*[a]	*Average individual payment*[b]
La Salina[c]	637	59	9.0	204	3 p 4 r
Sativa Sur	649	10	1.5	31	3 p 1 r
Covarachía	1,251	11	0.9	38	2 p 4 r
Iza	1,130	18	1.6	33	1 p 6 r
Cuítiva	297	4	0.3	6	1 p 4 r
Mongui	921	8	0.9	8	1 p
Tópaga	1,041	12	1.2	23	1 p 7 r
Güicán	943	22	2.3	87	4 p
Espino	1,240	15	1.2	34	2 p 6 r
Zetaquirá	907	26	2.9	60	2 p 2 r
Teguas	764	26	3.4	38	1 p 4 r
Canton of El Cocuy	20,220	293	1.5	795	2 p 6 r
Province of Boyacá	143,115	2,576	1.8	6,928	2 p 6 r

Sources: AGN, SR, Intendencia de Boyacá, leg. 2, fols. 454–632, June–Aug. 1823; Agustin Codazzí, *Geografía física y política,* 3:306; census of 1835.

[a] In pesos

[b] In pesos and reales.

[c] La Salina's 1823 population might have been closer to 900 than to 637, the lowest figure given for the century, but in either case, La Salina returned a substantial per-capita sum for this tax. I use the later, lower figure on the grounds that the town's population often dropped during periods of extended warfare.

> cant *resguardo de indijenas*) and permission to cut wood for construction, and permit residents to have small gardens, which Valderrama now forbids. [They] declare they will never take any action to obstruct the interests of the government as is proved by their liberal and spontaneous gift. Reflecting, over time, the advantages for the state in [controlling] the land with its hills and woods that supply salt production instead of leaving them in the hands of [these] monopolists, [Valderrama and Alarcón,] . . . who have only tried to devour and take from the community that has given them such wealth.[101]

There is no record of any response to this offer, and three of the petitioners were later named in the list of those who had left La Salina in 1847, but the timing could hardly have been worse. The proposal was sent at the very moment when the republican state was moving against the sort of communally owned lands these residents were attempting to preserve. The effort to divide and distribute the indigenous population's remaining common lands in the eastern highlands had strong support from both the national and provincial governments. Antonio Flores, the governor of Boyacá at the time, expressed his frustration with delays in the effort and blamed "the negligence of some minor officials, the personal interests of certain people, and the worries and stupidities of the Indians";

he added, "As I am persuaded of the advantage to the public good of this law, I have made carrying it out a priority."[102] Though the lands in question in La Salina were not a resguardo, the vecinos' petition went against the spirit of the day. The community's effort to maintain access to land and resources, crucial elements in staving off poverty in La Salina, was doomed by the logic employed in establishing direct administration.

Correspondence from Diego Davidson, who was the first manager employed by Ignacio Morales, documents both the declining local standard of living and the role of outsiders who ignored or denied the reality of local life. Writing in 1836, Davidson claimed that direct administration had stimulated the local economy, that the new construction projects had meant jobs, and that the rising prices for foodstuffs benefited locals.[103] This last claim would have been true if residents in La Salina had grown their own food or sold produce locally, but they depended on traders who brought foodstuffs to the town.[104] In 1839 a letter to the Interior Ministry explained, "In the parish of La Salina nothing is produced, and no human hand cultivates. Everything is brought in from outside and therefore can be obtained only at very high prices. The town's population has now been reduced to employees of the administration and workers in salt production."[105] Any increase in local prices hit residents hard.

A litany of reports on local life confirms the portrait of a monoeconomy that did little to provide for the majority of the population.[106] Provincial figures for *primicias* (an offering to the church based on crops that can serve as a rough indicator of agricultural production) in 1843 show that contributions in La Salina totaled forty pesos, the lowest sum for any town in the province.[107] Agustín Codazzi's livestock survey from the 1850s reported almost no animals in La Salina, and only five mules, a remarkable absence considering how crucial they were for transporting salt (see table 2.3).

In a special report in the *Memoria* of 1859, Vicente Gomez explained, "In La Salina there is no agriculture."[108] There is no mention of La Salina in Anibal Galindo's 1875 statistical survey of agricultural production, suggesting that it had no production worth recording.[109] A provincial livestock survey from 1883 offers a final example. There were only thirty-six cows and fifty mules in La Salina, but no horses, sheep, goats, or pigs. Even with the assessment of these fifty mules (finally making an appearance) as worth 3,750 pesos, the total value of livestock in La Salina was surpassed in all but three of the forty towns surveyed.[110]

Various factors contributed to this poverty. Manuel Ancízar reported on the practice of limiting residents' use of wood for any purpose. Timber was reserved for salt making, and Ancizár explained that even a simple house fire was viewed as evidence of illicit salt making.[111] When the Finance Ministry purchased the "bosques de la Iglesia" in the 1840s, it followed and then surpassed Valderrama's policy of limiting access to the woods and prohibiting agriculture. The administrator at the time of the purchase wrote that one renter who owned some cattle should be permitted the continued use of his plot but that another with a small

TABLE 2.3. *Livestock in La Salina and comparable towns,a canton of Cocuy, 1859*

			Livestock					
	Population	*Altitude (in km)*	*Cows*	*Sheep*	*Goats*	*Pigs*	*Horses*	*Mules*
La Salina	1,230	1.440	20	0	0	1	5	5
Espino	1,897	1.995	100	250	200	25	50	40
Panqueba	1,112	2.258	100	400	1,000	50	25	10

Source: Codazzi, *Geografía física y política,* 3:306.
[a] Comparable to La Salina in size and elevation.

farm on the same land should be allowed only one final harvest, adding, "in the future all labor that interferes with the regrowth of the forest should be prohibited."[112] One provincial official summed up the result: "The residents are so poor that they barely manage to eke out their miserable existence. They have no property, and the only profession in the district is [the salt making] that the government has established. He who does not get up at dawn to carry [wood] does not eat lunch."[113] Of course, the wages to be earned carrying wood were relatively low when compared to the high cost of living in La Salina: "Goods have become so scarce, and they are so costly, that the unhappy laborer works a week and hardly earns enough to satisfy his needs for two or three days[; workers] . . . have gone to other places where the cost of living is not so high, leaving the industry without the labor which is necessary for its existence."[114] Records of tax levies from later in the century show that these reports were not merely anecdotes (see table 2.4). By 1869 La Salina's returns were slightly below the regional average, and by 1898 the decline had become even more pronounced.

While there are no figures on income distribution, the percentage of La Salina residents with wealth sufficient to tax in the latter part of the century would likely compare unfavorably to the almost 10 percent of the population that fell into this category in the early 1820s. In seventy-five years La Salina had been transformed from a town where some of the wealth from its remarkable resources circulated among the general population to one of the poorest in Boyacá, itself a relatively poor province.

The same dynamic of impoverishment was in play in Zipaquirá and Chámeza, other places that housed Finance Ministry saltworks. Until the seventeenth century Zipaquirá had been a prosperous pueblo de indios with a stable population, but by the end of the eighteenth century, that indigenous community was impoverished; it had resorted to renting pieces of the resguardo but still had trouble paying its tribute.[115] Eventually the town could not afford even the timber and ceramics necessary for production.[116] Thus, even though Zipaquirá had legal access to a source of great wealth, the increasingly heavy hand of the royal treasury and other factors made communal ownership and management untenable. As in La Salina, the indigenous community was removed from its res-

TABLE 2.4. *Tax levies in La Salina (in pesos)*

	Levy of 1869		*Levy of 1898*[a]	
Town	*Total wealth*	*Taxable assessment*	*Total wealth*	*Total taxable assessment*
Corrales	80,000	100	262,460	1,180.07
Chita	98,415	525	879,600	3,958.20
Espino	27,500	185	303,800	1,367.10
Güican	90,911	314	689,020	3,100.59
La Salina	79,996	330	111,325	500.96
Panqueba	52,900	195	317,120	1,427.04
Sativa Sur	53,680	119	190,050	855.22
Zetaquirá	32,075	150	164,300	739.35
Cantonal average	69,532	363.70	521,918	2,348.60

Sources: Memoria de Presidente de Boyacá de 1869, 50–51; *Memoria de Ministerio de Gobierno 1898,* "cuadros."
Note: For this table I have used the towns included in the previous tables. The borders of the canton containing La Salina was redrawn in between these two levies. In 1869 it was the Canton del Norte, while in 1898 it was a slightly different area called the Canton of Gutiérrez. I have included the towns that were in the legally recognized canton at the time of the assessment in the averages.
[a] The higher overall figures from 1898 reflect the inflation of the era.

guardo during the resettlement campaign of 1778 and settled on the resguardo of Nemocón, a smaller salt-producing town nearby.[117]

Chámeza, which like La Salina lay between the highlands of Boyacá and the plains of Casanare, had a saltworks in the vereda of Guativita. A relatively small affair, the town was not targeted during the resettlement campaign of the 1770s.[118] At that time, officials reported no abuse of the indigenous population, which consisted of seventy-six households, by the local vecinos, who accounted for thirty households. Chámeza's economy was more varied than La Salina's, with typical highland agriculture and textile production (making *ruanas* [ponchos] and *mantas* [blankets]) accompanying salt making as prominent local industries.[119] In 1826 the Finance Ministry extended its involvement in Guativita with minimal impact on local life. Only after the division of the indigenous common lands, the *repartimiento de resguardo,* in the 1830s was the old order of life destroyed. In his 1833 report, Francisco Soto noted the decline of the local economy, while a letter to the governor of Tunja explicitly linked the land division to the existence of the saltworks: "It is not a coincidence that in the parish of Chámeza one of the results of the exploiting of the salt water springs was the removal of the indigenous population."[120] The former owners of the resguardo squatted on nearby *baldios* (state lands considered vacant), where they were harassed by the contractor. Later descriptions of Chámeza report no agriculture or textile production, explaining that residents worked in production, trafficked in contraband, or generally made trouble. During the 1840s production ceased in Chámeza altogether.[121] In 1858 the "Indians of Chámeza" were still battling the ministry contractor over the legal rights to their former resguardo, apparently without success.[122] When a saltworks was incorporated into the national fiscal system and communal land—whether owned by indios or by vecinos—was divided, the results were always the same: an economic decline. It happened earlier in Zipaquirá, because of its size and proximity to Bogotá, and later in Chámeza, but it was the same process.

The Community after 1836

While La Salina's economy declined markedly after the implementation of direct administration in 1836, the Finance Ministry employees' frequent comments that the town was inhabited only by transient laborers and shady operators were demonstrably false. For example, numerous documents refer to schools and to women's roles in local markets and as defenders of traditional gleaning practices.[123] While the references to schools are of interest because many small towns in the region lacked them, the references to women's involvement in salt making are another example of the way state building changed local life. Direct administration brought together discourses of patriarchal sovereignty and modernization, a move that simultaneously marginalized women's local economic privileges and elided the fact that women had previously enjoyed these privileges. In fact, women frequently participated in public disputes over land and marketing in the early decades of the nineteenth century. For example, more than a dozen women signed the petition to the government to take control over the town's common lands.[124] Of course, these women lost their rights to this land, and the attendant benefits, with the rise of Valderrama. The similar waves of dispossession that followed tended to run along the same lines, and the new landowners were always men. While the assault on women's traditional rights was a hallmark of republican Latin America, the concentrated effort on state building in La Salina accelerated the process.[125]

The changes that marginalized women in part reflected the transition of salt making from an occasional cottage industry, or at least a communal activity done by families, to a process centered completely in the fábrica.[126] While there are only passing references to salt-making methods before direct administration, references to a gleaning practice, the "repartimiento de agua salada," clearly indicate that women were involved in, and may have controlled, certain aspects of production.[127] Under this repartimiento, or apportionment, residents were allowed a share of water from the springs, ostensibly for household use.[128] This practice was a legacy from the era of communal access that residents defended vigorously. Almacén employees generally opposed the practice because it violated the spirit of the ministry's absolute control, but on occasion they supported it on humanitarian or practical grounds.

A review of the challenges made to this practice profiles the beleaguered survival of communal rights. The first complaints about the water share came from Diego Davidson, who oversaw production after the implementation of direct administration. In an appeal to the Finance Ministry, he argued that the practice violated a contractual clause granting him, as the contractor's representative, full control over all elements of production in La Salina. The judge in Tunja who heard the case ruled against him.[129] Despite ministry rhetoric and the larger issue of absolute property rights (a touchstone for those with visions of a market society) versus communal privileges, this court was reluctant to abandon traditional practice.

The next documented challenge to water sharing came in the town rather than in court. In 1841 a new administrator reported his surprise at seeing residents pass through the center of town carrying containers full of brine they had taken from the springs. He ordered the alcalde and guards to seize the containers and destroy them. Residents, including "the wife of the alcalde Isidoro Camino, who with her daughters live[d] separated from him," complained that they could not afford to buy salt.[130] The episode hints at a tumultuous world, a society where domestic disputes flowed into broader social conflict. It is also a rare documentation of a Finance Ministry employee exercising clear authority over a local government official. There was no immediate report on subsequent challenges to the practice in the 1840s, but during the 1850s and 1860s residents defended their interests with great success.

A series of notes in 1863 from the administrator Vicente Vargas and Medardo Cuadra document this repartimiento's difficult survival. Vargas wrote to the Finance Ministry to protest Cuadra's practice of forcing people to carry timber in return for their weekly allotment of brine.[131] At one time Cuadra was named as a contrabandista, and at another, as the leader of a threatening crowd outside the almacén, but at this moment he was the agent for Reyes y Niño, in which capacity he had transformed the water share into a privilege granted by the contractor. He defended this action by claiming, inaccurately, that the practice had been abolished in Zipaquirá.[132] He then echoed Diego Davidson, arguing that the production contract superseded residents' rights, and dismissed Vargas's complaints:

> Though your attempts to help the inhabitants of this town are noble, I must limit my applause. I cannot allow that a favor granted *by the contractors* to the laborers as an incentive for working in this business be converted into a source of conflict and the ruin of this business.
>
> An official who has been in this saltworks only a few days and who has more than enough work to occupy him cannot possibly know the inhabitants of this district or know the size of each family.[133]

In his view, not only had Vargas intruded on the contractor's rights, but his ignorance of local society and willingness to allot each resident two arrobas of water a week would have dire implications for the industry. If a family of five received ten arrobas—and various reports pointed out that residents pressed their full claims in such matters—they would use their share to cook two arrobas (twenty-five kilograms) of salt. Cuadra estimated that each family would have at least an arroba of salt to sell. Because the price of a single arroba equaled a week's wages, no one would work, and the industry would fail. Worse, officials such as Vargas would be duped by exaggerated claims of family size, and illegal salt would flood the market. Vargas replied that the water share had survived in Zipaquirá. Moreover, the share helped make local conditions bearable and therefore ensure a future workforce, which was more important than forestalling the threat of illegal production.[134] While Vargas's outlook was not common, it clearly had some adherents, because the water share endured.

Instructions in 1869 on amounts of salt residents should be permitted to buy at the almacén mentioned that they were "permitted" salt water.[135] In 1873 twenty-thee residents sent a petition to the Finance Ministry protesting the actions of the administrator, complaining, among other things, that they were mistreated when collecting their weekly share of water, "as [had] been the custom."[136] The administrator defended his actions on the grounds that he was not harassing people in this manner on Sunday, the traditional market day and, in his view, the proper time for such activities.[137] In 1885 an official listing the problems of La Salina explained, "In addition, there is a long established custom in La Salina de Chita that leads to contraband. Every Sunday they open up the springs to the residents so they may take pots of water for the week, but everyone takes as much water as they can to make it into caked salt."[138] Despite all the challenges mounted, there is no reason to think that the gleaning practice was ever stopped.

Other topics suggest that the history of local society in La Salina was not simply one of decline or maintaining a threatened status quo. Most notably, at the end of the century the Finance Ministry eased its restrictions on local agriculture by renting out plots located on its land.[139] The process of reversing a half-century of guarding land began in 1889 when the administrator Jorge Gaitán requested permission to rent land to squatters, revealing that the prohibition on local use was not quite absolute. Apparently the administration's view on residency and local farming had changed. The squatters-turned-renters remained in residence without challenge for the next three years; then, in 1892, when the administrator General Terán sent the ministry a list of the squatters, he argued that they were overrunning a nearby forest that he called "the hope of this salt works." Terán was also concerned about clashes between traders, who let their pack animals wander and graze freely while they were in town, and squatters, who were cultivating more land.[140] He recommended either allowing these people to use land already cleared or opening up land in Sácama and suggested stipulating that renters agree to raise specific crops, including coffee.

Working in the opposite direction, the squatters pushed for legal residence of the land they inhabited. With the backing of the parish priest (making a rare appearance in the record), they petitioned for and received permission to stay. In 1893 the Finance Ministry began to formally rent plots at fifty centavos a year for 100 square meters in the town center, with the condition that no one interfere with growing trees. Renters who did not cultivate their plots were to be evicted, a full reversal of the land-use policies implemented after direct administration. Throughout the next year, even though most of the rentals were registered in July and August, 178 people (including 18 women), rented 252 different pieces of land that, on average, measured 125 square meters.[141] In terms of the business done in La Salina, the rentals were a minor affair. To get a sense of proportion, the same year that the administration rented 31,424 square meters of land for a total of 137.65 pesos, the almacén brought in 164,496 pesos of gross revenue. But these minor transactions were a dramatic reversal after decades of banning any agriculture. By 1897, however, a letter from an administrator mentioned that 98 people were renting plots. The decline may have reflected a lack of enthusiasm on the part of officials, disappointment on the part of the renters, or the impact of the 1895 civil war. This was life in La Salina, a town showing more impulse toward change than did many others in the nineteenth century. It was a place where old traditions endured, if under siege, and at times life turned full circle. This fluctuating community, this small and increasingly impoverished town, was the setting where the Finance Ministry oversaw the construction of an unwieldy infrastructure in the hope of controlling the workings of a regional salt market. It had been made and remade by ministry policy, but a community of some sort endured.

It is not really possible to conceptualize this town without reference to the salt industry. The problem lies in the temptation to do the opposite, to see only a saltworks and let the community disappear from view. Certainly the voluminous correspondence generated in the almacén and sent to Bogotá emphasizes this perspective. But people lived in this town, even if they did so in a dysfunctional symbiosis with the ministry's saltworks. Through these people, this community, the ministry's project touched society and made history outside the confines of an institutional bureaucracy.

CHAPTER 3

Making Salt in a Ministry Works

Salt is now a humble substance. Industrial advances in drilling and chemistry have long rendered salt easy to make and cheap to buy. If anything, we now consume too much salt rather than too little. This was not the case for most of human history, during which salt ranged from relatively expensive to prohibitively costly. Thus, the contemporary experience of "sticker shock" at the price of a small bottle of hand-crafted gourmet salt has more historical resonance than do all the prepared foods whose high salinity threatens consumers' health. Nonetheless, both the high cost of salt in the past and its preponderance in the present are based on the same thing: humanity's universal taste for the substance.

As a species we cultivate a craving for salt that surpasses our biological needs. Beyond that, salt is important for preserving meat and fattening cattle. As a result, salt production forms a long, winding thread in the tapestry of human history. It is rarely visible at first glance, but a careful examination shows that it is a constant presence. If pulled, the thread of salt production unravels whole sections of trade, taste, infrastructure, and the state. Salt was carried in caravans across the Sahara Desert, along China's Grand Canal, between the city-states of Mesoamerica before the Conquest, and across northern South America in the region that would become Colombia. These histories of premodern industry and long-distance trade flow into the history of state monopolies, as governments including republican Venice, prerevolutionary France, and imperial China struggled for control over salt trades. In general, regimes sought uniformity as emergent bureaucracies worked to homogenize production and discourage local initiatives that impeded standardization.[1] In counterpoint to this tendency toward homogeneity, salt making is quintessentially a product of local knowledge developed with reference to specific conditions. Different saltworks tend to feature different techniques, or at least local variations on certain established practices. La Salina was both typical in the general scheme of salt production and distinct as a site where salt making was based on unique local conditions.

Cooking Salt

Making salt at the works in La Salina involved an updated version of the method established by the region's pre-Columbian indigenous inhabitants. Production techniques changed over the nineteenth century, but anyone who had made salt in the preceding centuries, including the indigenous Lache inhabitants, would likely have recognized the process.[2] Salt water was collected from the thermal springs, given an initial boiling to remove impurities, and then slowly heated in ceramic containers similar to those used across the Americas in producing sugar, rounded cones with a flattened end. When the salt was completely crystallized, it formed a hard cake in the shape of the container; the result, a salt without appreciable iodine that could be used for fattening cattle, was called *sal compactada,* or caked salt.

This millennial pursuit of salt was possible because nature had endowed Colombia's eastern highlands with banks of subterranean salt. Whereas Zipaquirá had been transformed into a mine in the colonial era, La Salina remained a surface works.[3] The springs generally provided enough brine for production. Various inventories listed a dozen or more springs in and around the town.[4] The varying estimates for the number of springs may have reflected the changing criteria employed by those doing the counting, but they also reflected physical changes. Any number of factors, including seismic activity, shifts in the Casanare riverbed, or even an ill-advised construction project, could diminish a spring's capacity or produce a new vertiente.[5] Generally, in La Salina the springs were near the banks of the Casanare, which flowed to the east, or in the vereda of Córdoba on the banks of the Chinivaque River, which flowed into the Casanare east of the town.

For all the many lists of multiple springs, the industry was based on only a few. Early reports mention a main well in the center of La Salina, called either Bravo, Pozo Grande, or simply the Pozo Principal. In Córdoba the main spring was called la Paloma.[6] As explained by Ignacio Caicedo, the main well was almost useless during the winter rains, which fell from March until July, when the flooded Casanare would rise above its banks, diluting the springs and ending production in the town center for four months. La Paloma was better situated, so salt was made in Córdoba all year, though that operation was more or less independent from the works of La Salina until 1836. One of the infrastructure improvements introduced by direct administration was a new road that linked

Córdoba to the town center, bringing it, theoretically, under the authority of Finance Ministry officials.[7]

In 1849 the administrator in La Salina wrote to the Finance Ministry explaining that the water in the old spring had diminished and that a new one, the Pozo Azul, should be used.[8] While there were no other reports that the main well was failing, work continued on this project intermittently, but not until 1869 was the Pozo Azul successfully incorporated into the production system.[9] At the end of the century there were attempts to exploit the Pozo Santa Bárbara, in the town center, and a second spring was used in Córdoba.[10]

Whereas Caicedo's proposal for a protective wall had gone nowhere, and Malo's petition to do the same in 1830 was denied, such a wall became a standard element in salt production after the implementation of direct administration in 1836.[11] Though these walls never quite matched the structure imagined by Caicedo, they were still impressive. An 1854 inventory measured the wall in La Salina's town center at thirteen meters long and the wall in Córdoba, at eight meters.[12] Of course, building a wall was only the beginning of the project, for one side was constantly buffeted by rushing water. Proposals for exploiting new springs always factored in the cost and trouble of both building and maintaining a new wall.[13]

Concurrent with the introduction of these walls was the practice of building larger wells or tanks to hold water collected from the spring. Caicedo explained that the spring emerged from a "rocky outcrop" and was collected in a "little well" near the river.[14] His drawing shows men with buckets kneeling to collect the water. Over time this open structure was converted into an enclosed affair. By midcentury such reservoirs had become structures where large quantities of water were held before being pumped through open channels or pipes into cauldrons for boiling.[15]

La Salina's abundant springs yielded water with a saline content ranging from 16 percent to 22 percent and a temperature ranging between twenty-five and thirty-six degrees Celsius. These factors made for relatively easy salt making. At least one observer reported that salt crystals formed naturally in the water.[16] Ancízar recorded the most detailed description of a spring in La Salina: "The saline springs emerge very close to the edge of the Casanare River so that they are submerged with the smallest flood. They are all thermal with, more or less, the same level of saturation. At specific spots nestled at the bottom of the hills a crystalline water flows out slowly." He also explained that the Tunebo Indians valued some of the crystals that accumulated around the spring for their medicinal value, the only reference to medicinal salts from La Salina I have found.[17]

On the rare occasions when there wasn't enough water, officials explained that a poorly planned oven had been built too close to a spring and that the radiating heat had affected the water supply.[18] In addition, officials sometimes observed that there was a problem with the manner of collecting water. For example, in 1895 Clímaco Silva declared that the current system for collecting water was a remnant of "primitive times," one that permitted "a great quantity of water to escape." By his estimate, in Córdoba up to a third of the water was lost when transporting it from the well to the ovens on the other side of the Chinivaque River.[19] While the system was probably inefficient, there is no evidence that more water was needed.[20] Although most other elements of salt production, such as wood for fires, were stretched to their limit, La Salina had all the brine it needed; in fact, it is not clear whether the infrastructure there could have handled any more.

Before direct administration, water was carried from the wells to the fábrica. After 1836 a system of pipes and *canoas,* open channels of wood or leather, was put in place. The complex network of tanks and channels that developed cut down on the labor necessary for moving water but did little to improve efficiency.[21] Once water was brought to the fábrica, it was boiled. Originally this reduction was done in kettles, but over time iron pans may have been introduced.[22] Of the very few references to these kettles or pans, only one stands out. In 1874 the Finance Ministry instructed a representative, Enrique Cortés, a resident of Seething Lane in London, to purchase larger iron pans and oversee their shipment to Colombia, yet the first record of such a transaction, the purchase of these items from the West India and Pacific Steam Ship Company of Liverpool and arrangements to ship them across the Atlantic, did not appear until 1881.[23] I have found no reference to their use in La Salina. The initial boiling, whether in kettles or pans, increased the brine's salinity until crystals formed. This process also isolated impurities in the brine, allowing them to be skimmed off.[24]

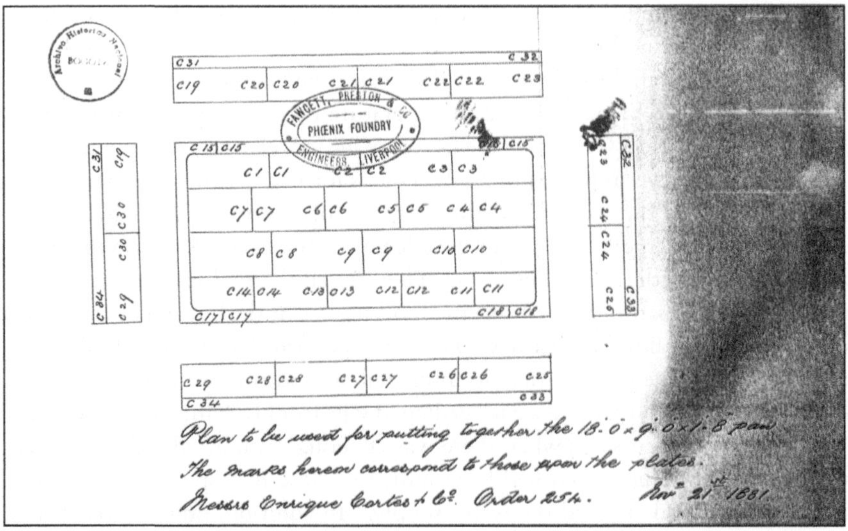

FIGURE 3.1. "Instructions for assembling a pan purchased in Liverpool in 1881." AGN, SR, Miscelánea de Salinas, leg. 4, fols. 35–36.

This reduction, though straightforward, was an art that demanded balancing a complex set of factors involving time, purity, and expense, and salt makers were skilled artisans. A relatively quick initial boiling meant less fuel used and therefore less money spent, but the reduction could not be done too quickly. Any use of catalysts to accelerate crystallization was considered suspect, for they left impurities in the brine and produced a poorer salt. Striking the right balance among these factors demanded local knowledge that defied the standardization inherent in state building. For example, the high saline content of the brine at La Salina confused outsiders who had worked in places such as Zipaquirá, where the initial reduction took longer.[25] One such administrator complained to the Finance Ministry after watching La Salina salt makers let crystals form quickly during this stage. He thought they were intentionally making an inferior salt and thereby defrauding both the public and the government.[26] He wrote again a month later explaining the misunderstanding: "The springs are so rich in their salt content that as soon as the water is put in the cauldrons it begins to coagulate."[27]

Salt-saturated water, the end product of the first stage, was then submitted to a second, more time-consuming reduction process. First, bell-shaped ceramic containers were arranged on the upper surfaces of the wood-fueled ovens; to prevent heat loss, salt makers used different-sized containers to cover as much of the ovens' upper surfaces as possible. The fires were generally kept low as the brine was reduced but were built up whenever new brine was added to the containers. When this stage was complete, the fires were put out, the containers were allowed to cool, and the salt was removed.[28]

These ovens were large affairs of stone and brick. Like other elements in the industry, they became larger and more complex over time.[29] The ovens were located in the fábrica, which transformed from an open shed to a completely enclosed structure. Brine, ceramics, and fuelwood were brought together at the fábrica, which was, predictably, the point where production bottlenecks were most likely to occur. Before direct administration, it took about a week to cook a batch of salt. After direct administration, there was at least one attempt to shorten this process by using a catalyst, though buyers refused to buy the salt made in this fashion.[30] There are no details on what happened after that experiment, but within a few years production time actually increased. By the 1840s it took two weeks to cook a batch of salt, with the ovens working twenty-four hours a day. Apparently the larger batches of salt served as compensation for the longer cooking time.[31] When managed properly, the operation produced thousands of kilograms of salt with each batch. When mismanaged, the process might take more time or use more wood, increasing the cost of production dramatically. It might also produce an inferior salt that would be hard to sell even at a reduced price. As a result, the ovens, which were often in need of repair, frequently became the object of improvement schemes.

The first detailed note about an oven, written by the administrator in 1843, mentioned that recent repairs had allowed it to cook 25 percent more salt with the same quantity of wood.[32] Eventually five ovens were built in the center of La Salina, though no report mentions that they ever worked at the same time.[33] Despite the frequent repairs and sporadic design improvements, keeping even a few of these ovens working was a struggle. In the 1860s, records begin to include more details on these efforts and their results. In 1869 the engineer Robert McCormick oversaw the addition of chimneys and grates to the ovens, intended to provide greater ventilation.[34] The administrator was optimistic about the improvements to the rebuilt oven 5: "It has been five days since we lit the fire, and all the people who know about these things report that, with the addition of a chimney, the oven is much better and saves fuel."[35] By May 1870, however, this enthusiasm had waned. Officials reported that oven 4 was not heating evenly, apparently because the recent construction had placed it too close to the giant oven 5.[36] Despite these reported problems, the ovens stayed in good shape throughout most of the 1870s. Notes on their capacity describe impressively large loads, with the potential for multiple batches of 20,000 kilograms or more in a month. After the late 1860s, though, no reports on production give any sense that construction projects had enhanced the fuel efficiency of the ovens.[37]

The increasingly enclosed sheds that housed the ovens were massive structures that must have dwarfed most of the town's other buildings; by midcentury they also included tanks for holding water. An 1854 description of the fábrica building in La Salina mentioned two doors and a covering "held up by 27 supports, nine of them in bad shape."[38] A report from 1874 measured one such structure at 36.3 by 19.7 meters and stated that it could house two ovens. As with every other element of the industry, the fábrica's sheds were difficult to maintain. Writing in 1892, one official stressed the need to totally rebuild a fábrica structure with durable materials, preferably stone and brick, explaining that repairs to wood and palm covers lasted only two years.[39] A few years later, however, a report mentioned a metal roof at the fábrica in La Salina.[40] After a visit in 1895 Antonio M. de la Parra reported that the new oven, called the "Libertador," was built on "a solid and ample base of stone."[41]

This entire complex of wells, tanks, pipes, ovens, and enclosed sheds formed the heart of the industry at La Salina. This infrastructure was both functional and precarious, a system that transformed the timeless act of salt making into something difficult to manage. For all the obstacles inherent to this cumbersome apparatus and the frequent reports of intractable problems, the production and sale of salt continued fairly steadily in the town. From the 1840s on production of caked salt in La Salina averaged about 114,000 kilograms a month. Sales of the same in the almacén averaged about 112,000 kilograms a month (see table 3.1).

While this average is useful as a benchmark for considering production, various factors prompted month-to-month and year-to-year fluctuations in both production and sales. For instance, a contractor might take advantage of clauses that elevated the payment for deliveries above certain quantities; the Fi-

TABLE 3.1. *Average monthly caked salt deliveries and sales in La Salina (in kilograms)*

	Deliveries	*Almacén sales*
Calculated by monthly data	114,019[a]	112,284[b]
Calculated by yearly data	111,480[c]	106,440[d]

Sources: Monthly reports sent by the administrator in La Salina to ministry offices in Bogotá, Tunja, Santa Rosa, and Sogamoso found in various collections of the AGN, Sección Republica, especially Intendencia de Boyacá, Gobernación de Tunja, Gobernación de Tundama, Correspondencia de Boyacá, Miscelánea de Salinas, Salinas de Boyacá, Administración de Salinas, and Ministerio de Hacienda; other sources include the *Memorias de Hacienda*.
Note: Each figure was arrived at independently, which explains the entries' differences in months and years. Yearly totals have been included in the monthly averages and combined with reports from individual months. The averages obtained by looking at only the months where both deliveries and sales were reported were 112,284 kilograms delivered and 110,760 kilogram sold, only a little lower than the totals in this table. I have attempted to correct for entries that included sales figures from the subordinate ministry works in Muneque, though this was not always possible.
[a] Data from 443 months.
[b] Data from 500 months.
[c] Data from 31 years.
[d] Data from 30 years.

nance Ministry typically paid a better rate for any salt above 125,000 kilograms in a month. A savvy manager, such as one of the Espinosas, might delay deliveries at the end of a month so that the following month's delivery would exceed that quota. Such incidents were probably rare, if only because few contractors could manage to produce this much salt while avoiding penalties for delivering too little in other months. Reasons for more protracted fluctuations in production included timber shortages or the chaos of civil war. The early 1860s and late 1890s were years of low production, in part because of the effects of political disorder. In contrast, the highest production levels of the century were reached during the 1880s and early 1890s, a heady period of stability and relative peace. In 1894 Finance Minister Pedro Bravo wrote that it would be possible to produce three million kilograms of salt a year at La Salina, an amount that breaks down to 250,000 kilograms per month.[42]

La Salina never produced that much salt for more than a month or two, yet officials and entrepreneurs continued to propose and even attempt to implement schemes for improving the industry. Much of the impetus for change and most of the proposals for improvements came from outside La Salina. Many proposals called for minor additions or alterations, while others were intended to remake some fundamental aspect of the production process. A common proposal for improving production at La Salina was to make *sal de caldero* (kettle salt) rather than caked salt. More generally there were proposals for introducing a system of open access to salt making referred to as *libre elaboración* ("open production") in various saltworks and for using steam engines to turn kettle salt into caked salt where appropriate, thereby eliminating the time-consuming second reduction stage.[43] These three projects were frequently linked. For example, some proposals called for a system of open production wherein independent workers would produce kettle salt. As an extension of this, some called for a system in which these independent salt makers would sell their kettle salt to entrepreneurs operating steam engines, who would then make caked salt to sell.

Most of these calls for reforms came between 1850 and 1885, the heyday of nineteenth-century radical liberalism in Colombia. During this period, the antimarket reality of the state's salt monopoly was particularly galling to the many powerful men who advocated sweeping reforms. Of course, these proposals were not offered in a vacuum. After decades of discussion, legislation in 1881 established open production in the Cundinamarca saltworks of Zipaquirá, Nemocón, Tausa, and Sesquilé. Things proceeded a bit more slowly elsewhere: "In La Salina de Chita, Muneque and Chámeza they will continue to make caked salt until they can install sufficient kettles to make enough kettle salt to satisfy public demand easily."[44] Open production had a profound impact on the type of salt made and sold in Zipaquirá, Nemocón, and the minor works of Cundinamarca (see table 3.2).

In the 1873–1874 fiscal year, the almacenes at Zipaquirá, Nemocón, Tausa, and Sesquilé sold more than 5 million kilograms of caked salt, almost 1 million kilograms of kettle salt, and more than 5 million kilograms of crude salt. A decade later these same works sold only 300,000 kilograms of caked salt (none in Nemocón), almost 2 million kilograms of kettle salt, 12 million kilograms of crude salt, and 2.5 million kilograms of brine (all in Nemocón). I have found little explicit commentary in Finance Ministry reports on the way this change affected regional markets. Given that the industry in Zipaquirá was dominated, whether formally or informally, by a few family enterprises, perhaps these numbers did not translate into more access to salt production for the general public. The lacuna illustrates the degree to which ministry commentary fluctuated between clear-headed, fairly accurate descriptions of the industry and the posturing for the public that followed such discussions.

In the broader context of salt markets and kinds of salt made in Finance Ministry works, comments about salt's role in public health are strikingly rare. Over time, populations that consume salt without iodine are prone to developing goiter and, eventually, developmental disabilities. In Colombia goiter—*bocio,* or as it was termed, *coto*—was a matter of some concern, but the discussion usually stressed the lack or high price of salt, when salt was mentioned at all. Codazzi mentioned the condition when describing Tundama: "Goiter predominates, an illness whose causes are still unknown. . . . [It produces] imbeciles after the third generation."[45] He made no mention of salt when considering the matter. The problem was taken seriously. For example, a law passed on April 20, 1857, established a prize for the best study on coto and its treatment; a report on the competition ran through six issues of the *Gaceta Oficial* in 1859.[46] In the 1860s this concern dovetailed with efforts to reform the salt monopoly. In 1868 Finance Minister Jorge Gutiérrez de Lara argued that salt shortages were the

TABLE 3.2. *Salt sold before and after introduction of open production (in kilograms)*

ZIPAQUIRÁ				
Year	*Caked salt*	*Kettle salt*	*Vijua salt*[a]	*Brine*
1873–74	4,002,844	466,025	3,402,650	—
1883–84	297,925	1,312,150	10,555,775	—
NEMOCÓN				
Year	*Caked salt*	*Kettle salt*	*Vijua salt*[a]	*Brine*
1873–74	1,015,800	—	1,911,625	—
1883–84	—	516,200	1,201,875	2,563,400
TAUSA				
Year	*Caked salt*	*Kettle salt*	*Vijua salt*[a]	*Brine*
1873–74	752,125	—	—	—
1883–84	437,200	9,675	—	—
SESQUILÉ				
Year	*Caked salt*	*Kettle salt*	*Vijua salt*[a]	*Brine*
1873–74	872,475	30,975	88,475	—
1883–84	184,837	—	—	—
LA SALINA AND MUNEQUE				
Year	*Caked salt*	*Kettle salt*	*Vijua salt*[a]	*Brine*
1873–74	1,228,431	83,943	—	—
1883–84	1,672,700	105,237	—	2,825,050
ALL OFFICES				
Year	*Caked salt*	*Kettle salt*	*Vijua salt*[a]	*Brine*
1873–74	8,109,859	580,943	5,697,356	354,675
1883–84	2,906,778	1,973,735	11,915,735	2,825,050

Sources: Memoria de Hacienda y Fomento de 1874 (Parra), 205; *Memoria de Hacienda de 1885* (Paúl), 39.
Note: Though the 1888 report did not include detailed information on the type of salt sold at La Salina, it included figures from the Cundinamarca works from 1884 to 1887 that confirm the figures in this table; see *Memoria de Hacienda de 1888* (Felipe Paúl), xxxix–xlii.

[a] "Sal vijua" was the term for crude salt; it probably referred to the vijagua plant, whose leaves were used to wrap salt. On this definition, see Donadíu y Puignau, ed., *Novísmo diccionario.*

problem and that it would be better for "public health" if the ministry allowed people to make salt freely, as would be the case under open production systems.[47]

Proposals for open production and the introduction of steam power provided a more attractive topic than the thorny question of public health. Less romantically, such proposals often advocated making kettle rather than caked salt. Calls for open production resonated with dominant attitudes about the inherent value of economic competition, while steam power evoked industrial modernization. Neither proved easy to bring to La Salina. In 1856 the Bogotá newspaper *El Porvenir* published a rejoinder to a piece in another Bogotá newspaper, *La Esperanza,* that had advocated for both the production of kettle salt and the use of steam power in La Salina. Bernardo Espinosa was the publisher of *El Porve-*

nir, and even if he was not the author of the unsigned piece, his influence and desire to protect his family's interests in La Salina were clear. The article argued against either innovation. With respect to the former proposal, it argued that caked salt was easy to transport and that regional consumers preferred this form. Responding to *La Esperanza*'s claim that the need for bags to transport kettle salt would promote a bag-making industry, the author derisively replied, "the economic axiom that new needs create new industry is not certain."[48]

In making this argument *El Porvenir* was going against elite consensus. In the third quarter of the century, almost any kind of technological innovation met with great enthusiasm. In his annual report for 1866, Finance Minister Tomás Cuenca included a special section on changing salt making in government works. He described traditional salt making as "brutal" and argued that steam power would act as a tonic for the industry, favor buyers, and produce a purer salt.[49] In the report for 1867, Minister Alejo Morales went further, emphasizing that steam power would lower production cost, for it would allow the use of fewer workers and eliminate the need to purchase ceramics; as a bonus, the lower prices would prevent fraud.[50]

The enthusiasm for bringing steam power to La Salina endured. In 1883 the Finance Ministry advanced 6,000 pesos to a company to introduce this system, a necessary aid for overcoming the "inconveniences" of innovation.[51] Lorenzo Marroquín, the head of this company, explained that problems with introducing steam production in Zipaquirá had stemmed from worker opposition.[52] This statement would have alarmed anyone with experience in La Salina, where worker intractability was a perennial challenge to reform.

Most commentators considering the use of steam glossed over this unpleasant fact and focused on the issue of finding a local source of coal.[53] On two occasions people reported that they had found coal deposits, but neither discovery amounted to much. In one case, the find was too far from La Salina. In the other, the deposit's quantity and quality were inadequate.[54] Of course, if a suitable source of coal had been available, it would have sparked a completely different and possibly violent drama. A report sent to the Finance Ministry in 1859 explained, "In general the town fears that they will be unemployed the day that a coal mine replaces fuelwood, and as they are now accustomed to [the work of carrying timber,] they judge it impossible to earn a living in any other way."[55] This report was written in a decade when local defiance to ministry-initiated plans had reached its zenith. There is in any event little evidence that steam-powered salt making was ever logistically feasible in La Salina during the nineteenth century and every reason to doubt that it was possible to introduce any change that residents perceived as a threat to their livelihood.

The attempt to introduce steam power to La Salina reflected the broader desire for industrial progress held by most Finance Ministry officials. The widely held frustration with La Salina's intractability, particularly when it involved steam power or open production, fed the general inclination to treat the saltworks as a failure. Viewed from the perspective of the present, however, and against other modernizing projects of that era and region, the La Salina saltworks constituted a surprisingly functional industry.

If this claim about La Salina's success seems exaggerated, consider the story of the ironworks at Samacá, which illustrates how high hopes might soar and how dramatically efforts at industrial modernization could fail.[56] The village of Samacá lay southwest of Tunja, in a valley with rich deposits of coal and iron. In the 1850s and 1860s, the government of Boyacá sponsored various attempt to build an ironworks there, but the efforts yielded only a few useless buildings.[57] In 1874 the national government backed the creation of another company, promising that it would produce "the necessary equipment for making all of the machines and apparatuses used in agriculture and the arts[,] as well as rails for tracks, cables for bridges and telegraphs, and the other articles of iron needed by the country."[58] This effort failed as well. Then, in 1878, the president of the near-sovereign state of Boyacá, José Eusebio Otálora, who had been reelected in part because of the excitement over the ironworks, issued an executive decree to purchase the buildings in Samacá and establish yet another company to build the manufacturing facility.

Having been awarded the contract, Charles Otto Brown and Levi D. York, formerly of the Passaic Rolling Mill Company, in Paterson, New Jersey, received 75,000 pesos from the government of Boyacá and further funds from the national government.[59] For a while, it seemed as if the ironworks might actually be built. Some of the funds from the national government came, predictably, from salt receipts, a point that received some criticism.[60] But criticism was overshadowed by enthusiasm for the project, as is shown by a jubilant letter to *El Boyacense* in 1879: "Now begin to appear in the north the first rays of light from the splendid sun of progress that will illuminate with its beautiful face the darkened sky of Colombian industry."[61] With a bit less poetry, government reports were equally enthusiastic, predicting that, with additional funding, the ironworks would provide finished material for modern transportation infrastructure across the country.[62]

The heady allure of the project was reflected in its anticipated scale. An 1878 contract called for the construction of a furnace rising sixty feet from the floor and reaching twelve below, with a diameter of fifty feet. This contract called for a furnace that could make 2,000 tons of iron a year.[63] By 1880 little progress had been made toward these goals, but some felt these plans to still be possible. A subsequent contract was negotiated, and more than two hundred workers were employed. Nonetheless, in only a few years this effort proved to be another failure.[64] There were reports that the main building had been damaged by floods; that there were problems with the skilled workers from the United States; and that, in a final indignity, the very ground beneath the works was so unstable that the foundation was sinking.[65] Drifting decisively into comedy-of-errors territory, Brown and York defaulted on their contract. Brown had proved a particularly poor choice. He would not work with a government inspector whom he accused of spying, and U.S. workers refused to work under him. When he left Samacá, he took the plans for the ironworks with him.

The praise heaped on the project by the Radical Liberals during their heyday was replaced by scorn. Felipe Paúl, who would serve as finance minister a few years later, critiqued the project in a report on wasted revenue in the *Memoria de Fomento* of 1881:

> If the President of Boyacá had controlled the impulses of patriotism, he would have limited [the project] to the practical, starting by establishing a modest industry capable of producing indispensable agricultural or domestic equipment. For example, it is quite possible that today more than a few towns would be purchasing these products at lower prices than they now pay for them. . . . Tragically they tried to establish a huge industry immediately, . . . [one] that will be inscribed in the catalog of [national] frustration.[66]

That same year Thomas B. Nichols, in a report for the Ministry of Development published in the *Diario de Cundinamarca,* provided details supporting this assertion. The construction was ostentatious rather than practical; the main furnace did not work; no one employed in the project had the expertise to build such a facility; and most damning, the ore was too poor to support such a works.[67] In his 1882 report to Congress, Antonio Roldán argued for abandoning the project because the unstable foundation had damaged the furnace, which could not reach the temperatures needed to make iron.[68] A congressional commission in 1883 determined that despite government investments of 500,000 pesos, the criticisms of both the furnace and the ore quality were correct.[69] The president of Boyacá attempted to save the project, explaining that the construction was almost finished and that Thomas Southan was bringing twelve workers from England to complete the construction.[70] This was an illusion. The works closed permanently in 1884. By 1887, however, the government of Boyacá, forced to cover its debts by Federal Law 40 of 1884, sought to recoup some of its investment by converting the remaining buildings into a textile factory.[71] In the light of this adventure in industrial development, La Salina was a modest success.

Resources for Salt Making: Ceramics, Timber, Land, and Labor

Behind the relatively steady rhythm of production at La Salina were the crucial support industries of ceramics and timber for fuel. Cooking salt required both.[72] While the large ovens embodied the quest for modern design that would reflect industrial progress, the use of ceramics highlighted the continuity between republican production and its colonial predecessor. During the second stage in production, the bell-shaped molds were arranged on an oven with their larger ends facing up. After the salt was crystallized, the containers were allowed to cool and then shattered.[73] New containers were thus required for every batch of salt, though I found no indication that this was an excessive cost other than the

TABLE 3.3. *Prices and sizes of ceramics in Muneque and Agua Blanca*

		Price (in pesos)		
Container	*Size (arrobas)*	*Muneque (fired)*	*Agua Blanca (fired)*	*Agua Blanca (unfired)*
Cacharitos	1	0.30	—	0.15
Madres	5	0.40	0.35	0.20
Asientos	6	0.70	0.60	0.35
Catorceros	12	0.90	0.80	0.45
Seiseros	20	1.00	1.10	0.60

Source: AGN, SR, Miscelánea de Salinas, leg. 12, fols. 22–23, January 1895.

previously cited proposal arguing for a steam system. The ceramics came in a variety of sizes, which allowed maximum coverage of an oven's suface.[74] This technique also produced variety in the sizes of salt cakes, which made sales easier to manage. A price list from the neighboring saltworks at Muneque, where ceramics were made both in the town center and in the vereda of Agua Blanca, gives a sense of the variety in the range (see table 3.3).[75]

Ceramics production at La Salina mimicked salt making, though on a smaller scale. The only comment about clay sources comes from Codazzi, who mentioned two spots, one on the road to Córdoba, close to the town center, and another close to Rodrigoque.[76] There was, he explained, an excess of clay in winter, and I have found no mention of any shortage in either clay or ceramics. There were sporadic, if steady, references to the independent nature of ceramics production, which was centered in the vereda of Rodrigoque, a resettlement site for an indigenous community removed from its resguardo (see chapter 2).[77] By the advent of direct administration, this small community was generally referred to as "the families" of Rodrigoque, though I have found no direct reference to their ethnicity or ancestry in republican correspondence.

Over time, the Finance Ministry's salt administration moved to take control over ceramics production by purchasing the kilns used to fire the containers. As were similar efforts in other areas, this effort was only moderately successful. In 1870, when the administration sought to introduce a system of wage labor, paying by a set workday rather than by the batch or load, the most determined defiance came from ceramic workers in Rodrigoque. Though the hostility to the attempted reform emerged across the area, officials specifically mentioned the "deeply held resistance of the families" of Rodrigoque.[78] Whether this was based in the skills demanded by ceramics production or the cohesion of a neighborhood with indigenous ancestry, Rodrigoque remained outside the control of the administration or any contractor of record. However, the attempt to assert control over ceramics was driven more by the institutional logic of the state than by need, and in general this part of the industry functioned fairly well.[79] Apparently, abundant supply and a lack of ministry control was a successful combination.[80]

In stark contrast to the abundant supply of ceramics, the supply of timber for fuel presented a constant and formidable problem. In fact, the demand for fuelwood stirred more comments and proved more nettlesome than did any other aspect of the industry. As were ceramics, timber was largely handled by subcontractors. All detailed accounts of fuelwood refer to *tercios* ("bundles"), but very few specify the size of these bundles. The most complete description was recorded by Humboldt when he visited Zipaquirá. There a tercio was a "medium-sized bundle" of twenty to twenty-five poles between 6.4 to 6.9 centimeters thick and 2.7 meters long.[81] The most detailed description of a tercio in La Salina is in a complaint about Honorato Espinosa, who was demanding an excessive bundle in return for the standard rate: "each stick must be at least 1.8 meters and of a diameter that barely fits into a man's two hands." Espinosa insisted that this bundle contain "at least twenty poles."[82] A worker would need to make two trips rather than one to collect and carry this much wood. This complaint clearly indicates that a standard bundle in La Salina was more or less like the one described by Humboldt in Zipaquirá and could be collected in a single trip.

The high saline content of the brine in La Salina made for efficient production in terms of bundles burned to kilograms of salt produced. Humboldt's eighteenth-century report on Zipaquirá mentions that one bundle yielded 10 kilograms of salt. Most reports from La Salina claimed that the saltworks got a better ratio, with estimates ranging from 12 kilograms to 20 kilograms of salt per bundle.[83] An 1878 contract argues the opposite, noting that the normal bundles of wood in La Salina produced less than 9 kilograms each, and a bundle of the best quality produced only about 12 kilograms of caked salt.[84] Assuming a compromise production rate of 12 to 14 kilograms of salt per bundle, La Salina's monthly salt production (using the lower figure from table 3.2) of 111,480 kilograms of caked salt demanded between 7,963 and 9,290 bundles of wood.[85]

It is clear that thousands of bundles were used in the fábrica each week. Supplying this wood was an enormous challenge. Like the smoke of the ovens, the threat of limited or expensive timber hung in the air.[86] Early in the century laborers carried wood directly to the fábrica. The subsequent construction of a network of cables extended the range for timber exploitation, though this neither eased the labor needed nor ensured a steady supply of wood.[87] As forests were exhausted, laborers had to travel farther and farther to obtain wood, and costs increased. Almacén employees tried to prevent the harvesting of younger trees, for they burned poorly, which drove up costs and imperiled future production.[88] They also tried to make sure that previously exploited areas were allowed to regrow.[89] They never had much success with either effort.

For salt makers access to or legal control over forests was both the most vital and most tenuous element in the industry. A potential contractor with access to forests, whether through a lease, ownership, or strong-arm techniques, could set the terms for the contract. Finance Ministry employees complaining about their lack of control over the industry referred explicitly to the issue of forests.[90] Over time, the ministry purchased as much of the forested lands in and around La Salina as it could. Simultaneously, a series of notable individuals and corporate interests, which often held or sought production contracts, managed to wrest control of the forested lands in and around the town. In theory these dual efforts at control might have led to some stability, but in practice they rarely did. The process of concentrated land ownership began first. By the 1820s most of the heirs of the vecinos who had purchased the old resguardo had lost control over those lands. Lucas Wilches sold land in Rodrigoque to Valderrama in 1811, and, as discussed in chapter 2, Trinidad Alarcón sold most of the old resguardo's land to him in 1836.[91]

While there is no record of the response to the Wilches sale, which took place during the chaos of the Wars of Independence, Alarcón upset many by selling what they thought of as the town's commons lands to her *compadre*. A number of vecinos brought suit against Valderrama and Alarcón over this sale and wrote to the governor of Tunja offering to "donate" the land in question if the government would intercede on their behalf.[92] The petition had seventy-nine signatures, so a fair portion of the town's households were locked in an ultimately unsuccessful battle with Valderrama for control over the land. In his 1889 pamphlet *Los montes de La Salina de Chita,* Benjamín Novoa Zerda took particular interest in a notarized sale of this land by vecinos, writing, "there is a clear conflict between the rights implied by the power of the deed from 1845 and those that are based on the deed of 1836. [The documents] refer to the same lands and are based on the sales made by individuals who say they acted as heirs of those who bought the land at auction [in 1778]."[93] This conflict was never really resolved. Contractors and other concerns continued to control the land, though the dubious legal nature of the titles, which rarely met the exacting standards of legal documentation needed for formal transactions, often proved an obstacle to the ministry's efforts to control salt making.

The state's attempt to purchase local forests outright was the other part of this story. If the Finance Ministry controlled access to the timbered lands around town, it could ensure the continued viability of the industry. But this strategy was not easy to put into practice, and the difficulties offer a primer on the challenges to institutional state building as a practical exercise. The ministry's first purchase, in 1844, was of forests that had belonged to the church, though the identity of the actual seller is not clear.[94] Over the next few decades, Valderrama's heirs made various offers to sell their land to the ministry, but they tended to make such offers after the hills had been cleared of usable timber, so it would be a decade before the land could be worked.[95]

By the 1850s most of Valderrama's heirs had either sold or leased their lands to the Reyes y Niño Company. The company was owned by Rafael Niño, Joaquín Reyes Camacho, and Gabriel Reyes Patria, the last a notable figure in Boyacá politics and the son of the independence-era general Juan José Reyes Patria.[96] The details of the company's transition from renting to owning land are murky, and the process brought accusations of dishonest dealing.[97] In 1860 some of Valderrama's heirs filed suit alleging that the company had forcibly removed them from their lands.[98] A few years later the company was accused of paying only 12,000 of the 28,000 pesos agreed on for a land sale.[99] None of this hindered the company's progress. By the mid-1860s Reyes y Niño controlled the lands called Rodrigoque, Resguardo, Rionegro, and La Estancia de Quilambria, "a single piece of land that, traditionally, form[ed] a single community."[100]

In contrast to Valderrama, who was unable to translate his position of prominence and control over land into an official position after 1828, the Reyes y Niño Company leveraged its control until it was one of the most important contractors of the era.[101] In 1866 Minister of Finance Tomás Cuenca reported that he could find no one to rent the works despite repeated invitations. As a result he had been forced to offer a contract to Reyes y Niño with terms that benefited the company to the detriment of the ministry.[102] The problems inherent in doing business with a company of this sort were exactly those that undermined state authority. In 1869 the Finance Ministry found that lands it sought to acquire from the Reyes y Niño Company were claimed by others, impeding any further transactions.[103] The issue was revisited in 1876 when the ministry did manage to rent the lands from the company, though now they were referred to as the "Bosques Valderrama," which is ironic given the protracted legal battles Valderrama's heirs fought to claim them.[104] The ministry paid 200 pesos for the use of these lands, which were then to be given or rented to another contractor. The agreement included a clause that gave the ministry the right to purchase the lands outright for 40,000 pesos, though the challenges to the company's claim to them seem to have prevented this transaction.

The Reyes y Niño Company was dissolved in the late 1870s. Its holdings were divided among the partners and eventually acquired by Várgas Santos and Company. The most prominent members of this concern were Gabriel Várgas Santos, a rancher from Casanare and one of the most important Liberal military commanders of the era, and an Italian national named Pedro Cantini. When this company was dissolved, Cantini ended up with most of the lands around town. In 1885 he signed a contract with the Finance Ministry stipulating that the government would acquire his property on the contract's termination.[105] This contract brought a new round of suits from Valderrama's descendants, who argued that irregularities in various deeds were proof that Cantini had never legally acquired the land.[106]

The complaints against Valderrama in the 1830s and by his heirs against Cantini a half-century later were serious matters. Neither the two men nor the successive interests who controlled the land between their tenures could fully establish that they were the legal owners. Of course, such niceties had little to do with the daily reality of control over this vital salt-making resource. Aside from the Espinosas, every major contractor in La Salina exercised control over these forests at one point or another. In the final act of this drama, the government bought Cantini's land for an enormous sum, despite heated criticism in Bogotá.[107] The land transactions in La Salina offer a variation on a phenomenon common in the nineteenth century: the wealthy and powerful wresting land from smallholders, either communal lands that were being divided by federal and provincial legislation or the theoretically vacant baldíos, where settlers could rarely defend their claims legally.[108] The process in La Salina was a variant of the first trend. While there is little novel in noting that the wealthy and powerful have often seized land that was not legally theirs, two matters are worth considering. The first is the duration of the process in this salt-producing area of Colombia. Roughly a century passed between the original division of the common lands and the Finance Ministry's acquisition of them. Every decade brought a fresh dispute over land ownership. For generations of La Salina's residents, the conflicts were a regular part of local life. The second point is that by the end of the century, the ministry—the institutional state itself—emerged as the owner. By the 1890s to cut timber in the hills of La Salina was to walk on state property.

The final element of the industry to consider is labor. There is, rather remarkably, relatively little information on the identities of those who labored in the saltworks or on the conditions of their employment. For every administrator like the attentive Vicente Vargas, who in 1845 purchased timber even before a new contractor had arrived in order to assure laborers that they would have work in the coming weeks, there were several like the administrator Juan Gutiérrez.[109] In 1841 Gutiérrez wrote to his superiors in Bogotá about his attempt to find out exactly how many people labored at the saltworks: "I wondered, . . . as I usually see no more than 30 workers employed in production, if it would be difficult to get the [contractor's] agent to give me a list of their names. He responded that there were more than 1,000 workers [directly] involved in production and showed me lists he has received with more than 500 names."[110] This admission is remarkable both for its frankness and for the implication of ignorance. Perhaps it is unrealistic to expect an administrator to have known exactly how many workers were involved in production, but Gutiérrez makes it clear that he did not even have a sense of the scale of the labor needed in La Salina. Finance Ministry employees were rarely so forthright in displaying their own ignorance or so perfect in articulating the blindness inherent in institutional state building, but many were as clueless as Gutiérrez. In the end he is notable mainly for his honesty. In terms of historical reconstruction, the disjointed nature of the industry and extensive system of subcontracting that supplied timber means that Gutiérrez's report is the best source for estimating the number of people who worked in salt production or related jobs.

In contrast to this vagueness, there are precise, if sporadic, reports on wages. A complaint from the cabildo of La Salina about Bonifacio Espinosa's actions while serving as contractor claimed that where cosecheros had previously been paid from fifteen to sixteen reales for each carga, Espinosa paid only twelve or thirteen.[111] I have found no reference to a payment as high as that, though the matter was taken seriously. After Espinosa's contract had been revoked, the government decree inviting bids to serve as La Salina's new contractor clearly stated the need to pay cosecheros decently: "Only with this measure can the Executive Power favor the poor resident[s] of La Salina de Chita, who in the exercise of their only industry are naturally subjected to the exaggerated demands . . . of the contractor."[112]

The next reference to wages comes from 1876. The administration tried to phase out the practice of paying by the carga and introduce wages paid per ten-hour shift. This attempt generated a pay sheet listing 120 people either working in the fábrica, making ceramics in Rodrigoque, or collecting wood at a spot identified as "el alto del Costal." The pay sheet also lists payments to twenty-four subcontractors who together delivered 12,141 bundles of timber.[113] This rationalized system set the price for ten hours of work, day or night, at thirty centavos. The change was intended to give the administration greater authority over salt making and, to a degree, deskill the trade. Part of this involved breaking up salt making's various aspects into individual tasks. Most of the jobs listed indicate some sort of work with the ovens.

The pay rate of thirty centavos for a ten-hour day allows for a tentative comparison between wages in La Salina and those in other parts of Boyacá. Reports on daily wages in the provincial newspaper *El Boyacense* from 1871 contradict the frequent complaints that wages in La Salina were excessive. Officials from five towns around Tunja reported that wages for unskilled workers in their districts ran between ten to thirty centavos a day with meals, while skilled workers in Samacá and Tunja earned between thirty and eighty centavos a day.[114] So in 1876, laborers in La Salina were paid roughly what the highest-paid unskilled laborers or lowest-paid skilled workers earned in central Boyacá five years earlier. Regional differences in the cost of living may shade this picture, but the general claim that wages were high in La Salina was based in entrepreneurial frustration, not fact.

The attempt to introduce set wages followed Honorato Espinosa's tenure as a contractor. Various notes clearly show that Honorato, like his father, had succeeded in keeping wages low, and they claimed that he tried to avoid hiring residents. He practically admitted this last point in 1870. In a lengthy report on industry conditions for the Finance Ministry, he discussed sending his brother to obtain laborers from neighboring towns.[115] Espinosa insisted the problem lay in the attitude of the residents, who would not work for a reasonable wage. In 1874 he reported paying adult wages to children younger than fourteen years old, noting that they did not give a fair return for this sum.[116] However the matter appeared from his perspective, in relative terms Espinosa was able to control payment. The 1876 effort to introduce standard wages was the administration's attempt to replicate Espinosa's control of salt workers. The effort failed. The administrators and the contractors who followed could not replicate Espinosa's control or success. Managing labor, whether the families of Rodrigoque or the roughly one thousand men who cut and carried timber, was an enduring problem.

It was also difficult to find people with the technical expertise necessary for maintaining or improving the industrial infrastructure of salt making. Traditional tasks might prove difficult to execute, as occurred in 1844 when a contractor wrote that a lack of skilled labor was causing problems for mounting new cauldrons; "Agdon Paniagua, one of the few blacksmiths who [could] do this kind of work, died a few weeks ago," he reported, adding that no substitute had been found.[117] As the production infrastructure grew and incorporated modern technology, such problems and the need for outside expertise, including foreign engineers, increased. At about the same time when the elite of Boyacá became obsessed with the ironworks at Samacá, George Simpson, a "mechanical engineer," was working in La Salina to improve the design of the ovens.[118] He was followed by Robert McCormick, mentioned previously, who took a contract to oversee improvements there, in return for which the ministry paid him 100 pesos a month.[119]

Despite the occasional presence of such exotic foreigners, the outsiders who had the most impact on La Salina were Colombians. The promise of wealth, or at least wages, drew all sorts of people, from regional migrants looking for work to elites who arrived with official appointments and contracts. The majority in this last group often knew of La Salina only as a point in the Finance Ministry's fiscal network, seeing it as a place for their enrichment before they had even set eyes on the town itself. Some had worked in Zipaquirá, meaning they had some familiarity with salt making. Others, perhaps most, arrived in La Salina with no experience in the industry. One of the frankest portraits of the expectations among these fortune seekers, as well as the shock of managing a saltworks, appears in Rufino Cuervo's account of his family's arrival in Sesquilé in 1862:

> It happened that the Government opened the saltworks of Sesquilé, located on land that was our brother Nicolás's through inheritance[, while] he was pursuing his studies in England. We entered into this business without any foresight, the partners dazzled by the enormous profits gained by entrepreneurs in Zipaquirá, Nemocón and Tausa, without considering that their exploitation had been begun by the Spanish government and that though the materials . . . might be deficient, they did not have to make it all, and they relied upon experienced workers. In Sesquilé . . . we knew there was salt, nothing more.[120]

Despite this limited knowledge, the Cuervos arrived with high hopes:

> Nobody's hopes in this business adventure were more ill-fated than those of my two brothers, Angel and Nicolás. Luis, the oldest of us, had contracted serious compromises to sustain the legitimate Government defeated by Mosquera. In such low spirits it is very easy to lose your head and get carried away in crazy

> schemes. He calculated that in a few months the saltworks would buy a Potosí and that this treasure would undoubtedly compensate any sacrifice. With the aim of assuring one of his creditors . . . Nicolás had mortgaged his property.[121]

Though various circumstances were different in La Salina, the promise of wealth and the frustration felt by those seduced by that promise were the same.

The Cuervos' difficulties stemmed from their naïveté, but other factors were also in play. The Finance Ministry could be aggressive in promoting the potential riches of salt making when it sought investors. For example, a flyer intended to attract a renter for the saltworks of Galera Zamba, on the Caribbean coast, in 1895 had "Earn Money!!!" plastered across the top in letters an inch high. The text explained, "There all can peacefully go and earn money in the comfort of working with the Government. Those involved in some part of this effort are exempt from military services and their vehicles will be protected."[122] Advertisements such as this and visions of profits, in concert with a paucity of other opportunities, ensured that woefully unprepared men arrived in La Salina all the time.

Complementing this dynamic was a tradition dating back to Humboldt's inspection of Zipaquirá: the outsider who fancied himself the possessor of expertise that should be applied to a rustic saltworks. At times these individuals directed true improvements, such as the chimneys added to the ovens in La Salina, but on other occasions they were in charge of ill-fated projects, such as the attempts to introduce steam engines. These men tended to be cavalier about any potential obstacles to their plans and often disparaged existing practices. Consider Emiliano Restrepo Echavarría's comments on his visit to the works of Upín, in the Llanos. While Restrepo testified to the high quality of the salt there, he warned his readers, "Understand that nothing could be more backward than the system of exploitation used in the saltworks of Upín. The goal of the system is to cost as much as possible and to waste the most beautiful natural product. Everything there is rudimentary and insufficient for meeting the goal set by the Government in administrating that rich saltworks."[123] He described a process where laborers dug to a layer of salt and then used gunpowder to blast out raw chunks for processing. He suggested that a system of sluices like those he had seen used in Antioqueño gold mines should be used to expose the salt bed.

On paper Restrepo's suggestion was reasonable, but despite his confident tone, there is no reason to believe that it was feasible. Restrepo never mentioned whether there was a source of water to work the sluices. Moreover, showing a deeper societal myopia, he offered little reflection on the effects of replacing twenty workers with only two, as he claimed could be done. In La Salina, where eighteen unemployed workers were eighteen people who might cause trouble, this would have been a vital point to consider. Tellingly, Restrepo offered these observations after only a brief visit. A glance was enough to convince him he knew what was best. La Salina had its share of Cuervos and Restrepos, ambitious fortune seekers and harbingers of technical modernity. In their way, with their varying kinds of ignorance, these men were as important in local life as were those who had success in La Salina. Their legacy was a string of unmet contractual conditions, unpaid workers, and cynical opportunities for the many actors in La Salina who gained from others' misfortunes.

It is difficult to attain a balanced assessment of salt making in La Salina. With the benefit of hindsight it is tempting to characterize the works as an unqualified success. After all, despite ups and downs and attempts at every sort of change, salt making continued in a steady, if unspectacular, fashion. The various improvement projects may have met with variable success—some of these projects allowed workers to make more salt or to do so more efficiently, as was the case with the wall by the Casanare River and chimneys, and some seemed to accomplish very little—but even the most ill conceived or poorly executed did not permanently damage the productive capacity of the saltworks.

Along with the changes in production came the unidirectional process of institutional state growth, a steady expansion spurred by the actions of the Finance Ministry. Decade after decade the state was built into a greater presence. The ministry acquired bigger and more impressive buildings. There were more state employees, guards, and inspectors charged with executing tasks on behalf of the bureaucratic state. And finally, the state emerged as the town's major land owner. While the relationship between society and the state, particularly between individuals and the state, is often conceptualized in abstract terms, in La Salina the reverse was true. Residents of La Salina interacted with state officials regularly and lived daily lives patterned around a massive infrastructure of stone, brick, metal, and less durable materials that constituted the material state. So manifest was this material state in La Salina that it seems to mount a frontal assault on scholarly assertions about the absence of the state in this world. But following this logic all the way leads to an argument that the Colombian state was fully formed in this peripheral town, which was not the case. Here, the subtle seduction of salt making impinges on scholarly assessment as surely as it seduced would-be state architects in the nineteenth century.

The Colombian state was certainly present in La Salina—or more accurately, the process of state building was focused in La Salina because of salt making. But the saltworks both attracted the state's attention and frustrated its designs. At times these frustrations lay in state agents' failures to acquire the local knowledge necessary to their tasks. But this frustration also lay in the dissonance between the steady rhythm of salt making and the state's attempt to treat it as a process of industrial gestation that would culminate in the birth of the spirit of economic modernity in this far corner of the country.

CHAPTER 4

The Ministry Monopoly and the Market Monopoly

> The day laborer [who gathers] sticks is poorly dressed in rags, and stooped under the weight of the bundles of wood. And at night [he] rests in a poor hut, where only the smallest fire is permitted, as to freely collect [wood] is seen as a sign of clandestine salt making. These things, and others that are worse, are the direct, cursed fruits of the monopoly over an article that is indispensable for life.
>
> —Manuel Ancízar, *Peregrinación de Alpha*

The history of salt making in La Salina presents many ironies, among them the widespread perception that La Salina's salt-making industry was a failure even though, at least in terms of production, it was a success. As with contemporary scholarship that considers the state, reflections on the salt industry often pose the question as a matter of success or failure. Elites saw the failure to create an ideal industry, an ideal market, and an ideal society of entrepreneurs and consumers. Local and regional residents had their own criteria for measuring success or failure, but they were similarly pessimistic in their assessments of the saltworks.

The sense that La Salina was a failure, despite its steady production of salt, was based on the general feeling that the administration did a poor job of overseeing the sale of salt, that ministry employees failed to ensure an equitable distribution of salt to the public. The explanation behind this sense of failure is a complex story of profits, expectations, consumption, and the workings of a regional market, matters all revolving around ritualized salt sales, the market ritual whereby state action shaped life in the region. Considering these sales details how people in La Salina and the surrounding cantons engaged daily with the institutionalized presence of the state apparatus. There were many discussions of the state's actual and ideal nature, but they were always conducted with a clear bottom line in mind.

The story begins at the point where production ended, with the deshorno.

After the caked salt was finished, the oven fires were put out and everything was left to cool for several days. The ceramic containers were then broken open and the salt removed. This batch of salt, which might amount to more than 10,000 kilograms, was taken to the almacén, where the bookkeeper received it, weighed it, and had it stored. The almacén was a combination of office, warehouse, and store. It was a clearly demarcated space created and maintained by the state. It was also a place where conflicting visions of the state's proper role were articulated during public sales. Visions of appropriate state action from Bogotá, expectations from residents, and the differing interests between the two—all these met in the almacén. The disputes that took place there and the problematic resolutions applied were disseminated across La Salina's entire consumption radius along with the cakes of salt, which were often sold at inflated prices across the eastern highlands and llanos. Disputes over marketing practices and state authority were renewed with each batch of salt delivered.

The Ministry Monopoly

During the independence era many people assumed that the onerous fiscal practices of imperial Spain would be abandoned in the republic to follow. Writing in 1826, at the dawn of this new era, the first finance minister of note, José María del Castillo y Rada, critiqued everything about the Bourbon taxation system: "The Republic must not base its rents on the ruin of its citizens. Its fiscal laws, I repeat, are not just or productive when it puts the interest of the state in contest with those of individuals."[1] The following year he expressed his frustration with the difficult legacy, writing that the earlier system "had no other principle than the blindness of the government that established it, nor other object than to maintain these populations in colonial dependency and through this, in poverty, in ignorance, and in wretchedness." Nonetheless, he recognized the burdens facing the new republic: "When a colony breaks from dependence on its metropole and constitutes an independent nation, it inevitably multiplies its needs and it must cover greater costs than those [incurred] during its servile tutelage."[2]

While Castillo y Rada advocated abolishing taxes such as the alcabala as a means to foster industry and trade, he argued that the Bourbon monopolies, such as that on tobacco (which he called a "lesser evil"), would have to be tolerated because they provided revenue.[3] As for salt, Castillo y Rada oversaw the ten-year rentals of ministry saltworks that began in 1826, balancing the inchoate desire for a free market, or at least a freer market, with fiscal expediency. When it came to salt, expediency always triumphed.[4]

The reliance on Bourbon revenues was a source of dismay for Castillo y Rada's successors, too. In 1835 Finance Minister Francisco Soto referred to the "sad legacy" of Spanish rule and characterized taxation as a necessary evil. On the one hand, he wrote, "The administration of New Granada, since its establishment, has recognized that there is no tax that is not essentially bad."[5] On the other, he acknowledged that revenues from salt were among those whose elimination would "compromise the health of the state."[6] This quandary, the state's need to profit at the expense of the individuals it claimed to represent, was mentioned repeatedly by the finance ministers of the 1830s. In 1839 Juan de Dios de Aranzazu wrote that he "would not vacillate" in selecting the salt monopoly for elimination if he could. The rent was immoral, he argued, and weighed most on the "poor and indigent classes." He also sounded a perennial theme, noting that forward-looking nations did not rely on salt monopolies. But since customs revenues were not constant, the rent had to be maintained despite "all of its consequences."[7]

The urge toward fiscal reform gained strength in the 1840s, though the focus was on the tobacco monopoly.[8] In 1845 the Finance Ministry turned one of its plantations over to a private company that was permitted to export tobacco not sold on the domestic market.[9] In 1848 Finance Minister Florentino González enthused over the move away from state involvement, expressing his faith in the market's transformative powers: "The improvements made in our system of tribute have brought greater liberty in production and relieved the burdens that have discouraged work. The residents of the country have been enlivened. [The reforms] have awakened an entrepreneurial spirit and are the reason that the resources of the Nation are beginning to be used for its enrichment."[10] These visions were to become a reality in 1850. Under the direction of Minister Manuel Murillo Toro, the government passed a flurry of legislation intended to promote international trade and finally free Colombia from its retrograde past. One of the key pieces of legislation was passed on April 20, 1850. It was intended to remake the taxation system. The tobacco monopoly was abolished, and other colonial taxes, such as tithes, were handed over the state governments, where they too were often abolished. The revenue lost to the national government was to be replaced with a direct tax and increases in customs receipts that would be the natural product of freer international trade. The abolition of the tobacco monopoly succeeded in prompting Colombia's first export boom, but not an era of long-term prosperity.[11]

One of the early drafts of the Law of April 20 included provisions for ending the salt monopoly. Murillo Toro specifically cited the burden the tax placed on poor consumers: "The rent is indisputably one of the most productive and best run of those composing the national treasury, but it has the shortcoming that it weighs heavily on an article of primary necessity . . . and it does so with a great deal of inequality."[12] Yet when this legislation was passed, the provisions to end the salt monopoly were not included.[13] In 1851 Murillo Toro tried to pass control over the monopoly to individual provinces, imitating the decentralization of tithes. In return, the provinces would pay the national government one-fifth of the total receipts from these works. Congress did not adopt the proposal.[14]

The national impulse toward reform cooled in the early 1850s. There was political unrest, rooted in part in the reforms, and the losses in Finance Ministry returns engendered by decentralization were not offset by increased customs re-

TABLE 4.1. *National revenues, 1827–1851 (in pesos)*

	Revenue source			
Year	*Salt*	*Customs*	*Tobacco*	*Total*
1827–28	229,164	869,894	598,059	—
1828–29	—	—	—	—
1829–30	—	—	—	—
1830–31	—	—	—	—
1831–32	265,837	770,847	—	2,327,310
1832–33	254,002	676,494	547,985	2,485,015
1833–34	—	429,999	—	—
1834–35	—	688,955	—	—
1835–36	195,000	726,000	583,000	2,539,000
1836–37	—	673,467	666,447	2,192,572
1837–38	255,382	519,761	692,900	—
1838–39	—	—	—	—
1839–40	273,742	141,251	722,244	1,965,710
1840–41	237,921	34,152	587,923	1,173,439
1841–42	276,464	235,076	587,923	1,838,758
1842–43	286,644	1,134,107	784,694	2,972,549
1843–44	321,986	1,124,346	820,439	2,972,751
1844–45	393,179	827,893	638,890	3,057,437
1845–46	—	978,496	—	—
1846–47	472,000	688,000	839,000	2,763,000
1847–48	467,000	562,000	827,000	2,553,000
1848–49	496,361	694,186	—	2,775,508
1849–50	468,458	670,207	—	3,340,131
1850–51	479,000	700,000	100,000	2,189,000

Source: Memorias de Hacienda, 1829–52.

ceipts. A review of government revenues illustrates the impact of the reforms of 1850. Prior to that year, tobacco and salt had been important sources of revenue (see table 4.1). Between 1827 and 1851 the ministry took in, on average, 2.5 million pesos a year. Salt accounted for 335,785 pesos a year; customs, 657,256 pesos a year; and tobacco, 642,607. The rest came from a variety of colonial taxes.[15] The salt monopoly went from returning just over 200,000 pesos to 500,000 pesos a year by midcentury. This steady increase stood in contrast to receipts from customs, which fluctuated significantly. Predictably, these revenues were not enough to cover government or military salaries or to service the foreign debt. As a result, government projects to encourage economic growth or develop infrastructure were generally difficult to support.

In this atmosphere it was no small deed to abolish the tobacco monopoly, which provided more than 20 percent of total government revenue and, in the late 1840s, often returned more than customs. The success of the coffee economy that flourished in the final decades of the century makes it hard to appreciate the risk of restructuring even an enervated taxation system in this fashion. Hindsight tells us that Colombia was bound to succeed with one of its exports, but the quick cycles of tobacco, quinine, and indigo exports that plagued the nation during the third quarter of the century and the many schemes to develop commodities such as silk or vanilla were all tenuous efforts with uncertain outcomes. Salt, however, fell in a different category. It rarely inspired visions of progress through commerce, and most agreed that it was important to maintain the 10 to 15 percent of national revenue provided by the monopoly on it.

Though finance ministers toned down their attacks on the salt monopoly in the 1850s, the critic's mantle was taken up by the Bogotá press.[16] Serving as finance minister in 1856, Rafael Núñez acknowledged the published critiques of salt contracting in Cundinamarca, admitting that "a dispassionate consideration of the matter" showed "the activity of the press . . . [to have been] a symptom of the popular hatred of the salt rent."[17] The hatred, he explained, had two sources. First, the monopoly was resented because it slowed the production of an article of "primary necessity." Second, following decentralization, a "powerful democratic tendency" had swayed opinion against such taxes. He acknowledged the role of popular sentiment, writing that it was necessary to construct a fiscal system that created "harmony" because "the wealth of citizens and that of the state are one and the same." When he returned to this theme later in the report, he opined that the ministry should foster "new links of union between the Government and citizens" rather than make itself "the object of disgust, of resistance and even disorder."[18]

Such words did little to appease critics who were particularly appalled that after 1850 saline springs in many regions could be exploited without restriction. In Cundinamarca, Santander, Boyacá, parts of Cauca, and Tolima (a province newly carved out of Cundinamarca), where the springs tended to be richer, the existing monopoly was applied even to newly discovered springs. Such springs were either integrated into the Finance Ministry's system or they were sealed, even though the price of salt remained high.[19] These conditions endured into the 1860s, even after the adoption of the Constitution of Rionegro, the most radical of Colombia's supreme laws. In a pamphlet published in 1863, Rafael Pombo pointed out that forcing residents of the eastern highlands to pay more for salt than did buyers in other regions was inimical to the values of this document and the liberal ideas it sought to embody. He concluded, "The monopoly on salt and the persecution of contraband are constant violations of the most important individual guarantees [of liberty]."[20] A few years later Finance Minister Tomás Cuenca blended economic theory into his disgust with the monopoly and corruption:

> The monopoly on salt, enforced with excessive rigor, is the most oppressive resource that can be used.

> The high price of salt is, for the poor classes, perhaps more deadly than war because it degrades them. . . . The moderation of this tax should not only be considered a measure that favors industry; it also prevents the degeneration of the races.[21]

Yet the scorn heaped on the salt monopoly during a decade when doctrinaire political liberalism reached its zenith was not enough to end it.

The salt monopoly endured in part because, in the 1860s, it was more important than ever in terms of generating revenue, approaching the levels from customs. Customs' role as the top earner for the government was more pronounced after 1850 than it had been before that date (see table 4.2); from 1851 until 1909 the Finance Ministry took in an average of 7,667,254 pesos a year, and

TABLE 4.2. *National revenues, 1851–1909 (in pesos)*

Year[a]	*Saltworks*	*Customs*	*Total*
1851–52	400,457	714,978	1,360,636
1852–53	394,143	1,081,035	2,075,664
1853–54	426,592	701,091	1,790,962
1854–55	325,129	640,075	3,377,223
1855–56	542,475	1,096,211	1,880,213
1856–57	551,080	977,969	1,893,503
1858–59	610,680	860,766	—
1859–60	600,000	925,000	1,766,000
1861–62	723,885	641,009	—
1862–63	973,101	822,171	—
1863–64	766,626	636,961	—
1864–65	621,007	1,337,946	2,106,891
1865–66	657,391	1,372,331	2,139,651
1866–67	1,066,613	1,148,668	2,567,926
1867–68	728,183	1,544,587	3,065,720
1868–69	669,616	1,504,229	3,937,500
1869–70	758,329	1,575,904	2,883,758
1870–71	770,143	1,561,082	3,570,580
1872–73	799,213	2,775,450	4,000,000
1873–74	656,000	2,811,000	3,921,000
1874–75	754,293	—	—
1875–76	3,331	1,999,231	2,912,151
1877–78	1,451,529	4,025,112	6,872,295
1878–79	1,349,349	3,992,424	5,919,548
1879–80	1,500,428	3,488,218	6,025,835
1880–81	950,284	4,292,835	7,019,094
1881–82	1,063,473	4,304,905	5,810,890
1882–83	818,945	4,360,413	5,624,655
1884–85[b]	12,67,869	—	—
1887	1,795,234	4,795,266	8,149,926
1889	1,727,021	7,725,599	14,014,543
1891	1,187,070	9,522,388	13,557,646
1892	1,159,753	8,384,671	12,465,513
1893	1,831,159	9,259,910	14,009,329
1894	1,605,162	8,024,468	13,273,592
1895	1,656,263	1,0847,966	16,242,784
1896	1,702,061	13,919,239	20,444,489
1897	1,767,607	13,537,301	1,9519,431
1898	1,405,988	9,471,743	—
1899	1,407,940	8,832,404	—
1900	1,947,859	4,409,540	—
1905	350,330	6,179,738	9,149,012
1906	811,723	7,414,776	13,935,674
1907	1,155,983	6,600,690	15,829,237
1908	1,147,008	6,169,239	16,138,444
1909	699,776	6,541,831	14,437,100

Sources: Memorias de Hacienda and *Memorias de Hacienda y Fomento,* 1853–1910.

[a] Data was unavailable for the following years: 1857–58, 1860–61, 1871–72, 1876–77, 1883–84, 1885–86, 1888, 1890, and 1901-4.

[b] In 1886 the ministry switched from a fiscal year that ran from September 1 to August 31 to one that matched the calendar year. The last four months of 1886 were reported separately and have not been included in these figures or these averages.

customs provided 4,382,463 pesos (57 percent) of that total. The salt monopoly was the second-most significant earner, accounting for 990,394 pesos (13 percent) a year. But in the 1860s salt came closer to producing more than customs than at any other time in the century. In the fiscal years 1861–1862, 1862–1863, and 1863–1864 combined, salt brought in 2,463,612 pesos in gross returns, while customs garnered 2,100,140 pesos, though the lower cost of managing customs meant that the government probably earned about the same from each source.[22] This unusual situation was caused by an extended civil war, which limited foreign trade, and a move to assess import duties by weight rather than value. These years were anomalies, but they underscored the point that salt was a reliable earner that would continue to function during crises when custom houses on borders and in ports might be threatened or inaccessible. The opposite was also true—during some wars salt centers were of little use—but that point was not considered in these issues.

Though the Finance Ministry was not prepared to end the salt monopoly, it introduced reforms intended to render it less objectionable. In Boyacá this involved working to improve distribution by establishing almacenes in towns in the state's center and eastern portion. The plan was meant to reduce the region's dependence on La Salina. It also dovetailed with the attempts to introduce open production and to produce kettle salt rather than caked salt.[23] In 1870 the chief administrator of the monopoly for all of Boyacá was stationed in an almacén in Sogamoso that was supposed to sell kettle salt made in Zipaquirá.[24] In theory this kettle salt, priced at 1.10 pesos an arroba, would compete well with the caked salt from La Salina, priced at 0.80 pesos an arroba.[25] In practice, the new almacenes neither improved market conditions in Boyacá nor appeased the monopoly's many critics.

Perhaps the most notable of these critics was Salvador Camacho Roldán. Roldán's writings on the salt monopoly are notable for a number of reasons. First, he wrote and thought about the problems stemming from the salt monopoly more than any of the other major figures of this age did. He was also the only major figure who embraced the possibilities of the salt monopoly, seeing in the salt trade the kind of catalyst for economic transformation that most reserved for export crops. While serving as finance minister, he arranged for a representative to tour saltworks in the United States and Europe in search of techniques that would increase production and efficiency. In 1871 he spoke firmly against turning the monopoly over to the states, arguing that the "servants of salt" would stop being "servants of the Nation" and come under the power of these smaller, implicitly less trustworthy regimes. Further competition among state governments for profits would become "an apple of discord, a source of civil wars between the States, and even perhaps of anarchy and the disassociation of the Colombian Nation."[26]

Roldán was, however, quite critical of the monopoly as it was: "The [monopoly] on salt increases the price of an article of primary necessity to a sum seven times what it costs to produce it. It equals, among the population that pays for it, more than seventy centavos per person among the working classes, which represents the daily wage for twelve days of labor for the father of four children, equaling almost 5 percent of their total earnings."[27] Initially, as the finance minister, he proposed selling the great works at Zipaquirá outright.[28] The following year he presented a variation, a plan to rent Zipaquirá and its satellites for twenty-five years, earmarking the earnings to pay down Colombia's foreign debt.[29] I have found no indication that this audacious plan went beyond his pen. Finally, Roldán's writings on salt were remarkable in unflinchingly naming the ministry's control over salt a monopoly, with all the inherent negative connotations of the term. In 1872 he wrote, "The monopoly creates contraband, and this is a source of moral corruption in the population. The monopoly creates violence, and this leads to resistance against, and the weakening of prestige of, laws and the authorities."[30] In this argument, the state was at fault.

Roldán's successors were less bold. Following tradition, they vacillated between the allure of using the state to promote a liberal society and the security of prudence. In 1875 Finance Minister Aquileo Parra explained, "it is not possible to establish a tax that replaces that on salt."[31] The following year Finance Minister Nicolás Esguerra insisted, "The monopoly on salt is not in accord with the principles of economic science, nor fitting with the nature of our institutions, which are intended to provide industry with the greatest freedom."[32] In 1877 Minister Januario Salgar worked through various issues relating to the justice of this monopoly before arriving at a rather disingenuous conclusion:

> If the saltworks can be appropriated, they also are susceptible to being raw material for monopoly; and if these monopolies, on an article of primary necessity, are not legitimate in terms of social rights, the saltworks must not be subject to appropriation. . . . The saltworks cannot be communal. What is not communal must be subject to appropriation. What is subject to appropriation can be the basis of monopoly. And monopoly is a socially damaging institution. . . . The saltworks are, then, of the consumers in the final analysis, and the monopoly that they support they [the consumers] establish and regulate, and they use them toward the common service of their general interests."[33]

This conflation of the individuals (here, consumers) within a society and the actions of the state that sought to control that society may have made sense in some theoretical realm, but it had no bearing on life in Boyacá. Returning to firmer ground, in 1879 Finance Minister Carlos Rico simply dwelled on the tragic impact of this tax: "The high price of salt is one of the calamities that most profoundly affect the needy classes of the States of Cundinamarca, Boyacá and Santander, where misery touches frequently if not constantly the huts of the multitude. . . . To earn for salt is a painful expression that is often heard as the sharp gasp from the lips of these martyrs to labor."[34]

Through the next decade finance ministers continued this litany of complaints about the salt monopoly—a tax they controlled—but not without an important change. The first half of the 1880s brought a political transition that culminated in the passage of the Constitution of 1886 and ushered in an era termed "the Regeneration." The Regeneration is generally understood as the first era in which the Colombian state wielded real power. Simultaneously, the emphasis on doctrinaire liberalism faded. The earlier finance ministers, however, justified the salt monopoly in terms that anticipated the policies of the Regeneration. Salgar's intellectual contortions were refined and smoothed out.

Writing in 1881, Finance Minister Antonio Roldán condemned the use of state monopolies but did not call for an end to government control over salt. Instead, and in a marked deviation from a half-century of reports that reflexively condemned such monopolies as antimarket, he argued that the solution lay in the guiding hand of a wise state. Roldán still expressed a dour view of monopolies: "Opinion energetically rejects all monopoly, [including] that on salt, that affects the poor classes of society[, forcing] them to choose between depriving themselves of the nourishment of a substance indispensable for health or sacrificing for the acquisition of that substance a part of their small wage destined to pay for supplying their homes with warmth and bread."[35] Moreover, the self-serving powers of private capital could not be trusted to work for the greater benefit of society, for free markets benefited only a few:

> Today under the official monopoly it is the pueblo itself that administers the saltworks through its legitimate representatives. He who sells the salt at the price determined by the demands of public service [does so] without caprice of self-interest. . . . Tomorrow, with the liberty of the saltworks, [of] the privileged producer, the owner of fuel, land, and capital—those who are able to produce the most at the least cost—competence will be unimportant. [The producer] will dictate to the consumer and the great triumph of liberty against the monopoly will become a great disaster.

Nonetheless, he proposed a compromise, writing, "The monopoly on salt, examined with impartiality and calm, admittedly has its defects, but it is not a social cancer."[36] His solution lay in legitimacy and order rather than market forces.[37] Writing in the same vein in 1884, Finance Minister Anibal Galindo argued that government vigilance, not market forces, was necessary to prevent an "industrial monopoly" or a private monopoly.[38]

A great deal of time had passed since Castillo y Rada had overseen the initial organization of the Finance Ministry. In those six decades, ministers rationalized the continued reliance on the salt monopoly with a remarkable philosophical flexibility that justified the ministry's substantial institutional inertia. Such a story could probably be spun out of many different themes in the nineteenth century in Colombia. If this one is noteworthy, its distinction lies in the degree to which discussion of the ministry's monopoly was mirrored by accounts of and disputes in La Salina itself. While the political debates concerning policy were carried out in the publications of Bogotá and in other urban centers, the disputes in La Salina were more concentrated, occurring in and around the town, and most commonly in the almacén. There, in a space that the ministry had built and demarcated, the agenda of the bureaucratic state came face to face with the strategies of the people the state sought to rule.[39]

A Local "Monopoly"

The frequent critiques of the salt monopoly published by finance ministers and other writers were balanced by reasoned explanations that the time was not right for ending this distasteful tax. But ministry employees, whether in Bogotá or Boyacá, exhibited little of this sincere ambivalence when they considered the market practices that developed around salt sales. The gathering of interested parties (often viewed as too many interested parties) at the almacén after the delivery of a new batch of salt was inevitable for a few clear reasons. First, a great deal of money was in play. Also, and perhaps of greater significance, the demand for salt usually outran the supply; in concert with the ministry's inability to control the salt market, this led to regular and even rampant speculation and market manipulation. Salt bought at the almacén was often resold at a price deemed excessive. In time critics came to refer to all the related practices—speculation, corruption, smart business, or simply rational market behavior—as "monopoly."[40]

The earnest usage of the term *monopoly* was highly ironic considering the existence of an actual legal monopoly. To justify their claims about the state's role and its task of market regulation, officials offered an increasingly simplified description of La Salina. Mirroring the reductive attitudes that led them to claim La Salina had no people, employees reduced the multifaceted mechanics of the regional salt market to a stark dichotomy. People were either "for" the administration or they were "against" it. People either sought to help officials in their sacred and patriotic task of distributing salt at the legally fixed price, or they were members of the immoral, degenerate, and criminal group that profited at the expense of the public. A review of the regional salt market shows not only that this was inaccurate but also that the discourse of state primacy was itself a failed attempt at state construction.

La Salina provided salt for more than 100,000 consumers, though how many more is not clear. Estimates usually ranged from 100,000 to 150,000 consumers, though at least one went as high as 300,000.[41] Working from an estimate that one person consumed 4.5 kilograms of salt per year, La Salina's annual 1,350,000 kilograms could have provided enough salt for 300,000 people. But much of the salt was used for fattening cattle in Casanare, so this number seems highly improbable.[42]

The physical extent of La Salina's market was more carefully documented. In theory La Salina competed with Zipaquirá in Boyacá. In practice the Chicamocha River was a dividing line between Zipaquirá's market to the west and La

Salina's market to the east. The Chicamocha divided Tundama, running southwest to northeast between the cantons of Sogamoso and Santa Rosa, before proceeding north through the canton of Soatá, though this was never an absolute boundary between the two markets.[43] While transport costs contributed to determining the reach of this market, consumer preference for caked salt from La Salina was equally if not more important.[44]

To the north La Salina supplied Garcia Rovira and Pamplona.[45] It also supplied Socorro until 1850, when the midcentury reforms relaxed restrictions on importing sea salt from the coast.[46] In addition, La Salina sent salt to the ranches of Casanare to the south and east, though here its trade competed with that from Muneque and from the saltworks of Cumaral and Upín farther out in the llanos. Providing salt for ranching was considered important, and it was customary to ensure that buyers from Casanare were favored when supplies ran low.[47]

These disparate areas of trade functioned as a single market, with supply networks and other forces causing events in one area to influence others.[48] Of course, La Salina's importance varied tremendously within this market. On the edges La Salina competed with other salt suppliers, but in its immediate vicinity it was both the only supplier of salt for consumers and a major source of employment and economic opportunity. It was within this more confined area that the impact of direct administration was most keenly felt. A government official visiting Córdoba in 1835 observed that residents from the entire canton went to La Salina to participate in salt making, but in 1837 another reported, "the present [production] contract has ruined the towns that depend on salt from La Salina."[49]

The issues of roads and road maintenance are enormously important in this context. The poor state of Colombia's transportation infrastructure presented one of the primary obstacles to economic development and state control in this era. The explication of fragmented cultural geography and difficult topography included in most studies on the country is not simply a nod to scholarly tradition; it is an attempt to explain one of the consistent realities of local life and national conditions. In the abstract La Salina was integrated into the fiscal network that extended to Tunja and Bogotá. In concrete reality it was embedded in a cantonal economy through a precarious network of roads cut across the region's difficult terrain.

Manuel Ancízar described the road into La Salina: "[T]he cutting of wood and the spiraling of smoke announce La Salina. One sees the hamlet with its modest church, crowded in the narrow ravine formed by the river and leaning so closely on the hills that the edge of the road is next to the roofs of the last houses, entering the town almost perpendicularly."[50] Augustín Codazzi also described the entrance to the town:

> [I]t is all a descent between rocky crags with precipices and it resembles a ladder [with steps] carved, some very deeply, into alleyways. There is a lovely waterfall.

> [The road] goes up and down brittle, rocky small hills; the road is not very wide, but it has moderate stretches until it reaches [La Salina]. . . .
>
> [Leaving the town, the road comes to] a bridge [that] crosses the Chinivaque River; passing on the left the road that goes from La Salina to [Córdoba] and staying on the banks of the Casanare River, [it] continu[es] until it passes the Chitambría gorge. A little after it crosses the Casanare where there is only a small bridge in bad condition.
>
> When the river rises, it blocks this passage for a day or two.[51]

Maintaining roads was both vital and difficult. Writing at the end of the century, residents from five municipalities in Casanare were quite specific on the topic: "Our needs grow greater each day as we lack roads that would put us in steady and easy relations with the interior of the state of Boyacá. Because of the fault of roads we lack the basic articles that they produce in that department, and if occasionally we are brought flour or potatoes, we have to pay a very high price for them. But we suffer much more because we lack salt."[52]

When Finance Ministry officials expressed concern over the state of roads, they generally focused on the role of infrastructure in economic development.[53] Road improvements were one aspect of the general improvements made with the implementation of direct administration in 1836.[54] From that point on, the administrator was charged with keeping the roads open and repaired, though salt contractors often did the physical work.[55] As were most projects involving the upkeep of infrastructure, the task was difficult. One administrator reported, "The towns to the North and Northeast of this saltworks consume an immense quantity of salt despite the fact that there are no roads, only veredas that cross the mountains and paramos."[56]

Any number of obstacles made road upkeep difficult. There were perennial shortages of funds and corvée labor, as well as harsh winter rains, flash floods, and landslides.[57] One report claimed the roads to be so poor that mules were in danger of falling from them, but given that this was followed by a request for 10,000 pesos for repairs, the author probably exaggerated.[58] Comments about the bridges in and out of town struck similar notes.[59] On one spectacular occasion, a winter flood lifted a bridge from its foundations and deposited it downstream, destroying a house.[60]

Local attitudes were factored into the difficulty of maintaining roads and bridges in good conditions. First, there was a general lack of enthusiasm for corvée labor. For example, in 1889 the prefect of Sugamaxi observed:

> It will be a long time before we can have good roads for communication, a consequence of the general poverty of the municipalities that are still getting used to contributing personal service, which properly organized and applied to road work would be a source of riches for everyone, leaving a large surplus that could be applied to public instruction. . . .
>
> The character of this province's residents, with few exceptions, is indepen-

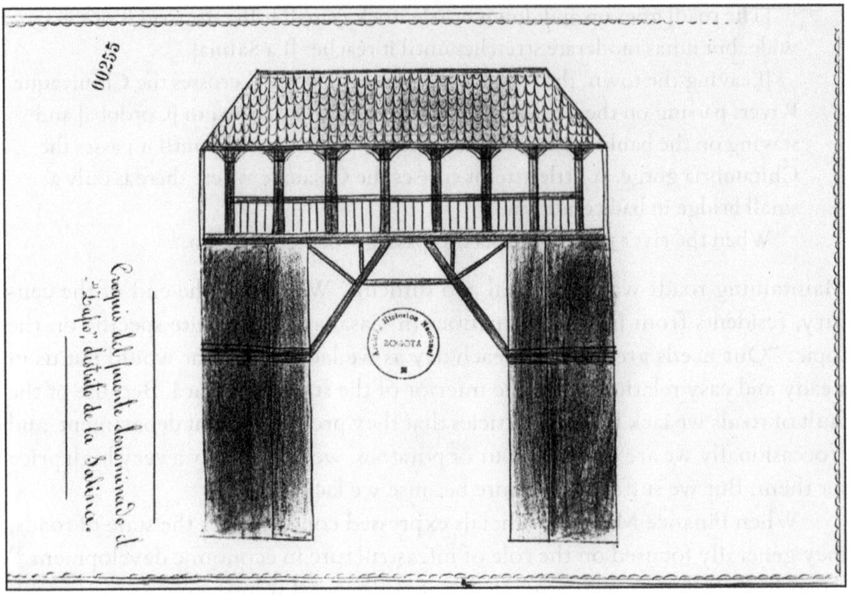

FIGURE 4.1. "Plans for the Zapo Bridge, 1865."
AGN, SR, Fondo Salinas de Boyacá, leg. 9, fol. 255.

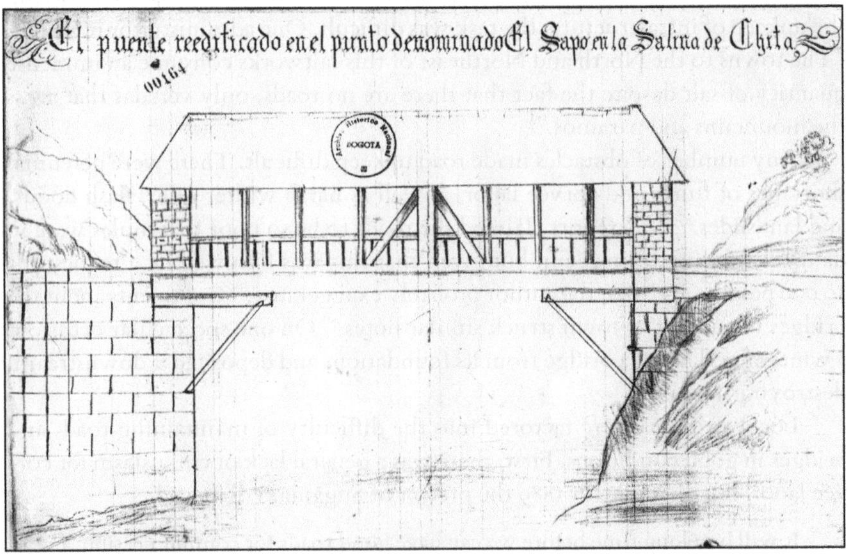

FIGURE 4.2. "Plans for the Zapo Bridge, 1880."
AGN, SR, Miscelánea de Salinas, leg. 3, fol. 164.

dent: they always seek to show disrespect to authorities, finding ways to mock their wishes.[61]

This reluctance to provide personal service could rise to open hostility when the project involved improving access to La Salina. As was well known, residents of Chita jealously guarded the profits that came from their proximity to La Salina and Muneque.[62] Evidence suggests that people in La Salina held a similar attitude. When a bridge on the road to El Cocuy collapsed, the alcalde of La Salina reported that he was paying workers for the repairs after he had "invited the populace to replace it with personal service but got little response."[63] Another alcalde reported that when he sought corvée labor for road work, residents "replied derisively that they would be happy to build roads so long as they were given the means to support themselves."[64]

The national profile of the Finance Ministry's salt monopoly must therefore be considered in this context. From 1855 until 1897 ministry works sold an average of 12,620,108 kilograms of salt per year, generating an average gross income of 1,680,118 pesos a year, with 1,173,971 pesos as profit.[65] Of course, Zipaquirá produced more than half this salt, selling on average 8,330,562 kilograms of salt per year and bringing in an average 607,990 pesos and 473,464 pesos of gross and net profit, respectively. In La Salina these figures were humbler but still substantial: 1,228,345 kilograms of salt sold, with returns of 103,757 pesos and 53,030 pesos of gross and net profit, respectively (see table 4.3).

Though the roughly 30 percent of the salt monopoly's gross returns needed to manage the monopoly might not seem like an excessive figure in a vacuum, it was far higher than the figure for customs, which needed only 10 percent of its revenue to function.[66] The relative expenses of managing the works on the Boyacá-Casanare border were much higher. When considering the question of resales and speculation in La Salina, it is useful to remember that many comments were made by critics who thought that the works were not as profitable as they should have been. In La Salina the point often came down to the more practical question of whether there was enough salt for all buyers to get what they wanted at the price set by the government. Generally there was not. My review

TABLE 4.3. *Average production and returns in different ministry saltworks, 1855–1897*

Site	*Average amount sold (in kilograms)*	*Average gross profits (in pesos)*	*Net profits (in pesos)*	*Costs/gross*
Zipaquirá	8,330,562	607,990	473,464	0.22
Nemocón	2,193,919	129,374	89,568	0.31
Tausa	477,618	45,543	39,572	0.13
Sesquilé	935,395	66,564	57,756	0.13
La Salina	1,228,345	103,757	53,030	0.49
Chámeza	267,428	19,935	7,162	0.64

Source: Memorias de Hacienda and *Memorias de Hacienda y Fomento,* 1856–98.

of monthly sales records in La Salina show five distinct periods when administrators reported significant quantities of unsold salt in their monthly summaries, but these were viewed as exceptions rather than the norm.[67]

Most of the time, demand outran supply. The practice of reselling salt bought at the almacén was the predictable result. Over time, the practice of reselling salt was termed "monopoly." Behind the overwhelming irony in the use of this term to describe speculation was the general understanding that any antimarket activity was essentially corrupt. References to the state's legal monopoly generally used other terms, such as *renta* or *estanco,* at least until the 1860s. At first the word *monopoly* was used to describe any activity in La Salina that impeded the absolute control of the Finance Ministry, and by extension, the contractor or his agents. Diego Davidson wrote about the entrenched defiance he faced as he attempted to manage the works after the implementation of direct administration:

> The authorities of this Canton have taken a step that any impartial person would see as part of a conspiracy against the industry. I can do no less than justly claim the protection that the government has offered to dispense to this enterprise as it is threatened with ruin. The step to which I refer is that they have named as alcaldes and as parish judges precisely the three people that control the production of ceramics used in this saltworks. It is notorious that traditionally these men receive advances to provide the necessary ceramics and that lately I have personally given them a little more than 110 pesos. . . . They have now attempted to renege on their commitment. I have no power to make them live up to the deal, as one is alcalde and the other two are judges. Furthermore two of them are closely related. . . . The old contractors, among whom are included the three ceramics producers, have formed a sordid plan of opposition so that the present contract cannot be completed. The contract would be rescinded and the monopoly over commerce would return to their hands. Money that belongs to the government would return to these eight or nine individuals who, by claiming they are protesting against the problems of the new system, had misled the peaceful workers so they will not work, even though their condition, though poor, is the same as always.[68]

In contrast, however, a note from an employee in the almacén in 1838 mentioned without fanfare or complaint that there were speculators in the town. Reselling salt was not yet a sign of the morally reprehensible capitalist monopoly. There was, to be sure, conflict stemming from the salt industry. For example, in 1838 Gregorio Murillo, the collector of ministry receipts for El Cocuy, accused the local judge, Juan de la Cruz Heras, of jailing the administrator on a false charge of not paying a debt solely to interfere with production.[69]

Other incidents in this era signaled that resistance to the organization of the industry under direct administration took various forms. In 1841 a contractor who had discovered workers selling salt directly from the fábrica expressed his surprise, remarking that he had "never imagined that one had to look for agents of [disorder] within the fábrica."[70] Of course, salt makers were the first to defy the Finance Ministry's control. A few decades later, Minister Jorge Gutiérrez described similar, if more flamboyant, practices in Zipaquirá and Sesquilé: "One of the methods used by the workers and laborers who steal salt is to throw, over the walls that close in the area of the fábrica, fairly large pieces of salt."[71] He considered and then rejected a plan to expand the exterior wall of the fábrica to prevent this practice.

For all the tension implied by these descriptions, a record of daily sales from May 1843 demonstrates that neither shortages nor resales were pressing problems. Though there were clashes involving the industry, they had not yet gravitated to the almacén, nor were public sales moments of constant conflict. The record lists 1,008 of the month's 1,161 numbered purchases, made on seventeen different days between the May 7 and May 31. At this point the almost daily sales were apparently low-key events (see table 4.4).

The specifics of the sales describe La Salina's market at the time, which was dominated by buyers from the immediate area who tended to make what were considered small purchases. Just over half the purchases (52 percent) were for quantities of less than 25 kilograms; another quarter (24 percent) ranged be-

TABLE 4.4. *Salt purchases at La Salina, May 1843*

Date	*Day*	*Buyers*	*Arrobas sold*	*Average purchase (in arrobas)*
May 7	Sunday	112	496	4.4
May 8	Monday	45	535	11.9
May 10	Wednesday	39	336	8.6
May 12	Friday	49	305	6.2
May 13	Saturday	29	145	5.0
May 14	Sunday	119	776	6.5
May 15	Monday	47	669	14.2
May 17	Wednesday	33	392	11.9
May 18	Thursday	18	103	5.7
May 20	Saturday	49	295	6.0
May 21	Sunday	118	635	5.4
May 22	Monday	36	292	8.1
May 24	Wednesday	29	245	8.4
May 26	Friday	65	443	6.8
May 28	Sunday	125	540	4.3
May 29	Monday	45	363	6.7
May 31	Wednesday	41	240	5.9
Total		1,008	6,810	6.8

Source: AGN, SR, Administracíon de Salinas, leg. 3, fols. 544–56. This record lacked details for the first week of the month.

tween 25 and 62.5 kilograms.[72] All but three of the remaining purchases were for less than five cargas, or 625 kilograms. Through the month buyers came from twenty-nine different towns to buy salt. The majority were from towns in the cantons of Santa Rosa, Sogamoso, and Cocuy—meaning Tundama—or in García Rovira, all within eighty kilometers of La Salina.[73] Individuals making over half the purchases, 545 of the 1,008, came from the towns of Soatá, Socotá, Chita, and Jericó, all within a one or two days' walk from La Salina.

The works at La Salina managed to meet regional demand in terms of both overall quantity and consistent availability. While the town did not produce excess salt during this era, the record shows buyers purchasing salt every day of the week except for Tuesdays and on seventeen of the twenty-five days covered by this record.[74] Significantly, there were sales on all four Sundays, La Salina's traditional market day. Together these four days accounted for 36 percent of the month's total purchases but only 29 percent of the amount of salt sold. The Sunday market was an important weekly event with a rush of buyers making, on average, the smallest purchases of the week. Despite this weekly burst of activity, there were no comments or warnings about shortages or limiting access and no complaints about small purchases. For example, the almacén was open on May 17 but made only eighteen relatively small sales. A few decades later opening the almacén on such a day would be unimaginable. Direct administration was reworking local life, but the transition was felt more in patterns of landowning and residence than in the salt market.

Bonifacio Espinosa's first production contract in 1845 catalyzed the changes waiting to occur. Even before he had arrived, the administrator Vicente Vargas warned that the outgoing contractor had stopped buying timber and ceramics, which worried local laborers: "It would be wise to take measures that would prevent the migration of laborers and would also be useful to the new business or the government. Possibly, from September first it might be permitted that timber be cut on the mountains owned by the government and purchased by the administration to be passed on to the new production company."[75] But, as discussed earlier, the stringent conditions Espinosa imposed on almost all with whom he dealt during his five-year tenure brought the tensions inherent in direct administration to a head. When the Finance Ministry removed him in 1850, his replacement, Mariano Sáenz de Santamaría, made extravagant promises, vowing to do a great deal for what turned out to be too little. After Santamaría left La Salina, he wrote to Antonio Prieto, the governor of Tundama, reporting that when it had become clear he would not be able to meet the terms of the contract, he had been approached "by an associate in a company formed to buy all the salt in existence in this saltworks and all of the rest that they [were] producing with the goal of monopolizing it."[76]

The term *monopoly* gained in significance and was applied with greater specificity after 1850. Both local events (e.g., the heightened tension that followed the tenures of Espinosa and Santamaría) and national events, specifically the liberal reforms of 1850, fostered the change. "Monopoly" was not simply an antiliberal abstraction but rather a practice that ran against the laws of the land—excepting the necessary laws that maintained this anachronism. Ironically, the term was now used to describe the practice of buying salt at the almacén and reselling it at a price higher than the one set by the Finance Ministry. In 1852 the administration reported that some sought to "monopolize" salt and that "the new monopolists" demanded a price higher than that legally allowed.[77] In fact, the early 1850s brought the first concerted efforts to prevent such practices. In response, the administration announced sales only a day before they took place, and the administrator kept lists of buyers; those requesting less than 1,250 kilograms of salt (ten cargas) were favored when the available salt was shared out.[78] As would often prove the case over the ensuing half-century, these measures did not make much difference. A letter to the governor of Tundama from the cabildo of Socotá in July 1852 named the ministry as the cause of a number of problems, including monopoly:

> La Salina de Chita, which used to supply this town with salt, produces less and has promoted a scandalous monopoly, in which, they say, government employees are involved. . . . This happened because the government has sealed off the fountains of life given to us by nature; although no government has this right. . . . They say that there is a monopoly in the saltworks of Zipaquirá too!! Therefore we raise our voice to the citizen president of the Republic asking that he open the saltworks of this province!! The government should stop hiding the gifts of God that are the riches of the people.[79]

An alcalde's report on almacén sales in June 1853 confirms that supply and demand were out of sync. During four days of one week, 813 people submitted requests to buy over 60,000 arrobas of salt when only 6,275 arrobas were available.[80] That same year, an official in El Cocuy reported that "monopoly" had reappeared, though he blamed a single individual: "He has enough money to buy the salt before it is weighed [and available to the public]. I have been told by a resident of Chita that in the last few days, Dr. Bernal was sold 22,500 kilograms, a whole batch."[81] The following year the administrator Juan Calvo was accused of holding all but a tenth of the available salt for Romualdo Levano, who presumably used his control over 90 percent of the supply to profit through resales.[82] Accounts where a single "monopolist" was named in this fashion, and accusations leveled directly at ministry employees, were relatively rare.[83]

More commonly, reports invoked unnamed agents of chaos without providing specifics, employees often describing a system where buyers in the almacén acted as agents for a powerful, shadowy figure. In 1852 one employee sketched such a scenario: "[A] trader comes to buy an arroba of salt. For a fee he requests two extra cargas, to which he is legally entitled and cannot be denied. So the monopolist hires fifty traders for a trivial sum and obtains 100 cargas without even appearing in the lists as a buyer."[84] Whether this level of organization was possible was beside the point; resales occurred, and they took place outside the Finance Ministry's control. One official admitted that though resales did not technically constitute a monopoly, he used the term anyway.[85] This is the only admission I have found that the term was inaccurate but fashionable. Others used the word without reservation, echoing the rhetoric of the time and positioning themselves on the right side of the general loathing for monopolies. For an example of this distaste, consider an item in the satirical Bogotá newspaper *El Loco* in 1856, a table of "Industrial Statistics" that looked like an official government report listing commodities and those who "monopolized" each one. According to the table, salt was monopolized by the "families of Zipaquirá."[86]

With a great deal less whimsy, the term factored into the tumultuous popular politics of the decade, specifically the artisans' rebellion of 1854. The rebellion, which was centered in Bogotá, ran from April 17 till December 4, during which time it seriously threatened the two-party system that ensured elite control over national politics.[87] The regime, led by General José María Melo, that ruled from April to December, along with the groups supporting the rebellion, all viewed their fight in part as a struggle against monopolists.[88] Article 14 of the decree that established Melo's new government included a guarantee of freedom in every profession, explaining that as a result, "all monopolies [would be] abolished."[89] Viewed with the benefit of hindsight, the loyalty of La Salina's residents to the anti-Melista constitutionalist forces was remarkable.[90] They were praised for their loyalty, though this did not affect the Finance Ministry's subsequent dealings with the saltworks.

The war seems to have made local conditions even worse. Reports from 1855 indicated rising food prices, in part because of increased speculation in salt, and anxiety that the contractor might not continue to pay locals for their work.[91] The increasing tension was manifest at the almacén during sales. An administrator reported that locals were particularly defiant when salt began running low during a sale in January 1855. In an attempt to distribute salt according to Finance Ministry protocols, he began to favor the requests of buyers who had traveled farther and come with mules:

> In sharing out the salt I distributed 100 cargas, giving to each one at least a third of what they had come for. . . . From the rest I separated 40 cargas to share among the residents of the district. . . . It may be supposed that among this crowd composed mostly of unfortunates, a part were very satisfied that they managed to purchase something at the official price. And those that are accustomed to come and buy and resell were full of complaints. For the forty cargas that were set aside for the district . . . 475 [buyers] were listed and this does not count small children[;] . . . each head of family had listed children, relatives, dependants, and peons. . . . The discontent among the notables of the town was so complete that we feared sedition, and Medardo Cuadra (who was ordered to leave the town in August) insulted me in an atrocious manner, and publicly defied me.[92]

This administrator concluded that the only way to prevent speculation by the district's "monopolists" was to set aside a portion of each batch for local residents, "excepting infants," and to divide it carefully. What that would achieve is not clear.

The degree to which residents in La Salina were influenced by the Melista rebellion is similarly unclear. Local changes engendered by events after 1845 had provoked enough discontent to produce widespread defiance whatever the state of national politics, but in 1857 the town rose in a genuine and fairly successful rebellion, forcing the Finance Ministry to abandon its plan to close the saltworks. Perhaps the rhetoric of artisans' societies, which had flourished in Tundama, or the town's active role in the constitutionalist triumph had emboldened the residents. Perhaps residents drew on both sets of experiences to justify their own actions. I will more closely examine the uprising of 1857 in chapter 5, but a quick description is necessary here, for it affected marketing practices and the rhetoric concerning monopoly in La Salina into the 1860s.

In 1857 the Finance Ministry moved to protect the official salt monopoly by shutting the saltworks in La Salina.[93] This policy, a Möbius strip of self-contradiction, embodied many elements of the process wherein an institutionalized salt monopoly—in other words, a state bureaucracy intent on producing markets and a market economy—limited productive activity because the bureaucracy objected to local market strategies. As the date for the closure approached, the administration lacked funds for timber deliveries, which meant that laborers went unpaid. The town reacted. With support from the surrounding towns, residents seized the works and began cooking salt on their own accord. In petitions the rebels insisted that the ministry maintain the role it had carved out for itself over the previous two decades. They proclaimed that their right to earn a living from nature's bounty was equal to the ministry's claims of primacy. In one letter, residents described themselves as "unhappy proletarians and *padres de familia*" and finally as "citizens and *padres de familia*."[94] These were notable—and, if not overly radical, then certainly very republican—twists on the traditional category. In the 1820s residents in La Salina had offered the state control over communal lands in return for continued access to them, only to be ignored. In the 1850s they successfully refused the institutional state's attempts to implement policy they deemed inappropriate. The saltworks stayed open.

The rebellion ushered in a decade of administrative chaos; contractors rarely lasted a year, production levels dropped, and Finance Ministry control fell to a minimum. Writing to demand greater institutional support while he served as the contractor in 1868, Honorato Espinosa reminded the ministry that it had made nineteen production contracts in the previous fifteen years and that all had failed.[95] National factors played a part in this instability. A network of powerful Liberals controlled many of the key salt positions in Boyacá after 1860 (Tundama having been dismantled as an independent province), and the federalism enshrined by the Constitution of Rionegro in 1863 provided little impetus for bringing the town under control. Resales thrived. In 1859 the administrator Vicente Gómez Maz. reported that a purchase made in La Salina for 8.75 pesos sold for 16 pesos in Chita.[96] Angel Galán inspected La Salina in 1862 and reported to the president of Boyacá that resales were frequent, food prices were high, and children were commonly listed as buyers in the almacén.[97] When he questioned the administrator about including children in these lists, he was told, "[T]his was the custom, as there is no ruling that prohibits it."[98]

In a country desperate for market activity, a state monopoly had produced an active commodity market, albeit a small and frantic one. But critics saw only the "monopoly" of private interests. Galán opined that private marketing practices in La Salina were "a hundred times worse" than those of the state in managing the salt trade, for the so-called monopolists added "a special monopoly that yield[ed] no benefit to the Government or society": "These residents do not sell any service, economically speaking." Galán's comments neatly describe the matrix of economic behavior, violence, and morality, as well as the corresponding calculations officials employed when considering life in La Salina. Galán even admitted that limiting sales to residents was a "violent" attack on their rights, but this solution was better than allowing the "additional monopoly" produced when people in La Salina could buy salt freely.[99]

In La Salina, of course, the violence was neither purely economic nor rhetorical. The administrator Vicente Vargas described a riot in the making in 1865:

> On Sunday the sixth of this month there was a mutiny in this district, caused by a fight between Eusebio Zuniga (implacable contrabandista) and the guard Francisco Nariño. It appears that [Nariño's] brothers defended him, and all of Zuniga's contrabandista fellows from Chinivaque took his side. The commotion was immediately so great that the urging of the local authorities, who were insulted, was not enough to calm it. . . .
>
> Some troublemakers, who are never lacking in these towns, exhorted the mutineers to arm themselves with clubs and rocks with the purpose of throwing out the national employees and taking control of the vertientes. The idea was taken up with enthusiasm and executed immediately. Fortunately I was at home and, though I was ignorant of what was happening, was informed and was able to reach the fábrica with six guards where we managed to block the doors.
>
> The mutineers appeared, numbering, between men and women, more than 50, but they found us sealing the entrance and resolved to meet force with force if it was necessary.[100]

While the crowd gathered outside insulting Vargas and the guards, "influential residents" arrived and managed to calm the crowd. Some residents tried to start more trouble the next day, but Vargas was ready with the guards and put it down. The threat of a rebellion like the one in 1857 hung over the town, though that uprising had been sparked by the Finance Ministry's threat to end production, whereas this one came from a single confrontation. All that was needed was the traditional Sunday crowd.

A complaint sent by Honorato Espinosa in 1866 described an industry where the use of force was standard practice. Espinosa was quarreling with the administrator about control over various production resources, and he was particularly annoyed that he could not use a road that had "always belonged" to the production company and provided access to the fábrica without going through the town center. He complained that the road was no longer open to him: "[T]his road has been usurped for three years by residents of the district who are now hostile to production. [They] opposed our attempts to open it on the 26th of last month with armed force." The response from the Finance Ministry was less than encouraging: the usual request for more information and the suggestion that if the road had not been explicitly listed in the production contract, Espinosa had no claim.[101] That same year, which saw numerous reports of thefts of ceramics and timber (presumably stolen to be used in illegal production), Espinosa cited an act that had a certain panache: "[A]s you know, they have robbed ceramics from the fábrica of Cordova, from the same place where the guards sleep."[102]

Espinosa, whose family exercised as much control as any contractor ever had in La Salina, sent a series of complaints to officials in Boyacá, the Finance Ministry, and even the "Citizen President of the Republic."[103] The letter to the president sent in 1869 and describing another Sunday disturbance has "urgent" scrawled in the margin.[104] He reported that on December 27, 1868, Braulio Toledo had insulted and then attacked the administrator in the almacén. Alcalde Estavan Barón neither prevented the action, punished Toledo, nor made an official report. Then, on January 6, 1869, a fire broke out in woods near the fábrica, an act of arson that Espinosa believed was intended to provide a distraction while people robbed the almacén.[105] The new alcalde, Daniel Estava (brother-in-law of Estavan Barón), did nothing to aid those fighting the fire and made no report.[106]

On Saturday, January 9, a group of forty men under the direction of Medardo Cuadra, an associate of Barón's, gathered in the plaza and attacked the employee Eduardo Flores and Lieutenant Francisco Mendoza of the guard unit. The "peaceful inhabitants of this place" were alarmed at the lack of order and the complicity of local officials. A contractor from Bogotá who had come to build a new fábrica left town because, he said, there was "no guarantee except that of brute force."[107] Espinosa closed the letter to the president by claiming that the administration had ended resale in La Salina, though he did not use the term *monopoly,* and offering praise for the workers from outside of La Salina. In his account, ending monopoly was the direct cause of these disturbances.

The end of the 1860s also saw a flurry of threats to impress salt workers and similar forms of harassment intended to limit production and enhance the prospects for reselling salt. Skilled workers, who were supposed to be exempt from forced military recruitment, were a particular target.[108] In 1869 the administrator complained that Estavan Barón, then serving as alcalde, had jailed the subcontractor Ramón Chapparro and threatened to impress his workers even though they were not to have been "recruited or taken from their job by any authority."[109] Barón wrote that he was powerless to intervene, for the district judge and the cabildo had ordered Chapparro's imprisonment. He also argued that the rule limiting military recruitment was "absurd" because it impinged on the town's authority.

Two years later, Medardo Cuadra again targeted Chapparro's employees for military recruitment, though the executive of the Departmento del Norte had warned Cuadra that this was illegal. Chapparro, his employees, and the employees of the contractor Ambrosio Orosco, who oversaw work on the retaining wall for the Pozo Azul, went into hiding, effectively ending these projects. While I am skeptical of many of the claims made by Finance Ministry employees in La Salina, one rings true: "This individual [Medardo Cuadra] has always been the principal monopolist of this place." Cuadra's career is notable for the fluidity with which the man moved from one role to another. In one incident Cuadra was at the head of a rioting crowd, and in another he held a government contract.[110]

This fluidity demonstrates that despite the inherent tension between the interests of the national state and those of the local society, there were no set patterns. Instead, life in La Salina revolved around conflicts between ever-shifting factions and commercial networks. Employees described an intractable opposition to the Finance Ministry because a dichotomized portrait of local life served their purposes and reflected their experiences. One employee summed up the feeling of most who worked in the almacén between 1857 and the early 1870s: "I am a spectator to the scandals which cause a growing awareness of my impotence. I resign myself to the daily mortification of wasting my time in a fruitless manner, writing letters that serve no purpose. . . . In this manner vagrants, contrabandistas, and fugitives from all parts come to La Salina to hide their crimes and to satisfy their twisted passions, inculcating honest and humble workers with corruption and vice." Responding to this reality, in 1870 the Finance Ministry made one of its attempts to distance itself institutionally from La Salina. Jacinto Corredor was appointed administrator for the salt monopoly in Boyacá, but he served in an almacén in Sogamoso rather than in La Salina. His report from 1871 included a review of conditions in La Salina and praise for Honorato Espinosa.[111] Corredor reported that monopoly had reappeared in La Salina, a statement that supports Espinosa's 1869 claim that it had been ended. Predictably Corredor recommended supporting Espinosa in most matters. From that point on, Espinosa sent few complaints. Perhaps he simply needed time to reestablish his family in the town, or perhaps Corredor's support was an effective counter to the machinations of Cuadra and Barón. Once again, the Espinosas dominated the industry.

If the exact reasons for Espinosa's renewed success can only be guessed, the impact is clearly documented. In 1871 the La Salina resident Ramón Toledo, a "native of the State of Boyacá," offered one of the most comprehensive critiques of any contractor, much less one of the Espinosas.[112] He explained that he was not moved by "hate or ill will" but believed that it was his duty "to favor the government when it is possible." How was it, Toledo asked rhetorically, that Espinosa had never been on bad terms with the Finance Ministry given his record of malicious acts? The answer was that Espinosa's true character was unknown. This "enemy of the Government" had succeeded in masking his "sinister intentions." Toledo accused Espinosa of fraud and named him as the party responsible for salt shortages that he exploited for his own profit. In Toledo's version of the situation, residents were the victims, rather than the perpetrators, of illicit and immoral practices:

> There exists then, in La Salina de Chita, the monopoly. The reason for this is very simply . . . because the Señor contractor, instead of protecting the old salt workers, has taken away their harvests and will not give them work in the fábrica. He has reduced them to carrying wood to be bought by him, but he has set an extraordinarily large bundle[;] . . . a laborer must now make two trips and the mountains are so far. With what one earns [residents] are not certain of a living and they have to leave because of hunger. This is the reason that there is monopoly in La Salina de Chita. . . . Señor Espinosa has said publicly that one finds workers only at the price of gold. This is not the case, and it is certain that there are the best workers, in part because of the experience they have, and also because they are perfectly acclimated. But Señor Espinosa has said that if certain families do not leave La Salina, he will have the pleasure of seeing them die of hunger. [He is] moved by a gratuitous hate[, which] is the reason that, even though the salt workers work with perfection and are paid two reales a day, they are rejected and workers from other places are brought in though they do not know how to move a single stone and ask for four reales a day.

Toledo explained that shoddy craftsmanship was responsible for a poorly working oven and that a retaining wall had been improperly rebuilt "by a worker who had no scientific knowledge and only earned three reales a day." All this occurred under Espinosa, a "cruel enemy of the government."

Toledo went on to critique the administration. His description of a sale in the almacén perfectly reversed the accounts of Finance Ministry employees. He blamed the administrator Faustino Barbosa for violence in the almacén and implied Espinosa was behind these affairs:

> [The employees] present a batch of 37,500 to 50,000 kilograms of salt and there are 100 or more buyers. But they do not sell the salt to them, they wait until there is a crowd of two or three thousand people; then they begin to write down the people there to the last one. After listing them they call them again

> for sales. In this disorder they abuse and mistreat, and [buyers] even lose the money they have brought to buy salt. Then in the presence of the administrator Sr. Faustino Barbosa, an individual beats the buyers who are nearest to the door until the blood begins to flow, [and] they begin to call those farthest away in the tumult, as if for amusement (because passing through is so difficult). The method of sharing out the salt is to sell an arroba to each individual who has come with ten or more beasts to buy cargas of salt, over a distance of three, four or up to eight days, having to return with the animals unloaded after having lost a week or two from waiting.

Further, Toledo wrote that Barbosa hired a schoolboy who distributed licenses illegally. He closed with pointed comments about the specifics of Espinosa's contract.[113]

Toledo was neither writing in a vacuum nor yet speaking for everyone in La Salina. He may have been related to the Braulio Toledo whom Espinosa accused of assault in 1869 or have acted as a spokesperson for the faction led by Estavan Barón and Medardo Cuadra. But even if he had some vested interest in discrediting Corredor, Barbosa, and Espinosa, his description of a sale in the almacén is a reminder that the traditional administrative strategies for distributing salt in moments of shortage had dramatic ramifications.

Espinosa's apparent control in La Salina after 1870 had no impact on the broader dynamics of the regional salt market or life in La Salina. In 1874, for example, when the official price stood between 60 and 80 centavos per arroba, salt was resold in and around La Salina for 1.20 to 1.80 pesos; purchasers in García Rovira paid 3 pesos for an arroba; and those in Pamplona, as much as 4 pesos.[114] La Salina saw more acts of arson than it had in previous decades. Most of these fires were set between February and April, right before the winter rains, when timber collection was most difficult. If the saltworks did not have a ready supply of fuel, production was threatened, and, by the logic of La Salina, the market would be easier to control.[115]

Altogether, these factors normalized the practice of buyers' requesting outrageous quantities of salt, first documented in the 1850s. In May 1874 a number of contentious sales transpired. An official described a salt sale on May 13, when a crowd of people, perhaps because they felt they had "preferential rights during sales," attacked two guards stationed at the entrance of the almacén. When local officials failed to come to the guards' aid, the administration closed the almacén and ended sales for the day. The administrator resorted to selling salt to outsiders and residents on different days, hoping this would calm the situation, but problems persisted; he reported, "Yesterday I found on the floor an anonymous threat that offered to assassinate me if I continued completing my obligations, which the discontented here call hateful and discriminatory against the pueblo."[116] Despite the threats, the administration held a salt sale on May 17 where the gap between the supply and demand, as represented by purchase requests, was absurd. Two batches of salt totaling 76,065 kilograms (a quantity that would have constituted more than half a month's salt at almost any time during the century) were delivered. Buyers requested a total of 1,400,000 kilograms, more than the administration might sell in a year.

The absurdity of these proportions was not some anomaly, as sales records from July 4 make clear. A batch of caked salt totaling 53,750 kilograms was put up for sale and attracted 1,138 buyers from eleven different towns in Boyacá and from Casanare. Collectively they requested 339,525 kilograms of salt (see table 4.5). Residents of La Salina requested, on average, 639 kilograms of salt, twice as much as the average for all purchasers and much more than buyers from outside La Salina sought.[117] These requests are particularly striking when compared to the sales records from 1843 (see table 4.6).

In 1874 buyers who were not from La Salina made requests that resembled

TABLE 4.5. *Salt purchase requests, July 4, 1874 (in kilograms)*

Town	*Buyers*	*Average request*	*Total request*
Casanare	26	827	21,513
Chiscas	29	36	1,038
Guicán	7	25	172
Santa Rosa	4	382	1,530
Boavita	16	470	7,520
La Uvita	108	107	11,558
Socotá	72	45	3,224
El Cocuy	182	71	12,991
Jerico	13	12	150
Chita	265	52	13,872
La Salina	416	639	265,957
Total	1,138	298	339,525

Source: AGN, SR, Salinas de Boyacá, leg. 5, fols. 646–54, July 1874.

TABLE 4.6. *Salt purchase request amounts, 1843 and 1874*

	La Salina buyers, 1843		*Other buyers, 1843*		*All buyers, 1874*	
Arrobas requested	*Number*	*Percentage of buyers*	*Number*	*Percentage of buyers*	*Number*	*Percentage of buyers*
1–2	13	3	309	43	522	52
> 2–5	41	10	179	25	246	24
> 5–10	48	12	158	22	80	8
> 10–50	190	46	61	8	143	14
> 50–100	94	23	11	2	16	2
> 100–400	27	6	3	0.4	1	0.1
> 400	3	0.7	1	0.1	0	0
Total	416	100	722	100	1,008	100

Sources: AGN, SR, Administración de Salinas, leg. 3, fols. 544–56, May 1843; AGN, SR, Salinas de Boyacá, leg. 5, fols. 646–654, July 1874.

those made by all buyers in 1843, when almost no one from La Salina attempted to purchase salt at the almacén. In 1874 purchases of less than a carga accounted for 90 percent of the requests from nonresidents; in 1843, they accounted for 86 percent of requests by all buyers. Only a tiny fraction of either group asked for more than five cargas. When residents of La Salina sought to purchase salt in 1874, they asked for much larger quantities. For example, only one-quarter asked for less than a carga, while almost a third requested more than five cargas. The point is clear: even if only a portion of La Salina's population actively backed people such as Medardo Cuadra, a large part of the town's entire population sought to participate in resales by purchasing large quantities of salt at public sales.

Facing these demands in July 1874, the administrator took the standard measure of allotting out-of-town buyers a larger portion of the salt they had requested than that allotted to locals.[118] This affected how sales functioned, but not exactly in the way the officials had hoped (see table 4.7).[119]

Subsequent sales in July and August suggest that nonresidents as a group were responsive to market conditions.[120] Whereas the July 4 sale of 53,750 kilograms drew 722 outside buyers, the sale of 17,375 kilograms on July 21 drew only 407, presumably because salt had been difficult to buy at the first sale. However, when 51,950 kilograms were put up for sale ten days later, after the administration began selling to residents and outsiders on separate days, 1,493 out-of-town buyers made purchase requests. An August 15 sale of 18,775 kilograms drew 906 buyers from outside La Salina, double the number who had come for the sale of a similar quantity on July 21. It is clear that information about sales—particularly the conditions set by the administration—spread rapidly throughout the region. Regional buyers came to La Salina when it might prove worth their while, but at other times they tended to stay away. The number of buyers from La Salina was steadier, for they attended sales whenever salt was available; 416 came to the first of these sales, whereas 292, 384, and 362 came to the succeeding sales. Outsiders were less consistent, but they also attended in large numbers if there was a good chance they could buy salt. Everyone participated in "monopoly" in one form or another.[121]

Another point evident from these records is the degree to which these sales dictated the rhythms of local life. Gone was the town of the 1840s, when a Sun-

TABLE 4.7. *Salt requests and sales at La Salina, July and August 1874 (in kilograms)*

			Requests by residents		*Requests by nonresidents*	
Date	*Total on hand*	*Amount requested*	*Number*	*Average amount received*	*Number*	*Average amount received*
July 21	17,375	94,750	292	6.25	407	37.5
July 31	51,950	369,500	384	12.5	1,493	31.25
August 15	18,775	192,188	362	6.25	906	18.75

Sources: AGN, SR, Salinas de Boyacá, leg. 5, fol. 644, July 1874, and fol. 672, September 1874.

day market drew people to La Salina, and many made small purchases of salt while, presumably, conducting other business. By the 1870s the sale of a new batch of salt *was* the market. Four decades of state policy had superimposed new patterns on local life, and its rhythms followed the sporadic delivery of salt rather than the weekly calendar. The other dynamic informing the complaints about residents' involvement in the salt market is its effect on the labor supply. Honorato Espinosa himself traced these connections: "Since monopoly has increased, laborers employed in the fábrica have stopped working there. They spend most of the month waiting for the sale of salt. . . . Working two or three hours [as buying agents,] they earn more than the production company could pay them for a day's labor."[122] Here monopoly was a threat not only to the legal functioning of the market but also to the viability of the industry.

Given the accounts from 1874, it is somewhat remarkable to read the administrator Félix Camargo M.'s 1877 report that in La Salina, "the monopoly, or the resale of salt, [had] been destroyed." He added that there were no inflated prices, and people received their salt almost as soon as they submitted their requests.[123] If this was true, it was an exception in a decade of rampant speculation and resales. The late 1870s were marked by pessimism among Finance Ministry employees in La Salina and fierce critiques from finance ministers in Bogotá, who began to emulate Salvador Camacho Roldán's writings from the beginning of the decade. In his comments on open production in 1877, Nicolás Esguerra condemned the legal salt monopoly as being contrary to "the principles of economic science" and wrote, "[P]ublic opinion demands the abolition of this monopoly as an urgently needed reform."[124] The next year Januario Salgar addressed the issue, explaining that there was a "public clamor" in part because of "the influence of that passionate word," meaning *monopoly*.[125] Discussing the effect of resales in the eastern highlands in 1879, Luis Carlos Rico addressed profiteering and claimed that the high cost of salt had "social ramifications," continuing, "People only consume what they themselves need and sometimes not even that. I have been informed that in some places there are families that prepare their food with sugarcane juice because they do not earn enough to buy salt."[126] But in one of the reversals that confirmed the deeper continuities of the salt monopoly, the 1880s brought a rhetorical turn wherein state control was deemed necessary to prevent the private monopolies that would take its place without ministry oversight.[127]

Complaints about resales in La Salina continued from the 1880s on, though a number of them named almacén employees as "monopolists" and mentioned that nonresidents played an active part. One employee defended himself from such an accusation in 1882, explaining that outside buyers were those most likely to resell salt.[128] In 1893 officials from the towns of Boavita and La Uvita made similar complaints, but they argued that Finance Ministry employees were complicit when outsiders engaged in resales. They sketched a scenario wherein seasonal laborers benefited from their formal status as outsiders, though they were residing in La Salina, to buy salt on days when sales were limited to out-of-town buyers, depriving honest traders of their rightful share.[129] Almacén employees facilitated the practice by blindly following the guidelines to favor outside buyers with more salt.[130]

A series of complaints from the next decade depict what an out-of-town trader might have faced in La Salina in the 1890s. In 1893 a trader from La Uvita, a town that had sent more than one hundred people to the sale of July 4, 1874, was accused of returning with only a small load of salt in order to charge high prices. He responded by explaining that the administrator in La Salina favored dishonest traders. He himself was "moved by patriotic enthusiasm . . . to end the monopoly on salt run by many in La Salina de Chita that victimize[d] consumers" rather than by the hope for illicit gains.[131] An 1896 letter from the alcalde of La Uvita describes a similar scenario. On his first trip to La Salina, a resident of La Uvita was forced to pay the inflated price of 3.20 pesos an arroba, though he managed to purchase three cargas of caked salt. On the second trip he

> arrived in La Salina, where he had taken seven mules to buy salt. After twelve days of waiting he was able to buy only fourteen arrobas at the official price, though he had hoped for fifty-six arrobas for his seven mules, not two arrobas for each one. He bought ten arrobas from Eliseo Neira at 2.20 pesos an arroba, from Asunción Tarazana he bought twelve arrobas at the same rate, and from Avelino Ferrer eight arrobas for 2.30 an arroba. What is more, the administration lie[d] to outsider[s;] every day for five days they announced that there would be a sale the next day.[132]

Given such circumstances, the propensity for traders to make quick profits by reselling was not that surprising.

The last decade of the century brought a number of elaborations on the traditions of reselling salt or trying to limit the supply. The most impressive was described by the administrator General Terán: "Last year, when we brought in a mason to make repairs on the main well, he found it plugged with straw and a mortar made of mud. . . . This impeded the flow into the well and naturally diverted [water] back to the river. The well lost almost half [the water,] which limited the supply of salt and increased its worth."[133] The repair had been carried out at about the same time another official had reported, "It is an ongoing battle with the inhabitants of this population, who believe they have more rights to the [springs] and mountains than the government [has]." Whatever the beliefs involved, the rewards for limiting the salt supply were worth the effort. In 1892, when the legal price of caked salt was 1.40 pesos an arroba, it sold for 4.50 pesos in Sogamoso.[134]

Local clashes were as charged as ever.[135] An inspector who visited La Salina in 1893 sent a telegram from Chita on March 14 reporting, "Against the general will, Octavio Sánchez, who has openly declared his hostility to the production company, has been named alcalde of La Salina."[136] He explained that he was forced to send the telegram from Chita because the telegraph operator in

La Salina was an ally of Sánchez. Whether or not Sánchez enjoyed widespread support, he wasted no time in employing the standard tactics of a local official opposed to the administration. In a matter of days he had imprisoned a timber contractor, jailed two skilled workers who tended the ovens, arrested twenty production workers and sent them to El Cocuy, arrested seven buyers from Socotá and La Uvita, and then jailed another five production workers. The twenty workers escaped while on the road to El Cocuy, but Sánchez had made it clear that he would control the market and the terms of salt making in La Salina. In May he was accused of impressing more production workers.[137] The Finance Ministry's solution was to wash its hands of direct involvement. In August General Terán took over the works as a renter, making these problems his and not the direct concern of the ministry.[138] As the decade drew to a close, reports of monopoly continued. One critic in Bogotá claimed that "monopoly" could not occur at the saltworks "without the consent of government employees."[139]

Representing a turnaround in the situation, in 1898 Finance Minister Manuel Esguerra reported that there had not been "one single complaint about [monopoly], probably because production [had] been gradually increasing."[140] Whether this moment of calm could have endured is not clear but also purely speculative, for the century's final civil war soon ended this sort of discussion. Thus, at the end of the century La Salina was, as it had been for six decades, plagued by issues of market access, control over profits, and the workings of local power structures in the face of the national, bureaucratic state. Before considering what this particular history means, it is useful to consider the other practice that gave regional residents access to the wealth of the salt market: contraband.

Contraband

Contraband was less contentious than monopoly.[141] Though the illegal manufacture and sale of salt were widespread, this trade in contraband was less obvious than were speculative practices. It was also less offensive to the spirit of the era and as a result failed to arouse much in the way of condemnation. It was simply the kind of illicit activity common in any society rather than anathema. The secondary status of the legal salt monopoly contributed to the seeming lack of concern about contraband. When the state monopolies of the early republic were established, the punishments for contraband traffic in salt were not as severe as those for dealing in contraband tobacco or liquor.[142] Contraband was a broad and slightly nebulous category. Traders were supposed to have a (legally obtained) guía, or customs authorization, for all their salt. Further, any salt made outside Finance Ministry control was illegal prior to 1850, except for legally imported sea salt.[143] In the eastern highlands these restrictions were in place even after 1850.

Contraband trade overlapped the resale market, but the two were rarely discussed together. When people criticized monopoly, they did so because it demonstrably led to inflated prices. There was no equivalent discussion of the way contraband affected the salt market, and it is difficult to determine how it affected the region. For a decade after 1836, accounts in La Salina included an entry for confiscated contraband salt. For the forty months with entries, the average figure was 695 kilograms, a relatively minor amount.[144] In contrast to these documents, an 1842 report claimed that between 125,000 and 150,000 kilograms of illegal salt was sold in Boyacá every year, about a tenth of the amount sold in La Salina and more than enough to affect legal sales.[145]

It remans unclear, however, whether this higher estimate referred to salt made legally and traded illegally or to illicit production. The latter attracted more attention than the former because theoretically it could be more easily policed. In reality, doing so was not that easy. The region had numerous springs that would yield salt for anyone with a few pieces of equipment and a modicum of experience, far too many for guards to police it effectively. Unused springs were supposed to be sealed, either by the guard unit or by contractors who agreed to perform the work, but seismic activity or mudslides caused by winter rains could always produce a new spring where it was possible to use a few kettles to cook kettle salt.[146] Generally such operations were small, but on at least one occasion a complex was discovered. In 1843 the guards found an illicit fábrica that included sheds, tanks, and ovens, the infrastructure necessary for the two-stage production of caked salt.[147] The commander reported that the contrabandistas had escaped because they had been warned that the guards were patrolling the area.[148]

The regional population seems to have accepted contraband salt with no complaint; it was sold in local markets and transported on public roads.[149] There were accusations of official complicity, but they never amounted to much.[150] On more than one occasion people simply began using abandoned equipment at sites that were no longer managed by the Finance Ministry.[151] Contraband thus demonstrates that while actors challenged the institutional state at the heart of the salt market, the issue of supply and unmet demand was felt throughout the region.

The Finance Ministry took very few actions that calmed life in La Salina, fostering a situation in which daily life was measured on a continuum of conflict. By the 1860s the threat of violence hung over every commercial exchange and many stages of salt making. Oscar Oszlak has argued that one trait in the historical evolution of state apparatuses is "the conflictive character that the very process of state expansion acquires as a consequence of the fact that its apparatus becomes an arena of negotiation and dispute."[152] In La Salina, where there was more dispute than negotiation, the state's attempt to use salt as an autocatalytic for state growth had the opposite result, creating instead an autoinhibitor. The ministry claimed the primacy of its rights and in doing so negated the claims of others. Ironically, those who opposed the state in La Salina did so with greater flexibility, deploying the rhetoric of the era and devising strategies to profit from the market that would not have worked without the legal monopoly on salt.

State action, however ineffective, was an enormous force in this region, though sometimes the influence came out of attempts to implement policy rather than any successes in these efforts. How is this process to be understood in the broader sweep of Colombian history? Were disputes in La Salina like some sort of self-contained eddy, a whirlpool on the side of swift rushing river where diverted waters clashed upon themselves with no impact on the broader current? In one sense the answer is yes. La Salina was important within Tundama and perhaps in Boyacá as a whole, but its direct impact beyond this region was minimal. In another sense, however, La Salina's story has broader implications. First, it tells a story of societal interaction with the institutional state, particularly the institutional state constructed around fiscal policy that was not unique in its general dynamics. Second, the comings and goings in La Salina, while overwhelmingly focused on profits and the salt market, played into a larger network of politics and influence that reached further than the radius of consumers who bought the town's caked salt. I now turn to this larger story.

CHAPTER 5

La Salina and Colombian History to 1857

Yes, the administration of Finances is the most important point, and the most interesting of a government, because with revenues everything can be done and without [revenues] nothing can be done. . . . The establishment of a new administrative system, even in countries where peace reigns, where there is a spirit, and a public opinion, is the most difficult task . . . a government can undertake.

—José María del Castillo y Rada, *Memoria de Hacienda de Colombia,* 1823

La Salina was unique as a town in the way that an outsized state presence—specifically, the state's economic policies—remade local life there and in the way that this effort generated systematic documentation. But the portrait of local state building has implications beyond this single municipality and touches on broader themes concerning politics in Boyacá and across the eastern highlands. Considering this history offers a portrait of Boyacá as a politically heterogeneous region that played an important role in national affairs, points rarely considered in the general historiography of republican Colombia.

A full reckoning of the nineteenth century must consider political eras and the uneven rhythm of eight national civil wars.[1] As the home of a saltworks, La Salina was almost always a strategic objective for campaigning armies during these wars.[2] Moreover, the regular administrative correspondence contains rich details on succeeding conflicts, the path of campaigns, and the general tumult of these wars.

Two immediate factors beyond its status as a saltworks shaped La Salina's role in the civil wars. First, the Casanare River valley, which defined the immediate geography of the town, offered a path, however precarious, between Boyacá and Casanare. Forces from one region often used this corridor to travel to the other.[3] Second, La Salina lay close to Chita, which was a breeding ground of unrest in general and often of regional opposition movements in particular. Lists of criminal cases in Boyacá from the early republican era generally mentioned crimes—frequently violent—in Chita, and correspondence from regional officials was often pessimistic when considering the town and its influence.[4] Late in the century, these issues were evident in a petition from residents of Chita, published in *El Boyacense* in April 1887, asking departmental officials to reconsider a plan to relocate the district superior court away from the town.[5] They conceded that among residents there it was customary to "obstinately resist the action of the law" but explained that the harassment of court officials had involved acts of only minor disrespect and was in any case typical of a regional attitude rather than anything particular to their town. More practically, the petitioners pointed out that Chita had the best jail in the circuit and that keeping the court where it was had benefits. The business of the court brought in much-needed revenue to Chita, and it was better to use this institution to guide and educate a rebellious population than to leave them to their own inclinations.

The Early Republic and the War of the Supremes

After independence, Colombia's economy was in dire straits. Though the new government ended up maintaining most of the existing colonial fiscal practices, the revenue it collected was barely enough to keep it running.[6] The man charged with managing this challenging situation as the first notable finance minister was José María del Castillo y Rada, a patriot who had already served as vice president.[7] Though he penned strong critiques of Spanish fiscal practices, he did not rush to dismantle the system he took over.[8] In considering the obligations of his new ministry and the task of bringing order to Colombia's fragmented regional systems of revenue collection, including a semiautonomous revenue center on the Caribbean coast, Castillo y Rada eschewed fundamental reform.

The task of building this economic infrastructure was as difficult in Boyacá as it was in any other place in the country; Boyacá had a large population, suffered from a stagnant economy, and did not produce much federal revenue.[9] Salt returns made up a fair portion of the federal revenue it produced in this era. Sporadic reporting from 1823 and 1824 shows that salt yielded revenues greater than those from direct contributions, sales taxes, the liquor monopoly, or the tobacco monopoly.[10] This trend is confirmed by totals for twelve of the thirteen fiscal years between 1832 and 1845. Of the 1,119,255 pesos of federal revenue collected in Boyacá during those years, salt returned 263,557, or 24 percent; tobacco, 230,916 pesos, or 21 percent; and liquor, 145,348 pesos, or 13 percent.[11] The importance of salt, which at this point returned only slightly more than tobacco, was enhanced because salt receipts were used "to satisfy immediate public expenses," such as government and military salaries.[12] In cash-poor Boyacá, salt could be relied on to pay for immediate needs.

This decade saw the fragmenting of the first Colombia, presently termed "Gran Colombia," into Ecuador, Venezuela, and the Republic of New Granada.[13] Francisco Paula de Santander returned from his exile, and a constitution was written for the new republic. This document, the Constitution of 1833, created provincial assemblies with the power to nominate governors and judges rather than have the president name them. This relatively decentralized structure helped ensure that provincial political conflicts did not take on national implications.[14] Under Santander's presidential hand, the government worked out compromises between the competing elite factions that in time would form the cores of the Liberal and Conservative parties. The split between these two groups has been defined in various ways: followers of Bolívar against followers of Santander, a promilitary faction against a civilian faction, and an "aristocratic" faction against a "provincial" one.[15] The comparative tranquility of this period is evident in the election of the serving vice president, José Ignacio de Márquez, a native of Boyacá, to the presidency despite Santander's preference for José María Obando. Boyacá was similarly calm in the 1830s. The provincial government was nominally allied with what Victor Uribe-Uran termed the "aristocratic faction," which supported Márquez's candidacy in 1837, and pursued policies that, without being doctrinaire, were aimed at economic development.

This political calm, during which the Finance Ministry's books sometimes ran in the black, ended with the War of the Supremes (1839–1842), also called the War of the Convents.[16] This conflict comprised a series of loosely affiliated regional rebellions sparked by the government's decree closing smaller monasteries in Pasto, in the southwestern part of the country. The war, which was a bitter blow to the church in the new republic, revealed the precarious state of federal revenues: in 1841 Finance Minister Mariano Calvo reported that it had been necessary to borrow 41,000 pesos from three Bogotá merchants to cover expenses. In discussing this loan and Colombia's foreign debt, he explained that the promissory notes that had been issued to cover salaries were redeemable for a share of future profits from salt returns.[17] Given the complaints about the market monopoly that were so pervasive after 1850, it is ironic that ministry policy fostered the conditions for speculation with a system of vouchers for salt futures.

As a military affair, the War of the Supremes may hold the dubious distinction of being the century's most confusing conflict. It began with a popular uprising in Pasto in protest of a plan to close monasteries in the region; from there the disorder spread haphazardly. Hence, a proto-Conservative regime in Bogotá, the party that would later be defined by its loyalty to the church, faced a popular challenge from a devout region because of the anticlerical tenor of its policies. The initial rebellion in Pasto was suppressed rather easily but then revived under Obando, a proto-Liberal, who proclaimed he acted in support of the church and federalism. By the middle of 1840, revolts had broken out across the country. Some of these movements advocated the program of the "Liberal Party," and all espoused federalism.[18] The leaders of these movements tended to award themselves the title of "Supreme Commander," giving the conflict its name.

The eastern highlands had their share of "Supremes," though this area remained a secondary theater during the war. One revolt broke out in Velez in February 1840, another in Casanare in July, and yet another in Boyacá in September. On September 17, 1840, Juan José Reyes Patria, whose son was a major partner in Reyes y Niño Company, led a successful rebellion in Sogamoso. From Sogamoso his forces marched to and occupied the plaza of Tunja, which they held until they faced an organized government force.[19] The plaza and the city were symbolically the political center of the province, but neither position offered a particularly easy defense against a superior armed force. Reyes Patria then joined with the army of Colonel Manuel González of Socorro, and together they retook Tunja. This victory left the rebels in control of Boyacá except for the cantons of Garagoa and Tenza. Under González's leadership, these forces set out for Bogotá. On that path, they stopped in Zipaquirá, from which Bogotá presented something of an open city, but they were unable to gain a decisive victory when they met government forces.[20] From a certain perspective, the leaders of the rebellion in the eastern highlands were proto-Liberals; for example, Manuel Murillo Toro, who would emerge as an important Liberal leader at midcentury, served as González's secretary. But rebel publications stressed that they opposed Bogotá's anticlericalism, a popular but not particularly Liberal position and, more important, that they strove to prevent the region from descending into chaos.[21] These leaders were cautious to strike a moderate tone, making much of the fact that they prevented a priest-led crowd from attacking prisoners in El Cocuy, though they added that the majority of the clergy had avoided inciting the passions of the "reckless and simple multitude."[22] The rebels held their position until December 1841, when an army led by General Herrán and Tomás Mosquera defeated González and forced Reyes Patria's forces out of Tunja and east through Boyacá. Herrán and Mosquera continued campaigning east and north, winning a series of victories in Socorro, Pamplona, and through the region that was called the North throughout the first half of 1841.

From September 1840 until January 1841, as the rebels and the government fought for control of this theater, rebels were relatively secure in their control of the cantons of Tundama, though it was not yet a distinct province. In La Salina a rebel force of forty men led by Tomás Martínez managed the saltworks and used the profits to finance their war effort in the immediate region. The government's first attempt to reclaim La Salina took place in January 1841. Cavalry led by Colonel Concepción Melgarejo, who had restored the government to power in Casanare, attempted to take La Salina. The advantage in the fighting between Melgarejo's cavalry and Martínez's forces shifted back and forth.[23] When the attack from the llanos came, Martínez's forces took up a defensive position in the town jail (the only reference to the jail I have found) and, though expecting reinforcements from the highlands, forced Melgarejo to retreat. The two sides then agreed to a weeklong ceasefire. During that week, rebel armies suffered defeats throughout Boyacá, leaving Martínez's position untenable. Melgarejo soon entered La Salina without opposition, reporting, "They have received me with the finest sentiment in favor of the government." He also requested that rations be sent for his forces because, he said, "in the treasury of this [town] there is not a single cent." Rebels had managed the saltworks, but the benefit accrued directly to them.

In January 1841 the governor of Boyacá, Jorge Hoyos, moved to establish order throughout the province.[24] He called for reconciliation—a somewhat contradictory effort given the simultaneous campaign to push opposition figures from public life—and issued proclamations after victories against the forces of González and Reyes Patria.[25] Active rebels and their supporters were exiled, imprisoned, or subjected to forced loans, a tactic intended to ensure docile behavior and replenish the government's diminished funds.[26] A list of the worst actors was drawn up in a document titled "A TABLE: That shows the individuals who took part in the rebellion," with information on the "perverse" and "dangerous" rebels from Boyacá. More than half those listed were accused of absconding with public funds, but the charges were also replete with moral condemnations. For example, José A. Prieto was a "youth of a perversity without precedent": "[T]oday he is in exile in the territory of Venezuela, where he must remain forever so that in the school of disgrace he learns to complete his social obligations."[27] The shock of the war was profound, for the rebellion occurred less than a decade after Gran Colombia had fragmented into Colombia, Ecuador, and Venezuela. Postwar prosecutions were common and conducted with some rancor.[28]

Both Juan José Lozano, the Finance Ministry's administrator in La Salina, and Juan Nepomuceno Solano, its bookkeeper, were among those formally charged for their actions during the rebellion.[29] Lozano, whose working relationship with Pedro Ignacio Valderrama could not have been seen as a sign of devotion to the ministry, and Solano had already compiled dubious service records.[30] In May 1837 Vicente de Azuenaga, who had overseen the implementation of direct administration the year before, accused Lozano of gross negligence and theft.[31] Lozano denied the accusation of theft. As for the charge of negligence, he explained that he did not keep regular hours in the almacén because the office was too small to hold all the salt and that he could not oversee sales on Sunday, the market day, because he and Solano needed to go riding that day, for they were "taking ill from the sedentary life" they were leading.[32] They survived this accusation and held on to their positions. When the war broke out, they stayed at their posts, even writing official reports during the months the rebels controlled the town. After the war, the government claimed that Solano was among the most venal rebels in the region, that he personally brought forces from the llanos to the highlands, and that when defeat was imminent, he had fled to Casanare with cash from salt sales.[33] Of course, this meant that he was gone before government forces took control of the town.[34]

The only other Finance Ministry employee charged was Jesús Urrea, who was accused of the "crime of rebellion" and "hostility to the government" because he had worked with the rebels when they controlled La Salina. Urrea defended himself on the ground that he had feigned illness and fled to Chita as soon as he could.[35] The contractor, however, had demonstrated sufficient loyalty to the government by abandoning La Salina as soon as it was taken, though he was unsuccessful in seeking 1,225 pesos as reimbursement for his timber that the rebels had used to make salt.[36] Governor Hoyos showed little concern in his response. He acknowledged the inherent justice of the request, admitting that the government had seized the salt made by the rebels who had used this timber, but offered no recompense. As the guards took up their duties in La Salina, they found caches of salt throughout the town. The ministry's new employees reported that people who had worked with the rebels were taking posts in the administration.[37] Melgarejo might have been well received in La Salina, but the population seems not to have really cared who ran production as long as it ran. The accusations against certain individuals faded into the white noise of mutual recrimination that defined life in La Salina.

Nationally the rebel strategy of targeting saltworks prompted comment. Now serving as the finance minister, Hoyos explained the role of salt in the "interior" during the war:

> This rent, I say, was also without a single doubt one that served as the basis for the maneuverings of the rebels, as much for the quantity of revenue as for the need there is for this substance in one of the provinces that initiated the rebellion. The revolutionary movement that in the province of Tunja went to work immediately on the resources of the national wealth sought in the yields of La Salina de Chita the means that would serve to buy soldiers in another province who would support the [rebellion] and threaten with all of the evil of anarchy the capital of the Republic.[38]

In La Salina this concern dovetailed with the general effort to strengthen the institutional structures put in place with direct administration. The two years after the War of the Supremes brought one of the first challenges to the residents' water allotment, the ministry purchased the forested land that had belonged to the church, the administration removed tenants from ministry lands, and a new oven was built in the fábrica.[39] The guard unit was reorganized under the authority of the Junta de Hacienda in Boyacá, and officials in both Tunja and Bogotá requested information about La Salina's population and workforce.[40]

These points suggest two things about the intersection of La Salina and national history at this point. First, the events of this war, particularly the targeting of the saltworks as a campaign objective, prompted a Finance Ministry effort to actualize the plans implicit in direct administration. Miguel Centeno has considered the dynamic of international wars and state growth in Latin America, arguing that civil wars weakened states. He maintains that specific kinds of rent seeking, particularly the "extraction coercion cycle," permit states to "penetrate their societies in increasingly complex ways" but do not force elites to sacrifice their share.[41] The salt monopoly was precisely this kind of measure, emblematic of the deeply contradictory nature of state building. In a technical sense, in the early 1840s the ministry succeeded in extending the infrastructure of the state.

There were modest gains in the industrial efficiency of the saltworks, though little evidence of increased revenue. But the 1840s also brought reports of increasing tension, complaints that the town was losing population and, in 1846, the first report of arson. The infrastructure of the salt monopoly was a blade with a double edge, and as it grew larger, it cut deeper.

In retrospect, it is easy to see the War of the Supremes as only one war out of many or as a prelude to those that followed. But at the time the survival of the Colombian nation seemed to hang in balance. Given that the 1830s had brought the dissolution of Gran Colombia, the fragmenting of the Central American Union, and the end of the Peruvian Bolivian Union, it is easy to understand the fear that this Colombia too would break into smaller parts. La Salina was a part of this threat, for it provided rebels with resources in the field, but if handled correctly, it could also be part of the solution. When he served as finance minister, Hoyos drew on his legal training to explain the main obstacle to an improved system in his 1842 report. He pinpointed the difficulty for judges in applying the proper penalties, specifically "the consequences of the absences of laws."[42] Properly administering the saltworks was part of a large program vital for the nation.

The second point about the intersection of La Salina's history and that of the nation is the degree to which the contractors of La Salina were notable regional figures during these wars. Here the town's wartime strategic value bled into the world of influence and contracts. The links in this conflict are suggestive rather than explicit but still worth reviewing. Antonio Malo, the first contractor in La Salina, belonged to an aristocratic family of Tunja and stood firmly against the Supremes and their loosely defined federalist agenda.[43] It is also possible that the Reyes Patria family developed their interest in La Salina during this conflict, the father's military command perhaps drawing the son's attention to the saltworks. The intersection of La Salina and notable regional politicians continued in the following decades.

La Salina and the Changes of Midcentury

The War of the Supremes was followed by nearly a decade of peace and relative continuity in terms of fiscal policy and the tax regime.[44] The presidential elections of 1849 set the stage for the end of both. The erstwhile *ministrales,* now the Conservative Party, split their vote between two candidates, and so the Liberal candidate, José Hilario López, won the election; the influential Liberal Manuel Murillo Toro served as finance minister. On April 20, 1850, López's government passed a law abolishing a number of Bourbon taxes and giving provincial governments control over others, such as tithes. By fundamentally altering the nature of taxation and liberalizing conditions to facilitate economic development, the law was intended as a catalyst for economic development. There had been attempts in this direction before, and there was some hedging on liberalization later, but this law was viewed as the key to breaking with the enduring colonial economic system.[45] It was anticipated that revenue lost by the reforms would be replaced by money from a direct tax and the anticipated increase in customs returns that would come with the anticipated growth of international trade. The abolition of the tobacco monopoly did lead to a brief export boom, though not to an era of long-term prosperity.[46] The fiscal reforms were one part of the larger attempt to break from Colombia's colonial past, a sweeping tide of change that was welcomed by all. The civil war of 1851 was prompted when Conservative landowners in the Cauca Valley, who were struggling to deal with the final abolition of slavery, rose up in rebellion against the government.[47] I have found no evidence that this war directly affected eastern Boyacá. This was, perhaps, the only time a civil war did not explicitly involve maneuvering around or fighting over La Salina. For once, the town did not play an outsized role in a regional conflict.

The optimism and urge for change that prompted the reforms of 1850 peaked in 1853 with the adoption of a new constitution. That year Finance Minister Juan Nepomuceno Gómez praised the enlightened legislation of 1850, marveling at its effect:

> The soil of New Granada has not improved its fertility, or its size, or its healthy climates, or its brilliant geographic setting. Its mines are as plentiful, as numerous, as varied as before. Its domestic animals are the same, the same is the plow pushed by the laborer, agricultural practices that have tilled the countryside since the conquest, guided by the weight of routine, without being altered by a single new idea. The plants that are grown are not different. The roads and navigation have not yet received the desired perfection, nor a great population, nor has much capital come from outside, and peace, the sweet company of the industrious man, was never as serene as under the Colonial Government.
>
> What is it, then, that has changed? The men [and] the principles that guide the destiny of the Republic. With them, a quantity of legislation has reenergized in a favorable manner the public wealth.[48]

The Constitution of 1853, the first of a series of national charters intended to liberalize the nation's political sphere, reflected this sort of enthusiasm. It implemented universal male suffrage, direct elections, and the separation of church and state and moved the country toward federalism.

Of course, not everyone was enthusiastic about these changes. In addition to dissent from the Conservative Party, there was popular opposition to these changes from a faction of the Liberal Party, the Draconians (so named in reference to their continuing support of the death penalty). The Draconian faction, which was distinct from the other Liberal faction, the Golgotas (a more elite faction committed to further change), drew strength from the artisans of Bogotá, many of whom had joined formal societies, some of which had been organized by Liberal politicians. While the artisans objected to the lower tariffs established by the Constitution of 1853, because cheap imports directly threatened their livelihoods, the extension of the franchise had emboldened them.[49] The most famous artisans' societies were in Bogotá, but similar groups existed across Colombia and in the towns of Boyacá.[50] The Draconians also supported military interests at a time when many sought to limit the size and influence of the armed forces.

In 1853 José María Obando, the Supreme of Pasto during the War of the Supremes and the popular leader of the Draconian Liberals, was elected president. Congress, however, was dominated by Golgata Liberals and Conservatives. Early in 1854 Congress moved to reduce the size of the armed forces. On April 17, after months of mounting tension and public confrontations in Bogotá, the commander of the army, General José María Melo, led the army and artisans' societies against the sitting government. Melo's forces easily seized control of the capital. They were surprised when President Obando declined their invitation to serve at the head of a provisional regime. He was placed under house arrest, and Melo assumed the presidency. The coup was opposed by most of the prominent political actors in the country; every living former president joined the self-proclaimed "Constitutionalist" opposition. But the uprising had support from various popular sectors, and many members of the artisans' societies also served in Bogotá's National Guard units.

Melistas eventually seized control of a region centered on Bogotá. There were also uprisings in other cities with active artisans' societies, such Popayán and Cali, and briefer episodes in smaller cities, such as La Cienega de Santa Marta. For convenience, the war can be understood as transpiring in a region defined by a radius of Melista control extending from Bogotá to Zipaquirá and other population centers in Colombia's eastern highlands. Over time the Constitutionalists reduced this radius, and Melo's rule became increasingly dictatorial. Constitutionalist forces retook Bogotá on December 4, 1854.[51] Studies treat the war as noteworthy primarily as a rare moment when class conflict took center stage, for a brief moment breaking the stranglehold that the vertically integrated political parties established over Colombia's formal political life. There is also a sense that the war was most significant in Bogotá and its immediate hinterland.[52] Without disputing the general truth of such ideas, a closer examination shows that in Tundama, the war itself was a serious affair and had repercussions that registered after December 1854.

The years leading up to the war had been marked by tumult and unrest in Tundama, which had been established as an independent province in 1851 and confirmed by the Constitution of 1853.[53] The new provincial government, under the direction of José María Lasprilla and Luis Reyes, moved ahead with plans to end church-related taxes, but it was ill prepared to maintain order.[54] In July 1851 a Conservative uprising took place in Sogamoso, the largest town in the province. The provincial government had no regular forces to deal with the revolt, so it rallied Liberals from the surrounding towns to establish order.[55]

Two years later trouble flared again. On the night of November 7, 1853, politics turned violent in the plaza of Sogamoso. When Francisco Moreno entered the establishment of Mariano Izquierda, he found a number of men who were drinking and insulting the president. He responded by attacking these men. The result of that fray is not clear, but later that evening Carlos and Severo Solano retaliated by shooting at Moreno and threatening him with a sword. Moreno, though hit in the hand by the gunfire, was undaunted and chased the Solanos, who ducked into a store near the plaza. Details on the rest of the night remain murky; rumors of a political murder circulated, though Moreno's injury was the only one mentioned. I have found no reports from Sogamoso until the following March, although the town was clearly ripe for further unrest.

A report from Sogamoso described the rising panic: "some of the inhabitants of Sogamoso have abandoned their homes, startled by the terror that such events have inspired." Officials sought to arrest those it deemed responsible for this atmosphere, meaning Conservatives. To do this, they brought in a unit of twenty-five Liberals from the town of Paipa, a move they thought necessary because they felt they had no trustworthy allies in Sogamoso. Residents harassed the interlopers from Paipa as they made arrests, leading authorities to request additional forces. Eventually men from Santa Rosa, Belén, Paipa, Corrales, and Duitama were called in to help arrest a dozen men.

Writing in November of that year, during the last weeks of Melo's regime, the Constitutionalist governor of Tundama observed that Sogamoso was the first town in the province to rise up in support of the rebellion of April 17 in Bogotá: "In that district no branch of public authority has enjoyed respect[;] . . . it is impossible to prosecute criminals who hide in their houses during the day and at night gather in gangs that molest honest and peaceful people."[56] The particulars of party politics, which were quite complicated during the Melista rebellion, did not apply to Sogamoso. A town full of troublemaking Conservatives was transformed into a regional center of a rebellion led by Draconian Liberals, pointing to the limits of these labels for understanding local politics. As events in Sogamoso make clear, cantonal politics, popular discontent, and sheer tribalism were often as important as the political programs generated in Bogotá. Equally, these events illustrate how the rebellion provided opportunities for disorder well beyond the savanna of Bogotá and the fragility of governmental authority in the small provinces, such as Tundama, created in the 1850s.

As the war progressed in La Salina, it appeared to follow a more conventional narrative. In the middle of 1854 the Melistas' control around Bogotá extended through Zipaquirá and into Boyacá, while Constitutionalists established a congress in the city of Ibagué.[57] Tundama, which supplied the majority of the Northern Army's troops and a large portion of the forced loans levied in the region during the war, saw heavy fighting.[58] In addition to Sogamoso, Soatá and Chita were Melista centers.[59] Gregorio Zarate led the uprising in Chita. Once he had established control over the town, he imprisoned Constitutionalists, robbed those who traded with La Salina, seized 300 mules used for transporting salt, and committed other acts later deemed crimes.[60] This posed an enormous threat to La Salina and, by extension, the Constitutionalist military capacity in the region. In June a military commander in El Cocuy reported that he had sent fifteen men to defend La Salina, explaining, "enemies of the government, making much of whatever small triumphs they might have, will take her."[61] These poorly armed troops were not enough. A guerrilla force from El Cocuy occupied the town on July 2, making and selling salt on its own accord.[62] Reports written after the war claimed that these guerrillas had damaged the roads and bridges into the town as defensive measure. If this was the case, the guerrillas were more circumspect in their treatment of production infrastructure; an inventory taken the next year listed four working ovens.[63]

Melistas held the town until July 27, when a Constitutionalist force led by Juan Nepomuceno Rico attacked from the llanos. Rico wrote that in retaking the town, he had procured a "source of funds."[64] He held La Salina until August 8 or 9 (accounts vary), when a Melista force numbering several hundred retook the town.[65] Retreating back to the llanos, Rico and his men waged a guerrilla campaign. They captured Sácama from the Melistas after three days of fighting that ended on September 7. Then, on September 11, Rico and his men entered La Salina without opposition. There, he reported to the governor, "We learned of the glorious campaign of . . . Pamplona. We were greeted by the population and the agreeable notice that the entire canton of El Cocuy, except for the district of Chita, had risen up in defense of order."[66] The final battle against Melistas in Tundama took place on September 22 near Páramo Sativa. Though many towns contributed troops, the governor saw fit to mention La Salina specifically: "La Salina de Chita also rose as a group against the dictatorship. There was no man that did not take arms [from this] . . . town of 1,150. A company of infantry was raised that was the base of the division of the East victorious in Sativa, with arms that they themselves took from the enemy; brave and long suffering are the inhabitants of La Salina."[67]

Following the fighting came the inevitable review: what had occurred, who had done what, who should be held accountable, and in what condition had the saltworks been left? Finance Ministry employees reported that Melistas had absconded with 5,351.95 pesos (4,620.20 pesos from July and 731.75 in August) and 150 cargas of caked salt (a quantity that implied possession of a significant number of mules, though that was unmentioned), worth an additional 1,500 pesos.[68] The acting administrator, Francisco Antonio Durán, was suspended and charged for the loss or theft of 1,238.56 pesos. The bookkeeper was also accused of working with the Melistas.[69] Durán claimed he had been kidnapped by Melistas and forced to travel with them into Casanare.[70] I found no reference to the outcome of these cases, but the parallels with the aftermath of the War of the Supremes illustrate the perils of ministry employment during civil conflict. While Uribe de Hincapié and López Lopera argue that, except for exiling several hundred artisans to Panama, the restored government of 1855 was less vindictive in prosecuting rebels than the triumphant government of 1842 had been, in fiscal matters the prosecutions were quite similar.[71] The political crime of rebellion might be forgiven, but money owed or lost was not forgotten.

In the aftermath of the war, the Finance Ministry moved to reestablish production in La Salina while coming to grips with the emergency contract it had made with Rafael Niño during the rebellion. The terms of the contract favored Niño, because in the midst of the war he was the only person who could cover the costs of assuming the contractor's role without the support of a loan.[72] The Constitutionalist commander Juan Nepomuceno Rico took over as administrator. This made perfect sense in the logic of the spoils system of Colombian politics. Whether Rico was an effective administrator was a lesser concern than his presence in an unsettled region.

The governor of Tundama wrote to the vice president about continuing disorder in the area:

> At the end of November just past, the provinces of Tunja and Tundama had submitted to the constitutional regime with the exception of the district of Chita. Remarkable for its rebelliousness, [it] still has some guerrillas in the mountains who constantly threaten La Salina . . . There is no other remedy to resolve this conflict but to send the forth division of the Northern Army . . . to serve under the salt administration, authorizing the administrator to arm the inhabitants of La Salina and destroy the guerrillas.[73]

Of this era Bushnell has observed, "The intensity of party competition created a potentially unstable situation; petty outbreaks of violence at the local level were a normal accompaniment of election campaigns."[74] In Boyacá two factors added to this instability. First, as evidenced by election results from the 1840s and 1850s, while Liberal politicians controlled the state, the population was inclined to vote for Conservatives.[75] Second, these Liberals were determined to maintain their hold on power whatever the popular will.[76]

Maintaining order in the province of Tundama was part of this equation. The Finance Ministry stationed troops permanently in La Salina, a force that safeguarded the saltworks and functioned as an element of political control in the region.[77] La Salina's importance as the lynchpin of the federal bureaucracy in Tundama was enhanced because the provincial government was weak: "The only notable source of revenue that exists in this province is La Salina de Chita."[78] Without La Salina, Tundama was not really a viable polity. The provincial government sought to introduce a direct tax, each parish contributing according to its population, with the explicit goal of collecting 200,000 pesos in 1854.[79] This effort was unsuccessful because, among other things, the government was unable to organize the cadastral survey necessary for the tax. During the constitutional reform of 1858, Tundama was reincorporated into Boyacá, an act of regional centralization necessary as Colombia moved closer to its era of extreme federalism.[80]

The problems of Tundama as an independent province were cast into harsh relief by events in La Salina in 1857. While the forces that led to a new constitution in 1858 were much larger than a minor uprising in Tundama, the episode exemplifies the challenges of the moment, specifically, the way the era's radical liberalism, still on the cusp of the dominance it would exercise in the following decade, was articulated in contradictory ways. In the 1850s Finance Ministry officials displayed their increasing frustration with La Salina, which had a rate of profit lower, in terms of gross to net returns, than that of the Cundinamarca saltworks. Apparently it was acceptable to maintain a monopoly that contradicted the economic philosophy of the day as long as that monopoly provided sufficient monetary returns. The solution, apparently, was simply to close the almacén and end salt production in the town. Vice President Obaldía explained that the move was part of an attempt to "put the tribute system in harmony with the principles of the Constitution. To conserve the richest rents that support national expenses and abolish the smaller sources[, which] . . . only help complicate the accounts of the treasury."[81] Of course, shutting the works in La Salina would undermine the cantonal economy, wreaking financial hardship on "the brave long-suffering inhabitants" who had fought for the government in 1854 and the regional consumers and traders who depended on this salt, but such considerations did not figure in these calculations. In his annual report of 1858, Minister Ignacio Gutiérrez Vergara justified the attempted closing by stating, "for seven years La Salina de Chita has been a theater of contradictions and losses for the national treasury."[82]

Despite such comments and despite a decade of disorder, La Salina had returned money to the federal government in the fiscal years 1854–1855, 1855–1856, and even 1857–1858, the year following the rebellion.[83] These were not La Salina's best years, but they were hardly its worst. Ironically, a few months before the rebellion, the government of Boyacá had submitted a proposal whereby it would take over the Finance Ministry's saltworks in return for a fixed payment.[84] Similar proposals were taken seriously in later decades, but this one was rejected out of hand.

Against the backdrop of such national concerns, the plans to close the works in La Salina brought matters to a head in August 1857. Unable to find a permanent contractor and with the threat of closure hanging over head, the Finance Ministry settled for a temporary substitute, José Dovieski, even as it worked to establish almacenes in Belén and Soatá that would be supplied with salt from Zipaquirá, reducing regional dependence on La Salina.[85] The political head of Tundama, Antonio Valderáma, expressed doubts about the efficacy of these measures and warned against ending production in La Salina. This move, he said, would damage the economies of Santa Rosa, Belén, Soatá, and Málaga, for the shortage could be only "imperfectly remedied with salt from Zipaquirá." He urged the government to make a new contract rather than close the works.[86]

People in La Salina sent a letter to Valderáma on August 31 with the signatures of "numerous citizens":

> We have no other industry or other source of subsistence for ourselves and our families than the little we earn from our arduous work of carrying timber, making ceramics and producing salt in this saltworks. Until a few days ago we were paid for this work by the contractor José G. Dovieski. But now . . . the government has stopped paying Mr. Dovieski what he is owed for the salt he had delivered. As proletariat and heads of families, [we bear the burden of the government's] unnatural and unjust act[ion]. Our daily wage goes to buy the bread on which we survive[;] without it we face the horrible possibility of starvation. We are overburdened and have no means with which to travel elsewhere to find employment and so we witness the pathetic and miserable state of our children.
>
> The government can remedy this ill by paying Mr. Dovieski what he is owed, which will provide work in production.[87]

Dovieski was a Polish national who first appears in the historical record after the artisans' rebellion of 1854. He was captured with Melista forces after the battle at Sativa, the very confrontation where residents of La Salina had distinguished themselves, and was listed as one of "the most important men" of the rebellion in the region.[88] While in prison in Bogotá, where he was taken and charged with the crime of rebellion, he sent a series of letters protesting his innocence and asking to be released until he could be brought to trial.[89] Producing written testimony from others to support his story, Dovieski claimed that though he had traveled with a party of Melistas from Santa Rosa to Sativa, he held no rank with the party.[90] Indeed, he had accompanied them only to safeguard his animals, which had been taken by Melistas. Further, he added, he had come to Colombia from Poland because he had heard its "Republican voice" but now felt he had been deceived, since this was a country that did not respect individual liberties. Government officials were unmoved by his words, and one noted that Dovieski might "use more moderate language when he had to address the authorities of a country that had welcomed him in his exile." Despite the disdain shown for his case, Dovieski was sent to Tundama to be dealt with by provincial officials, the transfer being part of a general effort to clear the jails in the capital and have provinces take over the task of deciding whether to prosecute rebels. The details of the judicial process in Tundama are not clear, but Dovieski was probably among the many accused who were granted an *indulto,* an act of clemency, in 1855.[91]

Not long thereafter, Dovieski, cowed neither by the warning to mind his manners while a guest in Colombia nor by his time in jail, was in La Salina holding the position of contractor. The mechanics of moving from a cell in Bogotá to holding a federal contract worth thousands of pesos are not clear. Dovieski had worked for Lucio de Hinestroza in Chámeza and must have still had dependable contacts and access to enough money to guarantee his contract, but I have found no details on this impressive transition.[92] He took a contract to begin delivering 125,000 kilograms of salt in August 1855. Almost immediately he failed to meet the conditions stipulated by the contract, particularly concerning the regular delivery of salt to the almacén, though it is impossible to say whether this was because he was unable to meet the terms of the contract or because he was manipulating production and engaging in shady dealing. In either case, the contract was rescinded by November.[93] In January 1856 Dovieski was fined for nondelivery of salt: six centavos for each kilogram short of the stipulated 125,000 kilograms.[94] Though such violations were routine, and many contractors found it difficult to deliver 125,000 kilograms, the amount Dovieski owed was enormous. The fine for September was 7,500.04 pesos and for October, 8,250 pesos; he was then assessed another fine of 1,200 pesos when the contract was terminated, bringing the total to 16,950.04 pesos. Later these fines were forgiven.[95]

Though he no longer served as contractor, Dovieski remained in La Salina, taking miscellaneous contracts for repairs.[96] He also made at least one gesture indicating his status as a man of substance. Of the 31 pesos collected to build a new road to Sácama, part of the improvements necessary for establishing a customs office, he contributed 6.04 pesos, the largest single donation.[97] In August 1857, when the Finance Ministry needed a temporary contractor even as it planned to shut the works, he was the only one willing and able to fill the role.[98] So by September the scene was set. As Valderáma forwarded petitions protesting the planned closure, a man with a checkered personal history oversaw production, and the ministry, ignoring the signs of discontent, opened an almacén in Santa Rosa that sold salt from Zipaquirá.[99] Residents expressed their view that the state had obligations to its citizens, but their appeal, including their description of themselves as "proletariats," provoked no reaction.

On September 13 the almacén ran out of salt.[100] Worse, there was no salt at the fábrica either ready to deliver or cooking in the ovens. Dovieski claimed he could not afford to finance salt making until he was paid what he was owed for previous deliveries. This was the last straw. On September 20, perhaps with Dovieski's prompting, residents seized control of the town and the saltworks. Predictably, the rebellion began on Sunday, the market day when residents were accustomed to their share of salt water, ensuring the presence of an anxious crowd near the fábrica. In the account of the administrator Pedro Franco Pinzón, the spark was a dispute between a group led by Canuto Vega and the almacén guards.[101] Vega threatened violence but explained that "if he fought, it was because he was hungry." As they wrangled, Pinzón interceded, attempting to explain that notices had been posted about closing the saltworks. Vega responded that all the springs should be opened for production. The alcalde, who was not named, also confronted Pinzón and accused him of acting beyond his authority, telling him that these people were not his to control.

According to Pinzón, all these people were prompted by a "hidden hand," perhaps Dovieski, who later referred to himself as the head of the municipality. Pinzón remarked that he had seen people in the town go without work for six months without this sort of trouble, apparently not considering how the threat to close the works was understood. Whatever the actual composition of the crowd he confronted, and whether or not it needed any leading or prompting, the mob was more than enough to defy the forces the state was able to muster. The market crowd, now rebels, chased out the guards and almacén employees, took control of the saltworks, and openly carried brine to their homes. Others took control of the fábrica and began to make salt.

In correspondence regional officials assured each other that Dovieski led the uprising.[102] One explained, "the foreigner José Dovieski has invited the town to produce salt but not for the government. He has given them wood and ceramics so they might add to the extensive contraband." Another wrote that Dovieski directed residents to tear apart government buildings for fuel and that he was openly allied with monopolists.[103] The former claim seems apocryphal; the latter, inevitable. On paper Dovieski sought to distance himself from the mutiny. In a letter to the commander of the guard unit, he explained that he was acting as La Salina's highest civil authority, not leading the rebellion, and he had accepted this burden because no one else was qualified to take the position.[104] He allowed that some salt had been made illegally but insisted that the town was completely calm.

Officials were skeptical of these claims. Letters to the Finance Ministry written between October 6 and 10 agreed that the government could reclaim the town only by force.[105] The captain of the guard reported that such an attempt would involve fatalities. He also explained that even outside La Salina, it was widely believed that the "Presidency" had acted "against the interest of the town" when it tried to shut down salt making: "Now it is generally accepted that the town is perfectly within its rights to rise up against the Government and continue in its appropriation of the saltworks." Others explained that the rebels had been convinced they would be pardoned for their actions—a fairly reasonable calculation based on the practices of the time and the limited power of the state. There had been pardons granted for far worse offenses, and Dovieski provided a fairly convincing reminder of this point.

The descriptions of cantonal support for the rebellion were accurate. In Chita the perennially recalcitrant residents held "nightly reunions and . . . decided to mount resistance to any invading force." In the town of El Cocuy, the cantonal seat, the "important people" had offered no aid in dealing with this brazen defiance of the government. Finance Ministry officials expressed concern that the rebellion might grow into a general uprising.[106] Neither the government of Boyacá nor the subordinate apparatus in Tundama could marshal forces to quell this rising tide of unrest. A letter to the governor of Boyacá suggested placing an embargo against salt from La Salina, for there was no money for a military expedition in the rugged terrain around the town. "The situation of rents in this State is not pleasant; Tundama barely covers its expenses; Casanare is notorious on the same count; and Tunja cannot be said to have rents, only very large debts."[107] In another letter to the governor of Boyacá, Ignacio Gutiérrez suggested that, since the state government was powerless and the works were ministry property, perhaps the national government might intervene.[108] A marginal note to a petition from Tundama suggesting such an action stated that this was an "insignificant" mutiny.[109] The regional government would have to raise a force of 200 men, dealing with this uprising as it had with the one in Sogamoso in 1854. But this time there was no ready regional support on which to draw and nothing with which to pay those who might serve in such a militia: the traditional source of emergency funding for regional campaigns was the center of the insurrection. The fruitless debate dragged on. Government officials in Tunja argued for national intervention, while the prefect of Tundama explained he could do little to suppress the rebellion because of his poor health.[110] In the meantime, people in La Salina made salt and, presumably, sold it.

Throughout October reports that the situation was settling down were contradicted by notices such as this one from Tundama's pessimistic political chief: "I lose hope that the conciliatory measures of the government will be sufficient to halt the impetuous machinations of its enemies."[111] The national government was the first to blink. On October 22 President Mariano Ospina Rodríguez signed a decree that the works would remain open, though the rebels maintained their control of the town until the end of the month, making salt as they pleased.[112] Then, in early November, Rafael Niño and Evaristo de la Torre took a production contract. The governor of Boyacá reported that La Salina "was perfectly calm and obedient to the authorities."[113] But on November 15 the administrator wrote with the news that a shortage of timber had alarmed the community:

> The residents have given me a letter saying that without work they and their families will starve. Two of them have told me that they had planned to seize the whole town and the saltwater springs tomorrow to produce salt themselves. They were counting on support from El Cocuy and other neighboring towns. The new contractors have not arrived to start work, and it is said that they will not arrive for the rest of the month. I have no funds and find the situation so critical that I have contracted with José Dovieski to prepare the ovens and I have supplied 200 pesos to buy wood. I hope this measure will calm the town and it does not hurt the state as the timber will be bought by the contractor whenever he arrives.[114]

Officials sent nervous reports through November, even after Niño arrived to oversee production, but no violence ensued.[115] Dovieski continued to play a crucial role in the industry. Quoting the new alcalde, Valderáma reported:

> There is an individual in this district (the Sr. Jose G. Dovieski) who constantly attempts to inculcate in [the people of La Salina his own spirit] of disobedience, and even hostility, to the Government of the Republic[;] . . . it is important that the local authorities take every appropriate measure to work against this

> dangerous and malicious influence, daughter of an ungrateful spirit, ignorant of the many favors received from the *patria*. Therefore proceeding with the greatest care in investigating these events, taking declarations from people [who were involved] in attempts against members of the guard unit and the seizing of the springs.
>
> The tolerance or pretense in this matter leads one to believe that the highest local authority of the district was complicit, if not actually involved in the criminal and scandalous maneuvers of the sworn enemies of order.[116]

Apart from the decrees forgiving the fine Dovieski had accrued before the rebellion, this is the last reference to him that I have found. It is also the last explicit reference to the rebellion in Finance Ministry correspondence. As an act of defiance, the rebellion was enormously successful: there were no subsequent attempts to close the works, and it appears that no one, not even Dovieski, was punished for participating. The ministry found a renter rather than a contractor to manage production, an action that distanced it from direct control over the industry. The move fit the liberal spirit of the age, but it was a pale substitute for the complete withdrawal that the ministry had attempted.

∽

State monopolies and the logic of state building dictated that the Finance Ministry could not simply abandon the operation at La Salina; unregulated production there would have presented too great a challenge to health of the salt monopoly centered in Zipaquirá. As the ministry found, however, neither could it close the La Salina saltworks; the people of the canton of El Cocuy, from local notables who were supposed to keep the population in check to the poor whose only livelihood was carrying bundles of timber, would not allow it. The strong feelings and deep interests there eventually led to insurrection, and while it is clear that La Salina was full of networks and groups that were often at odds with one another, there is no hint of anyone opposing the rebellion in any fashion. In this light, it is tempting to see La Salina as sui generis, but that would be something of an overstatement. There were currents of discontent across the eastern highlands and along the Casanare border; La Salina was only the place where they were manifested most clearly. There the pitch-perfect confluence of popular discontent and profit seeking, a product of state intent and state action, exploded.

If residents of La Salina had not already reconceptualized their relationship with the state after twenty years of direct administration, they did so after the Melista rebellion. Their use of the terms *citizen* and *proletarian* in their petitions underscores the degree to which the "long-suffering" residents of La Salina felt they had a right to protect their interests. At the same time, influential patronage networks continued to shape life in the town and throughout the canton. For every reference to the rights of the citizenry, there were comments that all these people simply did the bidding of some figure whom all knew but, with the exception of the foreigner Dovieski, who could not be named. Still, patronage networks and local notables could not stir an uprising without an army of clients to do their bidding, and in truth the entire population of the canton viewed the Finance Ministry's effort to close the works as illegitimate. With all these ambivalent messages and forces in play, La Salina stood poised to enter Colombia's heyday of radical liberalism.

CHAPTER 6

La Salina, Boyacá, and Colombia after 1857

> Marriage, the family, the tribe, the pueblo, the race; the district, the canton, the province, the State, the Patria, the Continent, the universal Republic, the spirit of a place, provincialism, patriotism, Americanism, cosmopolitism, theocracy, feudalism, absolute monarchy, constitutional monarchy, the centralized Republic, the federal Republic . . . express the perpetual aspiration to fuse oneself in a communion more perfect and more universal.
>
> —Salvador Camacho Roldán, *Memoria de Hacienda y Fomento de 1872*

In La Salina 1857 marked the zenith of local defiance to the Finance Ministry. In the nation as a whole, it was the year when Colombia moved from being a country composed of provinces to one formed by increasingly sovereign states. Formally this process played out in the ratification of two constitutions and a major civil war in the next decade; in practical terms, however, large states, often formed by pulling together smaller provinces, simply took on a stronger presence in national politics. The state of Boyacá was formed from the provinces of Tunja, Tundama, and Casanare and parts of the old province of Vélez. Compared to regions or states such as Antioquia, Cauca, or the Caribbean coast at this time, the state of Boyacá has received relatively little critical attention with respect to its role in civil wars despite the region's status as Colombia's most populous province for much of the century. This lacuna arises partly from the perception of Boyacá as a region solely of pious, uneducated campesinos whose dedication to the church translated into unequivocal loyalty to the Conservative Party and who provided unthinking foot soldiers in various conflicts. This understanding has been bolstered by retroactively applying twentieth-century history, particularly the partisan activity of Conservatives during la Violencia (1948–1957, in the shortest estimate), to the nineteenth century. Further, there is a tendency to lump all of Boyacá, despite its diversity, into a single category. This perspective is historically inaccurate but has had a noted influence on scholarship about the region.[1]

One way to correct some of these simplified understandings lies in ac-

FIGURE 6.1. "United States of Colombia, 1863-1885."
Taken from James William Park, *Rafael Núñez and the Politics of Colombian Regionalism,* with the permission of the Louisiana State University Press.

knowledging Boyacá's internal diversity, with a particular emphasis on the role of Tundama, and considering both Boyacense political autonomy and the limits of that autonomy. Boyacá occupied a specific and oddly static place in the Colombian political system. Studies of nineteenth-century politics tend to emphasize the role of the Liberal and Conservative parties: when they developed, how they were informed by popular sentiment, how regional factionalism played into their composition, and so on. It is generally understood that specific states were dominated by a particular party. For example, Antioquia was politically controlled by the Conservative Party and supported Conservative regimes, and that fact shaped its role in civil wars during the second half of the century. Similarly, Santander, created by merging the provinces of Pamplona and Socorro, was unequivocally Liberal. Generally speaking, Santander supported Liberal regimes in Bogotá. When it did not, it was because these regimes were not sufficiently radical.

When viewed in those terms, Boyacá is understood as a Conservative province. Nonetheless, many prominent Liberal politicians of the era were Boyacense and had political backing in their home province. This apparent contradiction is resolved if the national political system is understood in terms of regional balance rather than ideology. During periods of peace (or relative peace, for provincial rebellions were frequent), the logic of Colombia's internal geopolitics produced a regional balance. As a part of this balance, the political apparatus of Boyacá, as controlled by the state government in Tunja, was allied with the national regime in Bogotá, whether that regime was Liberal or Conservative. The internal politics of Boyacá were often shaped by the effort to ensure this support, which frequently produced tension between Tunja and population centers in Tundama, especially Sogamoso and to a lesser extent Santa Rosa, which had predilections for opposing the state regime.[2] This arrangement held both when the government in Tunja was a stalwart supporter of a midcentury Conservative regime and when it backed the Liberal regimes of the 1860s.

Events in La Salina document the dynamic nature of these intersecting influences, demonstrating that Boyacense "quiescence" was not static but rather dynamic and had to be managed and maintained. This local history, integrated as it was into national networks of influence and finance, also suggests that in Tundama the spoils system was remarkably flexible. The last forty years of the century are divisible into two periods. The 1860s and 1870s brought the enshrinement of a doctrinaire political liberalism as the guiding philosophy of the national government. Many of the most powerful figures involved in the salt industry in La Salina during these decades were regionally important Liberals, commanders in the field during war and influential figures in peace.

In the 1880s politics shifted. By 1886 Colombia had formally entered the period termed the Regeneration; it had a new constitution that granted the federal government more power than it had ever held before, ushering in an era that gradually turned into one of Conservative dominance. Marking the change by the 1890s in La Salina, explicit references to Conservative loyalty had replaced fretting over the immorality of market "monopoly." Despite this new orthodoxy, Liberal figures involved in contracting and production during the 1870s continued to work, hold contracts, and dominate the salt works into the Regeneration. This apparent contradiction suggests that in Tundama political peace was worth more than pushing out established figures in favor of Conservative clients. In this, at least, it seems that Finance Ministry officials had learned the lessons of La Salina: outside governments—whether in Tunja or Bogotá—had to choose how to wield their limited influence. As with the profile of the salt industry and the resident population, the underlying theme in this tale is that the era's political shifts overlay deeper continuities.

La Salina and the Liberal Era

The years following the rebellion of 1857 brought the nadir of ministerial control in La Salina. Yet that disorder and the extreme federalism of the Constitution of Ríonegro did not significantly alter the process whereby La Salina was increasingly integrated into the national political system. During these years, events in La Salina paint a portrait of the spoils system in Colombian politics. The dramatic developments of the late 1850s and early 1860s began with events in Cauca, which had its own mix of volatile politics and rivalry with its neighbor Antioquia. Cauca was controlled by Tomás Cipriano de Mosquera, who had moved from being a Conservative to becoming the leader of a Liberal faction identified with the Draconians.[3] In the eastern cordillera, Santander was dominated by Radical Liberals and at odds with the Conservative regime in Bogotá. The government of Boyacá at the time was dominated by Conservatives,[4] though they were divided into factions. The split, however, did not harm the state's standing as a stalwart of the national regime. Rather, in an example of the political geography previously described, Boyacá stood with Cundinamarca and Antioquia as one of "the three States in whose loyalty [the government] could trust in completely."[5]

As the decade ended, a number of states were racked by rebellion. The Liberals in control of Santander suspected that the Conservative rebels there were receiving aid from the federal government. The disorder that spun out of fighting in Santander morphed into a single war that lasted until 1862, some months after Liberal forces led by Mosquera took Bogotá. This was the only time when a rebellion succeeded in removing the sitting government and putting the opposition in power, Melo's brief regime excepted.[6]

In Boyacá the fighting, which began in 1859, involved the state government's efforts to put down Liberal revolts while seeking to stave off open warfare with Santander. By 1860 this was no longer possible. When forces from the two states met at the Battle of El Oratorio on August 16, Boyacá's troops defeated the Liberals.[7] In Tundama, Sogamoso and La Salina had emerged as trouble spots. Campaigns in 1859 and 1860 solidified the latter's status as a field resource and marked its emergence as a center of, if not popular liberalism, then at least Liberal caudillismo. Rafael Niño and Gabriel Reyes Patria, familiar figures in La Salina, played a role in the attempted counteroffensive following the defeat at El Oratorio.[8] On August 26 the administrator in La Salina, who was on the run, sent a report to the Finance Ministry:

> Yesterday at eleven in the morning, there appeared on the road to this saltworks Sr. Juan Nepomuceno Rico, associate of Lorenzo Valderáma and of Sr. Solon Wilches, and fifty more men including Baltasar Sandoval, Leonardo Estava and others from Sácama, Estavan Barón and Pedro Escobar of Chinivaque with various armed peons. They attacked the commander Juan de Dios Jiron,

> who had gone to Rodrigoque with the guards to see if they could prevent the entrance of Sr. Rico, who we were told was coming to throw himself on La Salina. Between all of these people and the commander and the guards there was shooting that killed the guard Rufino Vergara and wounded the Commander and the guard Fernando Peña and who knows who else. They entered La Salina and took control of the administration. Today they are running it as if [for the ministry] and are continuing to cook [salt] in the ovens that were lit, in order to sell it. I was informed that the employees of Sr. Rafael Niño, Srs. Javier Tejado and Medardo Cuadro, ordered the fires put out but later agreed to continue cooking them and to sell [the salt] for Sr. Rico and his partners. I have escaped and the funds I had in the strongbox in my possession are buried.

He closed by writing of his hope that government troops would come "and take from the Liberals the rent" they had seized before they could mount justifications for this action; "Until now," he wrote, "they have only said they have done this because they are Liberals."[9] Niño averred that he had been forced to comply with rebel demands, but this claim was greeted with skepticism, for he and Rico were well known as business partners.

With the saltworks at Recetor, Pajarito, and Chámeza also under rebel control, the government in Tunja responded to this threat to the nation's fiscal system. In early September Lieutenant Colonel Guerrero led 100 men in an attempt to retake La Salina, but this force was repulsed with heavy losses. Later that month the government sent a force of 800 men and took the town without a fight. Liberal forces abandoned La Salina rather than face this attack. Some retreated to Arauca, and others, led by Rico and Reyes Patria, fled toward Venezuela.[10] They were pursued by government forces that succeeded in capturing arms, prisoners, and thirty loads of salt but none of the leaders. Shortly afterward, Reyes Patria appeared at the head of another uprising in Sogamoso, which, according to government correspondence, targeted La Salina. The government sent troops led by Colonel José del Rosario, who defeated Reyes Patria's "guerrillas" on a road into La Salina on December 20, 1860.[11] This is the last reference I have found to fighting around La Salina. By 1861 the fighting in Boyacá was closer to Tunja, which Liberal forces captured briefly in February but could not hold.[12] Reyes Patria's forces participated in an April battle that brought much of Boyacá, though not Tunja, under the control of Liberal insurgents. From that point forward, the war in Boyacá was swept up in the history of Mosquera's successes as he advanced out of Cauca. Uniting with forces from Santander, his army took Tunja and in July triumphantly entered Bogotá. This settled the outcome of the war, though fighting continued through the year.

Mosquera assumed the presidency, and the successful Liberals convened a constitutional congress. The result, the Constitución Política para la Confederación Granadina of 1863, or Constitution of Ríonegro, provided for an extreme version of federalism where power was held largely by the nine sovereign states.

Despite the radical implications of this federal system, there was no follow-up to the fiscal reforms of midcentury. The anachronistic salt monopoly endured. After all, the early 1860s were the precise point when customs receipts reached their lowest ebb, when salt returns, which were always important for paying salaries and guarantying emergency wartime loans, brought in as much as customs.[13] The reform best capturing the spirit of the day was a reorganization of the Ministry of Finance, which was remade as the Ministerio de Hacienda y Fomento, or Ministry of Finance and Development. Adding development to the charges of this ministry would be a crucial element in the survival of the salt monopoly during an era dominated by politics that demanded its abolition.

In La Salina theses changes, coming on the heels of the rebellion of 1857, meant that neither the Finance Ministry in Bogotá nor the near-sovereign state government in Tunja had the inclination or ability to exercise much authority. Though Angel Galán inspected the saltworks and almacén in 1862, producing a portrait of an industry operating outside the ministry's control, no one sought to directly challenge the serving administrator, Juan Nepomuceno Rico.[14] The three preceding civil wars had been followed by concerted efforts to prosecute rebels; this war prompted a general amnesty. On May 6, 1861, Mosquera had issued a decree granting clemency to those who had fought against him during the war. The conditions of the decree were relatively forgiving and preempted the standard process of appeals and petitions.[15]

In La Salina the people who had sided, or at least worked, with Rico when he was a rebel commander, including the perennially appearing figures of Estavan Barón and Medardo Cuadro, were free to continue managing their affairs as they liked. Many of these men were a part of the political networks that could maintain Liberal dominance in the eastern highlands. La Salina was a key resource for this network in Tundama, one that could be used to ensure that Boyacá's frequently rebellious eastern cantons would not align themselves against Tunja. Ensuring political quiescence in Tundama may not have been inherently difficult, for popular politics there seemed to echo the radical inclinations of Santander, but it was important.

Military commanders across Boyacá and Santander sent requests for funds directly to the administrator of the monopoly in La Salina. In real terms this meant that instead of working through the formal networks of the Finance Ministry and the regional juntas de hacienda, former rebels wrote directly to their ally Rico for funds.[16] In this fashion, the revenue system was subverted by regional politics and partisanship, and there was no effort to include a critique of these practices in broader attacks on the salt monopoly. In his 1865 annual report, Minister Tomás Cuenca saw fit to observe, "In La Salina de Chita, for example, there are eight [springs] . . . and all are used for contraband," but he offered little in the way of a solution or further moral condemnations.[17] Cuenca's failure to elaborate on this critique may have been linked to Boyacá's place in the political geography of the country during the 1860s. It had been a key supporter of Ospina Rodríguez's Conservative regime of the 1850s, but now it joined Cundinamarca and Santander as one of the three states that could be counted on to support Liberal governments.[18] The government's salt monopoly was still suspect. The *Memoria* of 1864 included a supplementary report on salt written by R. Peréz, who argued that though the monopoly appeared antiliberal, it was actually dedicated to the greater tenants of liberalism and by extension to the Liberal Party.[19]

Despite such intellectual exercises, critiques of the federal monopoly grew. Some of the fiercest were voiced by the state government of Cundinamarca, which eyed the enormous revenues that Zipaquirá generated for the federal government with dismay. In May 1869 the Finance Ministry attempted to mollify the government of Cundinamarca by arranging for an amendment to the fiscal code so that each kilogram of salt sold in a government almacén put one-half a centavo into a fund that was then divided among the five states where the national monopoly was still in force: 35 percent for Boyacá, 30 percent for Cundinamarca, 22 percent for Santander, 10 percent for Tolima, and 3 percent for Cauca.[20] This plan was neither well run nor very successful. The money was not always sent in a timely fashion, and the intended recipients did not always get what was sent.[21] For example, in one note the administrator of La Salina claimed that because proceeds from the saltworks had "not been enough to cover urgent costs in the last few months," he had been unable to send the mandated funds.[22] As with La Salina after 1857, though the ministry's actions did not satisfy anyone, they did ensure the survival of the status quo. In La Salina the result was decidedly unfavorable for the ministry.

Rebellion in Boyacá, 1871

Rebellion once again broke out in Boyacá in 1871, a dramatic episode sparked by discontent and culminating in a battle where armed forces faced off before a backdrop of flaming buildings as the capital waited to hear whether this disorder had been contained or would spread. Tales of a conflagration that threatened the nation itself were perhaps exaggerated in Bogotá, but the rebellion is a noteworthy episode, for it reflects some of the challenges to Liberal control over Boyacá. It also perfectly illustrates Boyacá's place within the national political system. The events of this rebellion move this story away from La Salina, to the center and west of Boyacá, but they are worth considering.

Conservatives had a number of reasons for discontent after the expropriation of church lands in 1861 and the education reform of 1870, which established secular education.[23] The federalism of the Constitution of Ríonegro, however, helped buffer Conservative states from national trends. Indeed, many of the decade's political struggles involved conflicts between Mosquera and Radical Liberals. After his removal from office in 1867, a string of Radical Liberals were elected president, each serving a two-year term, pushing the politics of the era. The education reform of 1870 was a part of this process and is generally considered one of the causes of the civil war of 1876–1877, which began with a Conservative uprising. Both Liberal factionalism and discontent with the high-handed behavior of some Radical Liberals were evident in Boyacá at the beginning of the decade.

The rebellion of 1871 was born out of opposition to Governor Felipe Pérez, whose rule met disfavor from various Liberals throughout the eastern highlands. Viewed from one perspective, Pérez was a politician who was "dedicated to change, to advanced ideas, . . . and the advance of civilization in the name of progress."[24] In more concrete terms, this meant that his government showed a marked suspicion about the role of the clergy in public life and at times acted on that suspicion. A letter from the political head of the Northern Department to the Ministry of the Interior captures the sense of distrust:

> Mr. Secretary should be aware that the majority of the priests of Boyacá are plotting an action against the existing general government and that of the State. In this manner the priests of Soatá, Guican and Capilla in this department worked openly during the last elections, exhorting the residents from the pulpit and the confessional and handing out ballots.
>
> The openness of the priests is unbelievable. They have now alarmed the entire population and excited religious fanaticism to the point that the authorities are not obeyed and are offered public insults. If the present congress does not pass a law to control them, society will very soon feel the horrors of a fanatical reaction that they are preparing.[25]

The threat of a priest-led Boyacense mob was a concern for the Liberal establishment beyond Boyacá. An editorial in the Bogotá paper *El Tiempo* fretted, "It appears that [Boyacá] is the only State in the Union of Colombia where the majority remain, until now, immersed in a misguided dark cloud, due to the influence of the clergy."[26] But blaming Boyacá for regressive thought did not translate into support for Pérez. In assessing the rebellion of 1871 after the fact, the editorial page of *El Tiempo* condemned both Boyacense militarism and Pérez's government: "As long as those who govern have a bureaucratic army to win elections, as long as employment is considered something that is not linked to merit but [is] a prize of combat and the spoils of winning elections, it will be vain to think of consolidating this peace."[27]

Pérez had paid lip service to working across political divisions, proclaiming in 1868, "to live in peace [and] to live as brothers are the most important tasks of the moment. . . . Boyacá with its fertile soil and agreeable climate needs only peace and harmony among its children"; there is, however, no evidence he acted in this spirit.[28] Pérez not only had to deal with Conservatives and rival Liberals in Boyacá but also, according to some accounts, faced opposition from President Eustogio Salgar, who opposed Pérez's rule and even facilitated the rebellion by withdrawing federal troops from Tunja and refusing to sell arms to Pérez's government.[29] Aquileo Parra, president from 1876 to 1878, later wrote that one point of disagreement between Salgar and Pérez involved a plan to turn saltworks over to foreign interests as a means of paying the national debt (Parra, however, is vague on the details).[30] As always, in a small way salt figured into a larger story of political control.

Rumors of an imminent revolt began to circulate at the end 1870, and in Leyva and Sogamoso, the second week of the new year saw uprisings under the direction of Policarpo Flores.[31] By January 25 Bogotá had received news of a battle in the central canton, where 800 rebels had faced a force half that size led by Pérez. Events of the next few weeks are undocumented, but by the end of February the insurgents had installed themselves in Tunja.[32] Once in power, the rebels claimed their object was liberty, progress, reform, and popular suffrage: "Remember the public scandals that in less than two years have sadly distinguished the sovereign state of Boyacá . . . [including] the unmasked replacement of popular suffrage." They added, "It is now indispensable to demonstrate that this state has never consented to what has been implemented by fraud." The editors of *El Tiempo* were not impressed. An editorial in March argued that the rebellion was a scheme backed by Santos Acosta, a Boyacense radical who had served as acting president for Mosquera during the final year of his 1866–1868 term, and Sergio Camargo, of the Colombian guard, so that they could manipulate Boyacá's single state vote in the coming presidential election. This possibility was viewed with alarm. Though the paper claimed not to "doubt the patriotism of the assembly born of a triumphant revolution," one that raised "the banner of reform," it called for the new government to abstain from the coming elections as proof of its good intentions.[33]

Because federal law forbade the national regime from interfering in state politics, perhaps such a promise would have helped the rebel government survive. Or perhaps it was doomed from the start. In either case, the government in Tunja offered no such promise; further fighting was coming.[34] Articles and editorials in both *El Tiempo* and the *Diaro de Cundinamarca* argued that the Conservative groups that initially gave the rebellion tacit support were increasingly disenchanted by its rule.[35] This not only weakened the new regime but also meant that those in power had less control over the unrest they had let loose. The Liberal elite of Bogotá expressed alarm over the specter of a population deemed untrustworthy: "That State, the most populous, is the one that has the most Generals of the new crop. . . . [I]t is the most warlike because its Indian character serves as an inexhaustible breeding ground of machine-like soldiers. If it is understood that by only a few days' journey from the capital they could overcome the general Government, and perhaps the two neighboring states, what Cauca has been unable to achieve, the Generals of Boyacá might try."[36] Leaving aside the irony inherent in the fact that a general from Cauca itself had succeeded in overthrowing the national regime ten years earlier, the point was clear. Though he had his critics, Pérez was needed. Liberal interests, including

Manuel Murillo Toro, who would be elected president in 1872, offered "the indomitable Pérez" support in his efforts to retake his office.[37] In early April *El Tiempo* reported that armed parties under Pérez's command were prepared to cross into Santander and western Boyacá from Cundinamarca, an area where the rebel regime had little support.[38]

April proved decisive. The government in Tunja moved to suppress a challenge from Pérez's ally Venancio Rueda, the political head of the Western Canton, and fighting followed.[39] At a battle in Guateque, Rueda's forces defeated a rebel force of 400, killing 70 and taking over 100 prisoners.[40] Pérez issued a proclamation that claimed his peaceful overtures had been rebuffed but also announcing that he would soon be in Tunja.[41] His prediction was correct. He solidified his control of the areas bordering Santander and from this base announced the formation of a provisional government in Santa Rosa.[42] After two weeks of victories in minor skirmishes, Pérez collected his forces and led them into the field for a final confrontation. This battle took place on May 1 in Paipa, where his forces won a decisive victory.[43] Soon the paper of record, *El Boyacense,* resumed publication and announced Pérez's return to power. The first issue contained his proclamation from Paipa, which he had written immediately after the battle:

> Soldiers! Without any interest except the triumph of law, you have fought for twenty-one straight hours against a fierce enemy, superior in number. They encircled you in a ring of flame, and with the horrors of combat they wished to mix the shameless picture of a hundred families thrown from their houses by the cursed hand of fire!
>
> Paipa, loyal town, honored town, is today a smoking dump[;] the fire of the revolutionaries has embraced all except for the strong heart of [Paipa's] sons![44]

Pérez went on in a more conciliatory tone, commending "soldiers of both armies" for having done their duty. He resigned as governor by the end of May.[45]

Venancio Rueda led the restored government. Under his guidance, the regime balanced reconciliation with settling scores. The government offered amnesty to the rebels in 1872 but condemned their actions in Paipa, insisting that they had violated the unwritten rules of honorable combat.[46] The key charge was that the rebels had fired houses when they occupied Paipa, an allegation supported by testimonies published in *El Boyacense.*[47] One rebel soldier, a seventeen-year-old student from Belén, testified that General Jesús María Chapparo had ordered his soldiers to burn houses, though he claimed he tried to prevent his comrades from following the order.[48] A resident of Paipa, Andrea López, testified that rebel soldiers had threatened to burn down her house and that she had seen "a multitude of soldiers on the patio, the majority of them Indians."[49] With respect to burned-down houses, the testimonies from *El Boyacense* specifically mention only those of Andrea López and the Ochoa family, though others may have been destroyed as well. Mainly, it seems the soldiers had used the threat of arson to extort money, food, and liquor from residents.[50] Pérez's references to a ring of fire were an exaggeration.

The political head of the Central Canton ordered that the military victory should not be used as an excuse to abuse suspected rebels, but Rueda's restored government did not follow through on this conciliatory tone.[51] The assembly passed one law allowing the governor to jail people for up to a year for disturbing public order and another levying a forced loan of 40,000 pesos on opponents of the regime.[52] Though the editors of *El Tiempo* deemed these laws too harsh, they agreed that Boyacá needed guidance from modernizers such as Rueda. They praised his administration for bringing peace to the state that, "bloodied and impoverished," sheltered "in its breast the most corrupt and demoralizing element of all time, an ignorant clergy, full of superstition and fanaticism."[53] If the Liberal establishment of Bogotá was pleased with the spirit of Rueda's rule, it must have experienced a more profound and gratifying emotion at the national presidential elections of that year, when Boyacá cast its vote for Murillo Toro, the leading light of the radical republic.[54] The only connection between La Salina and this event I have found is tenuous: Rafael Niño's name topped a list of officers from the Army of the North in Tundama who wrote to congratulate the restored regime in Tunja and offer their support.[55]

Despite this successful effort to keep Boyacá's formal political participation Liberal, a letter to the president in September 1871 made it clear the state was not fully pacified. Enrique Cortés, who addressed President Salgar as "respected friend," reported on being waylaid after attempting to meet with public officials in Ubate to help them organize public education.[56] After failing to meet these officials, he and his ailing father were traveling from Ubate when they were stopped by a group of armed men who claimed they were part of a provisional government based in Samacá under the command of a Colonel Acero. His letter is full of the mundane details of an evening in the group's company, including their threats and his eventual release. Cortés dismissed these men, labeling them criminals, and their espoused program of Boyacense sovereignty. While they claimed to oppose the state government, "their hostility to the national Government and to the state of Cundinamarca" was evident in the circular they gave him:

FELLOW CITIZENS!

> We agree with the dignified sons of Boyacá in Pachavita who have taken up arms. Today we have the honor of following them. No more shame.
>
> The titled legitimate government of our State is each time more illegitimate, more fraudulent and oppressive; more and more insulting to the half a million free men in the Republic who carry the name of boyacenses. The Governor of Cundinamarca each time is increasingly given the air of the arbitrator of the

> destiny of our patria. And that of the nation, responsible by law for preventing the aggression of one State against another. . . . What remedy is there to end this internal tyranny?

I have found no other references to this guerrilla force or the uprising, but the episode illustrates the ongoing tension over the role of Cundinamarca Liberals and Bogotá in the state's politics.[57]

The Zenith of the Radical Liberals

Jacinto Corredor began his tenure as the head of the salt monopoly in Boyacá early in the 1870s. Instead of taking up his post in La Salina, as had been the practice, he directed operations from the almacén in Sogamoso. His report written in 1871, *El informe del administrador principal de salinas de Boyacá,* shows that he was a committed Liberal imbued with sense of drama. The plan to establish this almacén was another attempt to lessen the region's dependence on La Salina by introducing salt from Zipaquirá across Tundama. Echoing Finance Minister Salvador Camacho Roldán, Corredor argued that effectively administered, the salt monopoly would "awaken activity and enthusiasm" in regional residents as they witnessed the benefits of trade and the entrepreneurial spirit of those delivering salt to the new almacén. Further, the new system was "favorable to the common mass of Boyacenses."[58]

In response to unnamed critics in the Finance Ministry, Corredor listed his powerful supporters in eastern Boyacá, including Januario Salgar, who served as finance and development minister from 1868 to 1870. At the end of his report, he offered a defense of his service that turned as much on the favor of Liberal networks in the region as it did on any real evidence of success:

> On other occasions I have told you that the towns are satisfied and thank the Government for having alleviated the cost of salt, and, let me add, there have been repeated signs of this satisfaction. Among the respected people who have presented themselves personally to me [on this manner] let me cite the Sñr Montoyas, distinguished businessmen[;] . . . the elderly and virtuous General Juan José Reyes Patria and his two sons Gabriel and Manuel Reyes; the Sñrs Chaparros, Tavera i Montejo, the Sñrs Corteses La Rota, Rafael Niño, Narciso Tórres, Prietos, [and] Soler i Holguin; the Sñrs Escovares, Parras, Camilo Réyes, Archilas, Camargos, Tejadas, [and] Duran i Valderáma; and the Sñrs Abellas, Návas, Cárdenas, Diaz, Gutiérrez and a thousand more. And if I tried to list those from the middle and lower classes, they would not fit.
>
> Moreover, soon the Senators and Representatives will be arriving in the Capital of the Republic[, and they] are competent to judge what is beneficial or prejudicial for their constituents and the country. They will give the Government, and the Nation, the final word on the matter.[59]

Whatever political support Corredor had, this almacén did not succeed, whether in Sogamoso or in Santa Rosa, to which it was relocated in 1879.[60] The cost of bringing salt from Cundinamarca was too high, and consumers preferred the caked salt of La Salina to the kettle salt sold under this plan.[61] Regional politics could not override regional preferences.

Corredor's belief that the monopoly would be a catalyst to transforming society may have been quixotic, but it was in keeping with the thinking of Salvador Camacho Roldán, whose bold plans and predictions for the salt rent were presented in the early 1870s. Writing two decades after the economic reforms directed by Murillo Toro, Camacho Roldán offered a zealous vision of the way salt would vitalize the domestic economy and the nation's broader transformation:

> What a powerful boost it would be for the weakened internal commerce of this country if the price were lowered on this product of such wide and universal consumption! The salt of Zipaquirá, carried to Cauca, would bring in return tobacco, gold and excellent horses. The salt of [La Salina de] Chita would bring the hats, the tobacco, cacao, coffee and the sugar of Pamplona and Socorro [to Boyacá]. The farmers of [Pamplona and Socorro] would use this [salt] to improve the feed of their cattle and to raise their mules. There would not be, as there are now, Granadinos who eat without salt because they cannot afford to buy it.[62]

Roldán's ambitious plans, whether for development or for rental of Zipaquirá to cover the foreign debt, were never realized. But his writings are worth considering, for few others combined such expansive vision and precise thinking. Roldán's tendency to get carried away, despite his access to the details of the monopoly's history and intractable inefficiencies, testifies to the enduring power of the allure of the salt trade when it was imagined as a stimulus for development.

At this point, some of the energy that had been dedicated to critiquing the state monopoly was diverted into the process of earmarking money from salt for other projects related to development, particularly railroads. Once this link was established, proposals to end the monopoly were never as sincere as they had been in the 1860s, and powerful elites in the eastern highlands had much of their discontent assuaged, for they were often investors in these projects. For various reasons—some based in the extreme federalism of the Constitution of Ríonegro—the federal government had only limited involvement in railroad construction until 1872. Then Roldán helped pass legislation directly linking the national government to railroad construction.[63]

In his report of 1873, Finance Minister Aquileo Parra (who was president in 1876–1878) succinctly analyzed the state of affairs: "The result is that the fiscal situation of the Nation can be summarized simply and accurately in the phrase: THERE IS NO DEFICIT."[64] But his report, penned the very year when it might have been possible to abolish the salt monopoly, contained long sections on Colombia's poor transportation infrastructure and lack of industrial development in Cundinamarca, Boyacá, and Santander, the three eastern states crucial to Liberal political control. The solution, both political and economic, was to build a railroad that would link Bogotá, via Boyacá and Santander, to the Magdalena River and by extension the Caribbean and the world beyond. This project was the ill-fated Northern Railroad, which would be built, in part, with revenue from the salt monopoly.[65] Just as salt taxes had underwritten the construction of the Erie Canal in the United States, the hope was that salt would underwrite a transformative project in Colombia. Parra admitted that this dream took precedence over other concerns in his report of 1874, stating that though the "fiscal necessity" for the salt rent was disappearing, it would survive if "the increase of this rent" became "indispensable for the construction of the Northern Railroad."[66]

The Northern Railroad was a favorite of politicians from the eastern highlands. It was also a notable failure. An editorial in *El Occidente,* a paper published in western Boyacá, argued that the railroad would bring moral progress, while the secretary of the government of Boyacá wrote, "The Northern railroad is the most important project that can be completed for the country."[67] Parra also linked this project to migration. He contacted a London company to introduce "Hindu workers," who, he said, were excellent workers "when obliged" and would cope well with the heat of the lowlands.[68] Despite such hopes, Colombia's difficult terrain made constructing and operating profitable lines an enormous challenge. Only the Panama Railroad was an unqualified success, though there were numerous attempts to establish other lines. By 1885 work had begun on nine different railroad lines throughout the country, but together they comprised less than 225 kilometers of working track. A decade later that figure had increased only to 557 kilometers.[69]

Disillusioned Conservatives of the following decade were quick to note that "railroads had not progressed beyond the flat lands to conquer the primary objective" of linking the highlands to the Magdalena River and the markets beyond.[70] The insight that the Northern Railroad's "magnitude in relation to the economic elements that then existed . . . determined its failure" shows merely the benefit of hindsight.[71] At the time the project met with widespread acclaim, particularly among the highland states dominated by Liberals. Given that the partial completion of a transportation network of roads, trains, and rivers finally occurred only in the 1920s, during the headiest days of Colombia's coffee boom, it is easy to dismiss these projects as blindly optimistic, but in the 1870s they were the order of the day.

By the end of that decade the tensions of the Liberal rule were beginning to tell. Conservatives felt alienated by the oligarchic nature of the Radical Liberal republic and were bitter over the educational reform of 1870. The result was the civil war of 1876–1877. As with most insurrections aimed at overthrowing the national regime, the rebellion failed, but it also marked the beginning of the end of Liberal dominance. The rebels hoped that Liberal factions alienated by the radical nature of the government would stand aside as they battled the Radical Liberals who controlled the federal government.[72] Considering the rebellion,

Finance Minister Januario Salgar simultaneously dismissed the rebels' claim of religious motivations and defended the educational reforms that had precipitated the war: "The establishment of schools has been the basis of scandal and the ostensible motive for the civil war." He compared primary education to a light shining on the "masses" and argued that those who objected to these reforms were in fact defending their control over this population rather than fighting for a real cause.[73]

The revolt began in Tolima, with subsequent uprisings in Cauca and Antioquia. For the most part, the war remained centered in these states. No major battles took place in Boyacá, but its western cantons experienced substantial disorder. The economy, both provincial and national, suffered. Exacerbating the economic impact of the war in Boyacá, that state experienced a labor shortage when it supplied its typical heavy contribution of soldiers.[74] The eastern highlands were also supposed to help finance the war effort. Decree 472, of August 30, 1876, called for a forced loan across the region: 400,000 pesos in Cundinamarca and 200,000 pesos each from Tolima, Santander, and Boyacá.[75] Only in Cundinamarca and Tolima, where state governments levied special contributions, was any money gathered. The governments of Boyacá and Santander thought these measures were ill-intentioned and sought to control expenditures and finance military expenses with the rents at their disposal, meaning the customs collected in Cúcuta, Santander, and the saltworks in Boyacá—namely, La Salina.

Correspondence during the war mentions various conflicts and guerrilla action around Chinquinquirá, Moniquirá, Ramiriqui, and Samacá, towns that had been trouble spots for a decade.[76] Still, officials in eastern Boyacá remained wary of disorder, and they had some cause to feel that way.[77] In May 1877 an attempted uprising occurred in Tundama, and though it was easily put down, officials voiced their concerns: "The insurgents of the North, who in the majority are composed of a multitude of ignorant fanatics, were among the most hostile and tenacious in the fight against advances in the country." After praising the "Liberals of the North" and bemoaning assassinations and arson, the report added, with what passed for optimism, that perhaps the defeat of the rebellion would convince the population "that it was not religion they were defending, . . . nor any civilized principle nor reasonable purpose, [but] only the goals of the most vulgar part of the clergy and a political faction with the most backward ideas."[78] Insurgents succeeded in capturing La Salina for a short time, forcing the administrator to flee with salt receipts. Of course, during the weeks of the occupation, the rebels made and sold salt of their own accord.[79] The resulting shortage of legal salt forced the almacén in Sogamoso to close for six weeks.[80] I have found no postwar reference to prosecutions against Finance Ministry employees or contractors for collaborating with the rebels.[81]

Nationally the war hurt customs returns and threatened salt profits. The lesser Boyacá saltworks had been occupied by rebels numerous times, and the

Finance Ministry had increased the price of salt to cover war expenses.[82] In 1877 Januario Salgar wrote about the war's negative economic effects: "This time, the spirit, startled and weighed down under the nightmare of war, disgracefully must present for viewing a very different picture. No richness of the imagination could sketch this completely, without penetrating into the abyss of general devastation to wealth and individual desolation."[83] Still, when he considered critiques of the salt monopoly, Salgar said what finance ministers usually did, that open production would invite private monopolies; he added, "I would not vacillate in recommending [to Congress] the idea of dedicating the salt rent exclusively to the construction of a railroad," meaning the Northern Railroad.[84] The dysfunctional continuities of Colombian fiscal policy marched on in war as they did in peace.

The Liberal victory in the war of 1876–1877 proved a pyrrhic one, though the deeper ethos of developmental liberalism would survive. The winner of the presidential elections of 1878 was a hero of the war, General Julián Trujillo. Trujillo was politically aligned with Rafael Núñez, who was the leader and to a degree the creator of the Independent Liberal faction.[85] After Trujillo's term, Núñez was elected for his first term as president. He served two years and then, as mandated by the Constitution of Ríonegro, left office. His reelection in 1884 precipitated the Liberal revolt of 1885. Radical Liberals had been dismayed by Núñez's reelection in 1884. Seeing their grip on power slip away and the policies they had championed abandoned, they moved toward armed rebellion in 1885. "Prominent party members such as Felipe Zapata and General Várgas Santos joined groups of the ever-restive Santander Radicals in advocating recourse to arms."[86] While these rebels enjoyed some military successes, they ultimately lost the war and in doing so provided their opponents with an opportunity for making deeper change. After Núñez defeated the rebels, he had a free hand to remake the government, and the result was the Constitution of 1886. This constitution, Colombia's most durable to date, ushered in an era of centralized rule dubbed "the Regeneration."[87] For Núñez and his allies, the term embodied their effort at rebuilding Colombia's society morally and legally after decades of decentralized chaos.

In the eastern highlands the fighting began in Santander, the logical place for Radicals to contest the looming end of federalism.[88] Fighting between these Radicals and a pro-Núñez coalition in Santander sparked a broader war. Núñez sent troops to western Santander to prevent a campaign against Bogotá, and the stymied Radicals targeted Boyacá. In his memoir of fighting with the rebels, Foción Soto recalled, "we could count on the Department of Tundama to increase our forces by more than 5,000 men and with a multitude of resources in money and beasts."[89] Specifically, the hope was that aid from Santa Rosa and Sogamoso would be enough to mount an offensive against Tunja.[90] Camped in Santa Rosa, the commanders grew discouraged by their inability to raise funds for this effort. But then, on December 19, they learned that General Gabriel Várgas Santos had led 200 men out of La Salina, where he had financial interests, to reinforce the rebels, along the way "gathering men and arms in the friendly towns (Tasco, Socha, Jericó, etc)." Soto mentioned Várgas Santos's reputation: "The General enjoyed great fame in Boyacá for his indisputable patriotism, his honorable reputation, and his military service since 1854."[91]

Várgas Santos, "El Gran Viejo," was a native of Santander and a rancher in Casanare. He had fought with the Liberals in every civil war since 1860 and would serve as the supreme commander of the Liberal forces at the beginning of the Thousand Days War.[92] Bolstered by his forces and personal presence, the rebels set out for Tunja. They entered the city on December 22, with little resistance, though the government managed to remove thousands of rifles stored there before fleeing. Unable to arm their men for a further campaign, the Radicals sent a commission to negotiate with Núñez. While the commission was in Bogotá, the national government mounted a counterattack, pushing the rebels back into eastern Boyacá. The defense campaign involved fighting in and around La Salina, though Várgas Santos showed a marked aversion to taking risks in the field.[93] Soto wrote that after losses in central Boyacá, the Radicals marched to Chita, where they were disappointed by the lack of support. Fears of a similar reaction in El Cocuy helped shape their subsequent plans:

> It was suggested that we march to Casanare, via the road from Cocuy to La Salina de Chita, taking defensive positions before La Salina, and there wait [for news] of the movements in Tolima and Antioquia, [planning] to go into the interior of the Llanos if our friends in these States succumbed, or to return and invade Boyacá via Sogamoso. The General Vargas S., who was approaching his home [and] knew every foot of this area, and General Pedro Rueda of Boyacá were the most enthusiastic supporters of the plan. Though there were those who said that the road we would travel was detestable, [the generals] and others assured [them] that it was not particularly bad and that it was used by mules loaded with ten arrobas of salt. Also we would not want for resources in La Salina or in Casanare, where there were people so rich that they could give us 50,000 pesos if they wished.[94]

After a few days of maneuvering, the army camped in La Salina on February 16 and 17, 1885. Once there, they gathered enough money to pay some of the soldiers, though the forces under General Gutiérrez, camped in Rodrigoque, deserted in large numbers. Having heard no heartening news, the force of around 1,200 men fell back to Sácama, planning to retreat farther into the llanos. But the government forces in Chita and La Salina were then reduced to 500 men, apparently because they assumed the rebels had dispersed; when Várgas Santos learned of this, he changed course. On February 28 he took La Salina, frightening "everyone" and clearing a path for the rebels to consolidate their forces in Bucaramanga, Santander.[95] This success proved short-lived. On March 5, the

Diario Oficial published an account of the government forces' reoccupation of La Salina that focused on the town's strategic importance as a portal from Casanare to the highlands.[96]

As ever, La Salina proved a challenge to control on various fronts. Bank notes accepted in Cundinamarca were not received freely by people in Boyacá. For example, caked salt sold for 3.20 pesos in specie but 5 pesos in notes.[97] In July 1885, after the army had left and the Liberals faced defeat in Santander, the administrator wrote about continuing disorder. He claimed that he was happy to send receipts to support the army in the field but that the region was too chaotic, for the road was "infested with armed bands, true highwaymen," who would "attack and rob without pity whatever victim" entering or leaving La Salina. In one case a party of forty-five men attacked a group of soldiers and Finance Ministry employees on the road as they left town. The victims escaped after their money, clothing, and goods were taken. Further, Facundo Gamaldeo, a merchant returning to La Uvita, was badly injured when bandits robbed him of his mules and salt. The minister lamented his impotence in the situation: "it is impossible to spare a single man to chase these rebels . . . [and the guards are] barely enough to guard the town with a population that is completely hostile to officials and the government."[98] Peace, apparently, was a relative state of affairs.

The Regeneration

In terms of fiscal policy, the Regeneration began before 1886. The Finance Ministry had already moved to reduce its debts and the portion of its budget spent on public works projects, which had been vitally important in the 1870s and consumed as much as 40 percent of the national budget.[99] In 1884 Finance Minister Anibal Galindo submitted a report summing up the attitude of the day when he explained that it had been a fundamental mistake to "found a federal system on a basis of the sovereignty of the States," adding that the Constitution of Ríonegro had robbed the federal government of its "proper force."[100] Turning to politics in general, he quoted his own writing from a few years before to explain real liberalism:

> No, liberalism says to all, like Christ to Lazarus, get up and walk, I have broken with the powerful arm of democracy all of the obstacles, all of the chains with which despotism . . . [kept] you bound to the post of ignominious destiny. I am Liberty, but I am neither philanthropy nor charity. Thanks to me you have been born not a slave, nor servant, . . . nor vassal, nor plebeian by class or destiny, but a free man. But your luck depends not on me, only on you.[101]

Núñez's long career traced a path from liberalism to the search for a less-doctrinaire politics that would help build a stronger country. Along with the new constitution, policies early in the Regeneration were initially designed to undo the excesses of the radical republic. The Catholic religion and other orthodoxies were returned to their favored place in public life. The national government in Bogotá established more control over the country than any previous republican regimes had claimed. States were transformed into departments; state presidents were once again governors, and they were appointed by the president. Attempts to radically restructure the fiscal system were largely abandoned, and metaphors glorifying progress as the triumph of light over darkness were replaced by calls for order, control, and stability. The successful development of coffee as an export crop in the late 1880s gave Regeneration governments the funds to implement these policies, though Charles Bergquist has argued that this prosperity also sowed the seeds of further discontent, because the increased profits from exports provided for the revitalization of the Liberal Party in the 1890s.[102]

Tundama had its share of political tension even during the nadir of Liberal resistance. In January 1889 troops fired shots in the air to disperse protesting crowds in Santa Rosa and Duitama. "Friends of the government" reported that in Santa Rosa, people had gathered early in the day shouting "Long live the Radicals" and "Down with the Government."[103] Telegrams from various points in Tundama were sent to the Ministry of Government relating rumors of rebellion and tracking the suspicious behavior of known Radicals entering the area from Venezuela.[104] In May Judge Belisario Losada reported that, after being sentenced for an unnamed crime, the notary of Pescua, Justo Zuleta, publicly called members of the court of Santa Rosa thieves and bandits, insisting, "[They] have robbed me."[105] Worse, no one did anything to stop him. Zuleta offered the excuse that he had been drunk, but he continued to mock government orders. He was removed from his position, but there is no record of any other punishment. At a festival in Santa Rosa in September of that year, a judge was assaulted. When government soldiers tried to apprehend the assailant, "some Sogamoseños and Tunjanos" defended the guilty party, attacking and disarming the soldiers.[106] The departmental elections of 1891 brought further disturbances. "Radicals" in an unnamed town in Gutiérrez killed an alcalde and his son, there were minor disturbances in Sogamoso, and the elections were canceled in a few towns. Still, Governor Pinzón reported that the elections were carried out with "complete calm despite the enthusiasm of the different groups."[107] In 1893 artisans' protests erupted in Tunja, while others, in support of an artisans' uprising in Bogotá, caused a commotion in Chinquinquirá.[108]

As the Regeneration and its attendant changes registered in the region, the Finance Ministry reversed its sixty-year policy of discouraging local agriculture in La Salina and rented out small plots of land. Though the decision merely recognized a fait accompli, the acknowledgment and management of squatting rather than its denial and repression represented a profound shift in the way the ministry treated La Salina. At more or less the same time, the ministry achieved its long-standing goal of obtaining complete ownership of the timber-producing forests around La Salina, a major accomplishment given the perennial problems in procuring affordable fuel over the decades. The specifics of these transactions document the survival of Liberal networks of influence in Tundama despite the increasingly polarized atmosphere of the Regeneration. The forests in question had passed from the Várgas Santos Company to the ministry, only to be rented by Pedro Cantini (who had been a partner in the Várgas Santos Company) in an agreement that gave him complete control over this land as long as he ensured regular deliveries of timber.[109] There is no indication that these entrepreneurs suffered because of their Liberal identities, but that might not have been accidental, for retaining the goodwill of such men would have been an important defense against wider popular discontent.

Of course, there were still critiques of the state's salt monopoly—in this case, how it was managed. The most spectacular critiques emerged in a series of articles published in Bogotá under the pseudonym S. A. L. Ustio, a pun on *salt* and *usury*. Most of the articles appeared in *El Relator,* but some appeared in *El Diario de Cundinamarca*; the official sponsorship of this series hints that the government of Cundinamarca had never completely abandoned its desire to control Zipaquirá. The articles examined all the Finance Ministry saltworks and the system for taxing sea salt brought into the interior of the country, doing so without bombast but with a thorough and scathing rigor: "The salt rent, which has been increasing since 1879, has progressively declined without there being any visible cause. . . . [T]his important branch of public finance, a source in other times of considerable resources for the government, has entered into complete disorganization, and contradicting the laws on the matter, has turned into a source of great private riches."[110] Ustio's method was practical and effective. Through a systematic reading of the Finance Ministry's *Memorias,* he (assuming the author was a man) compared production levels, profits, and the assertions made by various finance ministers. Though ministry officials criticized the years before 1886 as an era of "organized anarchy," Ustio pointed out that in terms of returns to the ministry, they were neither better nor worse. In one example, he cited the findings of a Senate committee that found the ministry had squandered 464,000 pesos in La Salina merely by granting Cantini a terrible contract and then buying him out of it. The ministry subsequently made a contract with Manuel Jiménez structured in such a fashion that he could not "honorably" make a profit.[111] Then Ustio's assessment then took on a more traditional tone as he considered the issue of shortages and resales: "the history of La Salina de Chita convinces us that monopoly cannot occur there without the consent of government employees." General Terán's attempt to run an honest administration was seen as a hostile act by those who were "accustomed to enriching [themselves] with the national resources located there."[112]

The Ustio articles are noteworthy in a number of ways. The essays, which caused a stir immediately on publication, form a lengthy exposé even by the standards of the era, when it was common for essays and reports to be printed in pieces over several weeks. The later pieces in the collection include references to letters, responses from other newspapers, and questions concerning freedom of the press under the restrictive Regeneration regime. The government's salt monopoly was again at the center of public discussion, but there were no references to larger economic principles or any metaphors crafted to reveal the transformative powers of market activity. Instead, Ustio provided a stream of comparisons of profits and production rates using the Finance Ministry's own figures. Nothing concrete came of this critique, but it was perhaps the most thorough deconstruction of the government's monopoly penned in the century.

The policies of Regeneration governments were to a degree guided by economic liberalism, but the belief that changing economic behavior would catalyze a much-needed societal transformation no longer held sway. A strong government with proper respect for the church was now charged with improving society. Finance Ministry reports from the era reflect that institution's diminished role, in which the ministry was no longer in charge of Colombia's future.[113] Writing in 1894, Finance Minister Pedro Bravo reflected the more measured approach of the era: "I am satisfied with the employees of the Ministry, who in general show zeal for their honor and that of the Government."[114] Vice President Miguel Antonio Coro, when addressing Congress as he assumed executive power in 1894, though Núñez remained the titular president, made it clear that questions concerning fiscal policy were subordinate to rather than the engine for larger issues: "The economic question cannot be separated from the moral question: Gold does not improve races, or improve the blood; it cannot sustain itself with its own virtue."[115]

The next significant national event was the civil war of 1895. After almost a decade of Regeneration governments, now led by the Conservative Miguel Antonio Caro, Liberals were locked out of public office. They had attempted a revolt in 1893 without success, and the revolt of 1895 was almost put down before it began. With rumors of an uprising circulating in January 1895, Caro imprisoned Liberal ringleaders in Bogotá. Others pressed on, and there was action against the government in Facatativá, in Cundinamarca.[116] With a solid fiscal footing provided by booming coffee exports, which meant that paper money could be issued without excessive inflation, the government defeated the rebels and restored order throughout the country in two months.

Liberal rebels captured La Salina on January 31, 1895, and held it until March 18. The administrator Alejo Luque wrote that when in possession of the saltworks, the rebels made salt and were so carried away by the "fury of speculation" that they even tore apart buildings for fuelwood.[117] Though details are sketchy, government forces had to fight to retake the town.[118] In early April the contractor Guillermo Torres, who was in Bogotá, wrote to the Finance Ministry about the challenges to meeting his contract given that rebels had "destroyed all the production equipment there." He asked for an advance of 25,000 pesos so he could make salt.[119] Reports from Alcalde Octavio Sánchez R. support Torres's claim that the rebels had intentionally destroyed industry equipment. He estimated that this damage, together with looting, cost the ministry 160,000 pesos. He also reported that rebels had taken 26,000 pesos from the administration, of which only 8,000 was spent on their campaign, though how he could have known that is unclear.

Reports after earlier wars focused on the way estranged elites with experience in La Salina used rebellions to abuse government property for their own, illegal purposes, but reports of damages were rare. In 1895, however, the rebels stripped the town of all resources, including production equipment, not just cash. Perhaps the tension of Conservative control in a Liberal region had taken its toll. Sánchez proposed implementing open production to meet the challenges of salt making; "all the inhabitants of this municipality would be allowed to manufacture salt" on the condition that it be sold to the Finance Ministry for 0.80 pesos an arroba. To ensure continued production, he suggested that those who participated in the rebellion without committing any serious crime should be allowed to participate. Though clemency was a common practice, this suggests that the large number of people working with the rebels would have made it impossible for the industry to function without them.

Soon enough, however, the process of accountability took priority over reconciliation and reorganization. By early May Sánchez was requesting advice about dealing with locals who had taken salt from the almacén in Córdoba: "[It] was left by the revolutionaries when they retreated from the plaza . . . Afterward thirty men and women residing here shared the salt freely." He also claimed these were the people who took the production equipment. For the rest of the year, officials in La Salina sent a series of complaints that were traditional in that they described untrustworthy residents, only now the complaints were all couched in explicitly political terms. The question was no longer whether residents were opponents of the administration or worked with a monopolist but whether they were politically loyal. A letter from a commander of the guards, José Puentes, began with a pessimistic description of the works: "In the fábrica there is not even a piece of iron." His assessment of public order was equally dour: "As I made clear via telegraph, I have been here since the twenty-second of August exercising my duties as a soldier and Commander of the Guards. Until now I have not been able to correct the barefaced abuses that I encountered when I arrived."[120] Puentes dismissed those he deemed insufficiently loyal or politically untrustworthy and urged that the administration hire "friends of the government."

A series of letters written by and about the administrator Alejo Luque in early 1896 demonstrate the increased importance of political orthodoxy. In one, Luque wrote that the guards were of no use in controlling the town in part because the commander did not recognize Luque "as his immediate superior in matters pertaining to La Salina."[121] At the same time, two contractors overseeing repairs to the Zapo bridge reported that they could not finish the job on time because their brother had been imprisoned by Sánchez. The alcalde admitted he had imprisoned their brother but explained that they were all part of a "subversive movement against the government" and that "there had been signs that a Liberal plot was being hatched," though he offered no proof. He then ordered these contractors to appear before the prefect of El Cocuy for judgment or pay a fine of fifty pesos.[122] Though such harassment was a local tradition, and impressment a popular tactic—after all, Sánchez had been accused of both in 1893—using party affiliation to defend the behavior marked a change from the way such actions had been justified in the preceding decades.[123] As Caro's control over the government intensified, the Regeneration became an era of Conservative dominance, a shift that was reflected in correspondence from La Salina.

The next act in the drama of life in La Salina during 1896 came with a letter from the alcalde of La Uvita, who complained about resales. In addition to accusing almacén officials of controlling the market, a predictable claim from regional officials, he charged that they were politically disloyal. He then portrayed the local inhabitants as their victims: "All the residents of [La Uvita] are conservatives from the cradle, and loyal supporters of existing institutions. We have noticed the favors and benefits given to the radical revolutionaries of La Salina on the part of the administrator and bookkeeper, and our patriotic sentiments have cooled."[124] He assured the Finance Ministry that the commander of the guards could be trusted.

The same rhetorical tension surfaced in a dispute over the rental of Finance Ministry land. In June Julián Muñoz petitioned the prefect of Gutiérrez for aid, complaining that Luque was evicting him from his rental near the fábrica in order to turn it over to "certain Liberals." A Liberal had treated him, a "good Conservative," as a "*being of no importance.*" This was followed by a note from Sánchez to the ministry about Luque's actions: "Yesterday I was attacked by a gang of administrative agents, radicals, and resellers of salt. . . . The administration used its agents against me and three Conservatives, because we had revealed his scandalous conduct in the salt trade and [that he] sent a radical note threatening the head of the guard." In the margin Finance Minister Manuel Esguerra noted that the governor of Boyacá had been informed and that the prefect of Gutiérrez had been ordered to investigate these matters. Luque defended himself by insisting it was his critics who were Liberal scoundrels. Muñoz, who Luque said "does not merit mention," had claimed more land than he was permitted, including a number of springs that should be under government control. Muñoz's friend, Miguel Vega, was "a man of no opinion, a Liberal [who] changed his views and would have changed them again if the revolution had succeeded." In alliance with Sánchez, these men, the head of the guard, and the telegraph operator frustrated his attempts, "as a Conservative," to produce as much salt as possible. To corroborate his account, he submitted testimony from residents of La Salina, adding that the head of the guards worked closely with Sánchez to impede production.[125]

On July 17 the prefect of Gutiérrez made an eight-hour visit to see what was going on. He took Luque's side, reporting that though the infrastructure was in terrible shape, the employees in the almacén were doing their jobs.[126] Luque wrote that he had tried to make peace with the population. He also appealed to people in the neighboring towns for supplies and transport equipment and called a meeting of salt makers, "asking their aid in improving the desolate state of affairs of La Salina."[127] In August the prefect of Gutiérrez dismissed all charges against Luque. The commander of the guards continued to send complaints about the administration, calling for Luque's removal: "here radicalism controls the monopoly and those who sell food would deny the troops a bottle of salt water to provide supplies!"[128] Reports from the next few years tell the same story: there were accusations of monopoly, cases of contractors using immature wood for fuel, reports of deteriorating infrastructure, and disputes between different members of the administration.[129] Life was very much as it had been since the 1840s, if a little more anxious and described with pointed, politically appropriate vocabulary. The logic whereby S. A. L. Ustio critiqued the salt monopoly not by referring to economic philosophy but by examining inconsistencies in the way the monopoly was managed was echoed in the move from general condemnations of immoral behavior to specific accusations.

The Thousand Days War: La Salina Fades from View

Concerns about local disorder intensified as the century drew to a close and rumors of another Liberal rebellion spread. In March 1898 the administrator wrote to the Finance Ministry complaining, "Last night, on the orders of the commander of the Third Division, the garrison retired from this saltworks, leaving 125,000 kilograms of salt in the almacén and money in the strong box. The only ones to watch over [the almacén] are the employees of the administration and nine unarmed [guards] who would be incapable of facing an enemy should there be a conflict."[130] His concerns were neither idly speculative nor alarmist. Traditional interparty tension had been exacerbated by the decline of coffee prices on the world market. Lower export earnings meant reduced imports, and the corresponding decline in customs revenue left the government short on funds. The country was poised for another civil war. The result, the Thousand Days War (1899–1902), was the country's most devastating civil war of the century.

When the war began, in October 1899, there was little reason to suspect it would be significantly longer or more violent than its predecessors had been. The rebels were not well armed and enjoyed strong support only in parts of Santander, Cundinamarca, and Tolima. For the first six months, the war was a fairly conventional military affair, with commanders on both sides leading organized armies, a period Charles Bergquist has dubbed "The Gentlemen's War." This phase lasted until the battle of Palonegro, in May 1900. At this site in Santander, near Bucaramanga, a large Conservative force defeated a Liberal army in the nation's bloodiest battle of the century. Though various Liberal leaders continued to fight—for example, Rafael Uribe Uribe led successful campaigns on the Atlantic coast—the rebels never again seriously challenged the government's control over the country. With only guerrillas fighting on, the struggle took on implications of class conflict.[131] The war finally ended in October 1902 with the signing of the Treaty of Neerlandia, an act that ushered in peace and three more decades of Conservative dominance. There are few references to La Salina as a strategic objective during the war itself, though later accounts suggest that there had been extensive fighting in the area. The only mention of La Salina as a part of the war involves an occasion when Rafael Uribe Uribe met with Várgas Santos to convince him to assume command of the Liberal forces; in one account they met in La Salina, though in another version they met at the general's home in Tame, in the llanos.[132]

The war marked the end of La Salina's outsized presence in the historical record. In time the continued rise of coffee as a successful export reduced the importance of the entire salt monopoly. In the immediate aftermath of the war, however, the monopoly was still a matter of some import. Writing in 1904, Finance Minister Benjamín Calvo echoed his nineteenth-century predecessors as he considered the industry: "The exploitation of our saltworks is done today in the same primitive and absurd method as by our predecessors. Applying to the rich mines that we possess modern and scientific processes, production will be doubled and, simultaneously, the costs of production will be reduced."[133] Of course, revitalizing La Salina would prove a challenge. Calvo quoted an inspector who, after visiting the works in April 1903, reported that not a single piece of iron remained in the fábrica:

> It is difficult to convey how completely deteriorated La Salina and its population are. . . . Most of its buildings were partially or totally destroyed. The inhabitants . . . were forced to emigrate because of the absolute scarcity of means of subsistence, many of them going to Casanare or other places in the interior. Today some families, in poor shape, are returning, and now there is some movement and life in this skeleton of a town. . . . The only things that remain standing are things that were too difficult to destroy, or those things that would yield absolutely nothing of worth. There are a few pieces from some bridges intact and some stone work (that would be the walls built to contain the salt water springs). Of the ovens [there] remain only the merest traces, and of their shelters there is no more than a piece of the last one built in the place of the oven called Quinto. The administrative office is almost falling down, with its roof collapsing in some parts, and various doors and windows are missing. Of the house that was given to the nation as part of a contract with Pedro Cantini, the only thing standing is a small part of northeast corner, useless for habitation. Everything else was destroyed. Of the shelters that covered the springs of salt water, the only one that remains, in very poor shape, is that of the Pozo bravo. The kettles are aged and completely useless; not a single document exists in the archive.

Still, La Salina had its allure. Calvo ended his stark description with news of recent activity:

> The Sñrs. Nepomuceno Monroy and Jerónimo T. Rodríguez are building two ovens to make salt that are larger than the small ones that are there. For this reason I believe that within three months this saltworks will have increased its receipts to forty or fifty thousand pesos a month.
>
> I have not commented on the saltworks in Muneque, as they are and will remain closed. . . . The guards protect that saltworks constantly so that there is no contraband and to ensure that no one takes timber from the young forests that will be used later.[134]

The salt monopoly still provided revenue for the government in the first decade of the twentieth century (see table 6.1), but La Salina's lack of fuelwood, as always, undercut attempts to revitalize production. Over the next decade the town faded from prominence and the national stage.[135] The Finance Ministry report of 1919 noted, with a dispassionate accuracy uncharacteristic of the previous century, that La Salina had always "rendered something" but that difficult roads and the high cost of timber kept it from turning a real profit. By the 1920s La Salina was not even mentioned as a source of national revenue, but the revenue from sea salt and Zipaquirá more than made up for the loss.[136] More important, the wealth generated by the coffee export boom eclipsed such minor revenue sources. In this sense, Colombia escaped one of its colonial legacies—or perhaps simply outlasted it.

In the twentieth century the distant fiats of the Finance Ministry ceased to shape the daily life in La Salina. Residents of the town began to deal with the state in the same way that other residents of the region did, not as a local economic competitor. Networks of influence and contracting found new channels and objects. This change is reflected in the organization of the Archivo General de la Nación de Colombia. The collection *Salinas de Boyacá* ends with documents from 1895, and the collection *Miscelánea de Salinas* ends at 1900. Documents from and about La Salina in the early twentieth century are scattered with other miscellaneous correspondence and tell a story different from the one told here.

TABLE 6.1. *National revenues, 1905–1909 (in pesos)*

	Revenue source		
Year	*Salt*	*Customs*	*Total*
1905	350,330	6,179,738	9,149,012
1906	811,723	7,414,776	13,935,674
1907	1,155,983	6,600,690	15,829,655
1908	1,147,008	6,169,239	16,138,444
1909	1,403,647	6,541,831	14,437,100

Source: Memoria de Hacienda de 1910.

The Significance of La Salina in Colombian History

La Salina de Chita has had a remarkable history. A small population center, its saltworks supplied salt for residents of a fair portion of Colombia's eastern highlands. Though far smaller than the works at Zipaquirá, its fábrica produced salt that was consumed by at least a hundred thousand people and played an important role in the regional ranching industry. As the epicenter of a regional market, it was both a strategic object during Colombia's civil wars and the home of a bureaucracy that embodied the institutional state. Working in the ramshackle offices of that bureaucracy, employees of the Finance Ministry wrote innumerable reports on life in La Salina and the surrounding cantons. They were agents of the institutional state whose presence in this place collapses the conceptual gulf between Colombia's center and periphery. For all its many shortcomings, the bureaucracy of this inchoate state linked events and life in this town directly to the halls of Congress and the presidential chair in Bogotá.

After the environmental conditions that allowed for salt making, state action was the most important single force in La Salina's nineteenth-century history, not in some abstract fashion, but in material, physically concrete terms. This is why, although I have touched on the examination of political culture central to current literature on republican Latin America, I have focused on state action, a factor often elided from this scholarship. Such a history shows, first, that charting the course of state building with reference to fiscal policy offers fresh insights into the workings of Colombia's political culture. Second, and more generally, in conceptualizing state building as a process rather than working in a dichotomized understanding of success or failure, such an account opens up new avenues for understanding republican histories.

The administration in La Salina was the Colombian state in all its confusion and disorder. Party politics, regional politics, regional economic stagnation, patronage, and the chronology of developmental rhetoric constitutionally enshrined in Bogotá—the state provided means through which all these were manifest as important elements in this local life. Via contracts, subcontracts, inspections, patronage, and legislative reform, national forces made themselves felt in La Salina, where they intersected with the efforts of succeeding generations of residents to hold on to or regain control of local forests, to defend local gleaning practices, and to practice new methods of profiting from the salt market. Complicating this dynamic were vertical relationships built on patronage that extended from day laborers who carried timber, to local subcontractors, to entrepreneurs with national reputations. Rounding out the portrait, there was an additional tangle of connections and rivalries by which residents from across the canton of El Cocuy and Casanare were drawn into market negotiations and confrontations.

In this complex web of concerns and interests, the state was both a quotidian material presence and a local economic actor. Physical spaces constructed under the supervision of Finance Ministry officials were sites where local and regional actors challenged the state. For all the rich literature on the state and its fitful growth in Latin America, little of it focuses on the state as a daily economic actor in this capacity. In La Salina, conflicts between state and society were not abstract; they involved immediate profits. These contests for control were informed by ideas but were always rendered immediate and material.

Of course, the state was present as a local economic actor in other places throughout Colombia. A review of the Finance Ministry's annual reports suggests that there may have been other such locales beyond the dozen or so salt administrations in the eastern highlands, though none seem to have produced the public theater of that particular monopoly. La Salina was far from being the only place where the Colombian state, no matter how dysfunctional, was manifest during the nineteenth century.[137]

Viewed in this light, the history of La Salina and the Colombian Finance Ministry intersects with recent work on cartography and the state, particularly the points made by Raymond Craib in his study on Mexico. Craib describes "state fixations" with cartographic projects, particularly those involving surveyors and other officials in "a burgeoning bureaucracy," through which the state "saw" or "came to know the countryside."[138] These points certainly hold for Colombia, where the government had its own cartographic fixations, as revealed by its teams of surveyors attempting to measure common lands for divisions and its own mapping commission. There is also an energetic group of scholars considering this issue with rigor and producing work on Colombian cartography.[139] The story of the salt monopoly dovetails with these ongoing inquiries employing cartography, for the information gathering so crucial in the state-sponsored mapping projects described by Craib and exemplified in Colombia by the Comisión Corográfica led by Agustín Codazzi was a constant facet of the salt administration. In almacenes across the eastern highlands and into the Llanos, officials generated document after document that mapped society and fixed locations for the officials in Bogotá. Decade after decade, the government's official gazette contained more information relating to salt production, markets, and sales than any scholar has yet succeeded in tackling. In this sense, the salt monopoly was part of the mapping of Colombia.

More generally, the story of the state's material interventions in La Salina evokes the broader history of republican Latin America, for it allows a close view of the disjuncture between a state's rhetoric and its actions. On the one hand, there were finance ministers and a crowd of officials, economists, and polemicists who sought to implement the freedoms associated with political liberalism. On the other hand, there was La Salina and the territories where salt was managed through the ministry's monopoly, places where these practices were never implemented. There were also the publicly acknowledged fiscal needs of the state to consider and the less openly discussed though widely known interests of those who profited from the salt monopoly, or at least the machinery that had developed around the monopoly. The one real attempt to close La Salina was not simply a dismal failure; it was the kind of failure that preempts the possibility of a second try.

Not until the 1860s, after residents had prevented the closure of the works and just as political outrage over the retention of a backward colonial economic practice reached its highest pitch, were profits from the monopoly irrevocably wedded to the seductive developmental liberalism that came to dominate Colombian politics. Through trade, railroads, or some other developmentalist dream, salt would foment, to borrow the vocabulary of the time, modern economic behavior. Proponents of this brand of liberalism were sure that the general population would be able to claim freedom only after such behavior was widespread, whether diffused via regional salt traders, by highly skilled immigrants who would arrive with their families and the tools of their craft, or perhaps by the anthropomorphized Northern Railroad itself. Here this history articulates an observation by Elizabeth Dore that "Latin American liberals promoted freedom of property, not freedom of persons."[140]

In the variation of that theme at work in La Salina, state intervention circumvented both the freedom of people and the freedom of property. In their quest for progress and development, state architects offered the rationale that the sin of controlling economic behavior in one market was a necessary precondition for the more general spread of modern economic values in other areas. This basic dynamic endured from one era to another as this colonial legacy survived the waxing and waning of doctrinaire liberalism. Even when Regeneration-era finance ministers proposed a role for the state with regard to the salt monopoly that differed radically from the one favored by their predecessors, the way that ministerial action shaped life in La Salina remained fundamentally unchanged. For all the shifting politics of the era, there was no reimagining of the state's role; monopoly remained monopoly, and the community of La Salina remained criminal and untrustworthy.

Despite the intensity with which the institutional state worked its way into daily life in La Salina, the project was never deemed a success by those involved. This perceived failure was rooted partly in Colombia's dire economic conditions and partly in the fact that, for all the fine words concerning fiscal modernization, the state never obtained the primacy it sought. If one measured salt making in La Salina by the standards of Boyacá, it was an industrial success. But it was not understood in those terms. Instead, the shortcomings of the institutional state garnered the most attention, though they were projected onto the regional population. The bureaucratic edifice of the state grew in size and complexity, if not in moral vision or effectiveness.

Seen in this light, La Salina encapsulates some of the main currents of the era: faith in the power of modernization, fascination with technology from the United States or Europe, and belief in the transformative power of the market.

It is almost trite to discuss these practices as significant findings, for they were widespread at the time, but seeing them so clearly marked in this small town is novel. Roughly a decade ago, the scholar James Scott departed from his earlier examinations of moral economy and resistance culture to consider states' efforts to alter landscapes. In assessing state growth, Scott discusses how the "intellectuals and planners" behind modernist bouts of hubris were not always bureaucratic monsters: "their actions, far from being cynical grabs for power and wealth, were animated by a genuine desire to improve the human condition—a desire with a fatal flaw. That these tragedies could be so intimately associated with optimistic views of progress and rational order is in itself a reason for a searching diagnosis."[141] The litany of failures and frustration in La Salina offers some insight into the way the state operated locally, but not as searching a diagnosis as is desirable. My fascination with the salt monopoly is based in part on the way it survived without slipping into an ossified echo of the past. Rather, it exhibited a constantly renewed and carefully frenetic stasis, with a bureaucracy ever intent on modernizing a vital anachronism while always falling short. If on a certain level this characterization describes a great many state bureaucracies, it underscores the importance of conceptualizing both state building and state formation without referring to either success or failure.

Within the context of Colombian history, La Salina provides an opportunity to see how political disorder, regional history, and the institutional state intersected. Boyacense politics, the rise of political liberalism, the introduction of new technology, the waxing and waning of obsession with free market competition, and popular political rhetoric—all were absorbed into daily life. Each new wave ratcheted up the tension of local life. In response, residents went from petitioning the state as a guarantor of patriarchal beneficence to calling on it to protect their rights as citizens. The discourse of monopolies as antithetical to civic virtue and natural rights was also absorbed into local society and deployed for local means.

Except for works that emphasize Boyacá's role in producing important national politicians, current scholarship rarely places the department's history within a national context. To get a full sense of Colombia's internal political workings, it is important to understand Boyacá's place in the country's political geography, particularly its position as an ally to the national regime.

Though it was possible, fiscally and geographically, for the governments of Antioquia, Cauca, or Santander to oppose the national regime (whether as an opposing party or as a dissident faction of the ruling party), this could not happen in Boyacá. Keeping company with Cundinamarca, Boyacá had to be allied with the ruling regime in Bogotá. But this alliance or the means that ensured it did not result from a passive political culture; at times it took a struggle to keep Tunja and the rest of Bogotá in this alliance. The task was difficult in part because of Boyacá's intraregional rivalries, often articulated along cantonal lines.

Sogamoso and other parts of Tundama were loci of opposition to the government in Tunja, no matter the politics of the moment. Often, but not always, this opposition involved Liberals with links to Santander opposing Conservative regimes. The variations on this dynamic, and the implications inherent in charting cantonal rivalries, are endless. But even a brief consideration is enough to contradict the notion of a politically passive Boyacá.

National conflicts, then, did indeed play out in La Salina. Through participation in civil wars, as soldiers, victims, or refugees, residents of La Salina understood that the broader changes occurring around them affected their lives and governed how they would make and talk about their choices and loyalties. Residents adopted new attitudes about their rights regarding the industry in the 1850s, ones noteworthy in leading not just to novel declarations but also to actions that mirrored the changes in Colombia's political culture. In the same fashion the increasing demands of party loyalty articulated in the 1890s were important not because all involved were, as claimed, staunch Conservatives but because this emphasis shows the degree to which local life responded to national events. Residents of Boyacá were not isolated from Colombian history; rather, they were part of its mainstream, even in their physical isolation.

To a remarkable extent, the story of La Salina traces the broader sweep of Colombian history. The town housed the most important source of federal revenue in a densely populated and impoverished region, yet in an autoinhibitory process, state action in La Salina engendered resistance to itself. Residents were well versed in the practices that created this situation, and they proved both vigilant and creative in working to obtain a share of the profits produced in La Salina. As Finance Ministry officials championed the state's primacy in the name of societal good, locals insisted that they had rights relating to salt production, whether as citizens, property owners, workers, or community residents. The violence and daily conflict generated by this tension, exacerbated by party factionalism and clientalism, are not startling given the general workings of republican Latin America and modernization throughout the world. Indeed, the jostling in the almacén, theft, contraband salt making, and corruption that defined La Salina were mundane practices from a certain perspective. What place in nineteenth-century Colombia did not feature one or more of those elements? But the story of La Salina is unique and instructive for two simple reasons. The first is that officials extensively reported affairs there, documenting how the spaces created by the state became places where people challenged the institutional workings of that state. The second is that people in La Salina understood that even as they pressed state agents to their mission, they were competing with the state over who would benefit from the riches of the salt market.

꩜

La Salina hardly figured into fiscal matters in the twentieth century, and little has been written about its history in any era. A special issue of the journal *Cari-*

barare with articles about La Salina was published in 1992, but apart from that, there are only a few references to the town in regional studies or other media. Often enough, it is violence that prompts broader recognition. The special edition of *Caribarare* mentions that the town was a target for aerial bombing during the era known as "la Violencia," and national publications have noted it as a site of political violence.[142] Otherwise, the town attracts no more attention than does any other small municipality and perhaps even less than it should.

Some find the weight of this condition to be a point of concern. I was reminded of this when, as I was finishing this manuscript in late 2010, I visited La Salina's website to check on a few details. Colombian municipalities maintain official, relatively standardized websites as a part of Colombia's Gobierno en Línea program. La Salina's (lasalina-casanare.gov.co) has many points of interest. In addition to practical information about the present day, there is an explanation of the town's official symbol; a shield with images referring to its church, the Lache, salt, and coffee; and a poem by Egidio Pabón Jaúregui, "Nostalgia Salinera." So it was with great surprise that I read the brief sensational history section when I visited the site on November 29, 2010:

> La Salina was founded in 1527 by the Indigenous Laches. It still conserves its cobblestone streets and its houses with balconies that reflect the past as one of the oldest municipalities of Casanare. [It is remembered] as the historic command center of the army under the authority of General Santander.
>
> Through history, our glorious municipality has been advancing progressively on both the departmental and national level, in spite of the difficulties that it has suffered in the realm of Public Order, natural disasters, and worst of all, "THE COMPLETE ABANDONMENT AND LACK OF RECOGNITION BY THE STATE" that has been present in each and every process, impeding that economic, political, and social development reach this territory.[143]

This was, of course, a remarkable statement to see on an official website. When I returned to the site the next week, the text had been changed to something more politic that touched on the salt industry and the town's troubled history but assigned no blame: "The region has been affected by numerous demographic fluctuations and of permanent depopulation, [and] this factor has been, perhaps, the greatest obstacle to the population increasing [in terms of] the number of urban residents and socioeconomic development." This much more diplomatic version of the local history was still up in 2011.[144] The incident is a mystery to me, and the explosive statement could have been a response to any number of factors. In any case, someone in La Salina is clearly still searching for a working relationship with the state. Among the many differences between the nineteenth century and the present, that is an exceptional and notable continuity.

Notes

Chapter 1: The Salt Monopoly, the State, and Boyacá

1. For more notes from Caicedo, see Archivo General de la Nación de Colombia (AGN), Sección Colonial (SC), Fondo Salinas, legajo (hereinafter abbreviated "leg.") 7, fols. 297–301, September–December 1806. Caicedo's efforts were part of a broader program investigating Colombia's natural wealth. For an overview and context concerning the Royal Botanical Expedition, led by José Celestino Mutis beginning in 1783, see Jorge Orlando Melo, "La idea del progreso en el siglo xix, ilusiones y desencantos, 1780–1930," in *Revista de Estudios Colombianos* 36 (2010): 16–29.

2. "Informe relative a los datos estadisticos," AGN, Sección Repúbica (SR), Gobernación de Tunja, leg. 23, fol. 41, January 1837. Unless otherwise noted, all translations are mine.

3. Benjamín Calvo, "Informe del administrador de las salinas de Chita y Muneque," in *Memoria de Hacienda de 1904,* sección 2, p. 2.

4. S. A. M. Adshead, *Salt and Civilization* (New York: St. Martin's, 1992), 178. On the state and salt monopolies, see Kwan Man Bun, *The Salt Merchants of Tianjin: State-Making and Civil Society in Late Imperial China* (Honolulu: University of Hawai'i Press, 2001). For a very useful general study, see Mark Kurlansky, *Salt: A World History* (New York: Walker, 2002).

5. This tradition dated to the birth of the republic and lasted for decades. On the government's covering domestic debts by issuing promissory notes redeemable for salt in the saltworks of Cundinamarca, see Luis Fernando López Garavito, *Historia de la Hacienda e el tesoro en Colombia, 1821–1900* (Bogotá: Banco de la República, 1992), 20–21. On later examples of promissory notes issued by the government to be redeemed for salt at ministry works, see Anuar Herna Peña Díaz, *Sal, sudor y fisco: el proceso de institucionalización del monopolio de la sal en las Salinas de Chámeza, Recetor y Pajarito, 1588–1950* (Bucaramanga: División de Publicaciones UIS/Universidad Industrial de Santander, 2008), 212, 226.

6. I draw extensively on the *Memorias de Hacienda,* the annual reports that finance ministers submitted to Congress. These reports, published both as single volumes and serially in the official gazette, were novel in documenting the workings of the government to the public. The starting point for conceptualizing the role of print culture in nation making, though not necessarily state building, remains Benedict Anderson, *Imagined Communities: Reflection on the Origin and Spread of Nationalism,* rev. ed. (New York: Verso 2006). On the development of the public sphere in Colombia prior to independence, see Victor Uribe-

Uran, "The Birth of a Public Sphere in Latin America during the Age of Revolution," *Comparative Studies in Society and History* 42, no. 2 (Apr. 2000): 425–57. On the history of efforts by "modernizing elites," such as those who ran the Finance Ministry, to increase the state's "cognitive capacity," see chapter 2 in Lawrence Whitehead, *Latin America: A New Interpretation* (New York: Palgrave Macmillan, 2006). For an example of scholarship drawing on *Memorias,* see Malcolm Deas, "The Fiscal Problems of Nineteenth-Century Colombia," *Journal of Latin American Studies* 14, no. 2: 287–88. On the Finance Ministry in Colombia, see Luis Fernando López Garavito, *Origen y fundamentos de la Hacienda y la economía colombianas* (Bogotá: Universidad Externado de Colombia, 2000). A number of studies are useful for understanding the role of finance ministries: Paul Gootenberg, *Imagining Development: Economic Ideas in Peru's "Fictitious Prosperity" of Guano, 1840–1880* (Berkeley: University of California Press, 1993); Barbara A. Tenenbaum, *The Politics of Penury: Debts and Taxes in Mexico, 1821–1856* (Albuquerque: University of New Mexico Press, 1986); José María Dalence, *Bosquejo estadístico de Bolivia* (La Paz: Universidad Boliviana/Universidad Mayor de San Andrés, 1975); Linda Alexander Rodríguez, *The Search for Public Policy: Regional Politics and Government Finances in Ecuador 1830–1940* (Berkeley: University of California Press, 1985); and Javier Tantalean Arbulu, *Política económico-financiera y la formación del estado: siglo xix* (Lima: Centro de Estudios Para el Desarrollo y la Participación, 1983). For colonial predecessors to the *Memorias,* see Gabriel Girlado Jaramillo, *Relaciones de mando de los Virreyes de la Nueva Granada: memorias económicas,* Archivo de la Economía Nacional, vol. 13 (Bogotá: Banco de la República, 1954). For examples of similar reports published in Venezuela, see the following three works edited by Tomás Enrique Carrillo Batalla: *Hemerografía económica del siglo xix* (Caracas: Biblioteca de Academia de Nacional de la Historia, 2002), *El pensamiento económico de Francisco Aranda* (Caracas: Academia Nacional de Ciencias Económicas, 1993), and *El pensamiento económico de Santos Michelena* (Caracas: Academia Nacional de Ciencias Económicas, 1993).

7. Fernando López-Alvez, *State Formation and Democracy in Latin America, 1810–1900* (Durham, N.C.: Duke University Press, 2000), 24–25.

8. Stacey Hunt, "Languages of Stateness: A Study of Space and El Pueblo in the Colombian State," in *Latin American Research Review* 41, no. 3 (2006): 90. Of course, the challenges to stability that were a central part of Colombia's history at various points during the twentieth century feed the impulse to consider the state in terms of failure; see, for example, Harvey F. Kline, "Colombia: Lawlessness, Drug Trafficking, and Carving Up the State," in *State Failure and State Weakness in a Time of Terror,* ed. Robert I. Rotberg (Washington, D.C.: Brookings Institution Press, 2003), 161–82.

9. Oscar Oszlak, "The Historical Formation of the State in Latin America: Some Theoretical and Methodological Guidelines for Its Study," *Latin American Research Review* 16, no. 2 (1981): 3–32 (quotation, 12). Recent works useful for framing the state within a political context include Jeffrey D. Needell, *The Party of Order: The Conservatives, the State, and Slavery in the Brazilian Monarchy, 1831–1871* (Stanford, Calif.: Stanford University Press, 2006); and Roger A. Kittleson, *The Practice of Politics in Postcolonial Brazil: Porto Allegre, 1845–1895* (Pittsburgh, Pa.: University of Pittsburgh Press, 2005). A touchstone for examining the state over the last few decades remains Peter B. Evans, Dietrich Rueschemeyer, and Theda Skocpol, eds., *Bringing the State Back In* (Cambridge: Cambridge University Press, 1985).

10. Oszlak, "Historical Formation," 7. I favor the study of state building over that of state formation, for the former describes the labor involved in the process, whereas the idea of state formation implies, inaccurately, an organic process. For an overview of the multiple processes involved, see Nils Jacobsen and Cristóbal Aljovín de Losada, "State- and Nation-Building Projects and Their Limitations," in *Political Cultures in the Andes 1750–1950,* ed. Jacobsen and de Losada (Durham, N.C.: Duke University Press, 2005), 69–73. On the state, see Charles Tilly, *Coercion, Capital, and European States, AD 990–1990*

(Cambridge, Mass.: Basil Blackwell, 1990); see also Tilly, "Reflections on the History of European State-Making," Gabriel Ardant, "Financial Policy and Economic Infrastructure of Modern States and Nations," and Rudolph Braun, "Taxation, Sociopolitical Structure, and State-Building: Great Britain and Brandenburg-Prussia," in *The Formation of National States in Western Europe,* ed. Tilly (Princeton, N.J.: Princeton University Press, 1975), 3–83, 164–242, and 243–327. Other significant works include Philip Corrigan and Derek Sayer, *The Great Arch: English State Formation as Cultural Revolution* (Oxford: Basil Blackwell, 1985); see also Corrigan, "State Formation," and Sayer, "Everyday Forms of State Formation: Some Dissident Remarks on 'Hegemony,'" in *Everyday Forms of State Formation: Revolution and Negotiation of Rule in Modern Mexico,* ed. Gilbert M. Joseph and Daniel Nugent (Durham, N.C.: Duke University Press, 1994), xvii–xix and 367–78. For general points on this process and the value of *la longue durée* when examining state formation, see Richard Bonney, ed., *The Rise of the Fiscal State in Europe, 1200–1815* (Oxford: Oxford University Press, 1999). Useful works on the state in Latin America include Jeremy Adelman, *Republic of Capital: Buenos Aires and the Legal Transformation of the Atlantic World* (Stanford, Calif.: Stanford University Press, 1999); David Rock, *State Building and Political Movements in Argentina, 1860–1916* (Stanford, Calif.: Stanford University Press, 2002); Miguel Centeno, *Blood and Debt: War and the Nation-State in Latin America* (University Park: Pennsylvania State University Press, 2002); Robert Holden, *Armies without Nations: Public Violence and State Formation in Central America, 1821–1960* (New York: Oxford University Press, 2004); and Robert G. Williams, *States and Social Evolution: Coffee and the Rise of National Governments in Central America* (Chapel Hill: University of North Carolina Press, 1994).

11. On political culture and related scholarly divisions, see Nils Jacobsen and Cristóbal Aljovín de Losada, "The Long and Short of It: A Pragmatic Perspective on Political Cultures, Especially for the Modern History of the Andes," in *Political Cultures,* ed. Jacobsen and Losada, 1–24. See also Lara Putnam, Sarah C. Chambers, and Sueann Caulfield, *Honor, Status, and Law in Modern Latin America* (Durham, N.C.: Duke University Press, 2005); Florencia Mallon, *Peasant and Nation: The Making of Postcolonial Mexico and Peru* (Berkeley: University of California Press, 1995); and Ulrich Muecke, *Political Culture in Nineteenth-Century Peru: The Rise of the Partido Civil,* trans. Katya Andrusz (Pittsburgh, Pa.: University of Pittsburgh Press, 1998), which is in part a response to Mallon.

12. On the state resurgence after this shift, see Fernando López-Alvez, "The Transatlantic Bridge: Mirrors, Charles Tilly, and State Formation in the River Plate," in *The Other Mirror: Grand Theory through the Lens of Latin America*, ed. Miguel Angel Centeno and Fernando López-Alvez (Princeton, N.J.: Princeton University Press, 2001), 153–76.

13. Pablo Piccato's recent work on honor in the public sphere in Porfirian Mexico is interesting in this context. While Piccato focuses on public and political culture at several key moments, the conflicts he documents, such as the Dueda inglesa, were explicitly prompted by matters of public finance; see Pablo Piccato, *The Tyranny of Opinion: Honor in the Construction of the Mexican Public Sphere* (Durham, N.C.: Duke University Press, 2010). Useful works on the state in Colombia include David Bushnell, *The Santander Regime in Gran Colombia* (Westport, Conn.: Greenwood, 1970); Christine Hunefeldt, "Reinventing Old Power: State, Tribute, and Indians in Puno," *Fronteras* 4, no. 4 (1999): 11–53; Fernando Botero Herrera, *Estado, nación y provincia de Antioquia: guerras civiles e invención de la región 1829–1863* (Medellín: Hombre Nuevo Editores, 2003); and Alfonso Múnera, ed., *Ensayos costeños: de la colonia a la república: 1770–1890* (Bogotá: Colcultura, 1994).

14. Recent scholarship has shown that institutional approaches provide a means of examining interactions between national elites and provincial actors, vital components in shaping later state practices. See Hillel David Soifer, "The Sources of Infrastrucutral Power: Evidence from Nineteenth-Century Chilean Education," *Latin American Research Review* 44, no. 2 (2009): 158–80; Soifer, "Authority over Distance: Explaining Variation

in State Infrastructural Power in Latin America" (PhD diss., Harvard University, 2006); Lina Del Castillo, "The Science of Nation Building: A History of Geographic Sciences in Colombia, 1821–1921" (PhD diss., University of Miami, 2007). See also Meri Linnea Clark, "Education for a Moral Republic: Schools, Reforms, and Conflict in Colombia, 1780–1845" (PhD diss., Princeton University, 2003).

15. Cecilia Méndez, *The Plebeian Republic: The Huanta Rebellion and the Making of the Peruvian State, 1820–1850* (Durham, N.C.: Duke University Press, 2005), offers an excellent look at the way issues regarding the state and taxation played into the broader sweep of politics.

16. On the relative weakness of the Colombian state, I am guided by Hillel Soifer's comparison of Chile, Peru, Mexico, and Colombia in chapter 4 of his unpublished manuscript "Authority over Distance: Institutions and Long-Run Variation in State Development in Latin America," which the author generously shared. Considering Colombia in an Andean context, Jeremy Adelman states, "While Andean societies share some features that contrast with other regions in Latin America, there are significant differences among them as well. In some senses, Colombians acknowledge the legitimacy of their constitutional framework—what they bemoan is its inability to territorialize itself" (Adelman, "Unfinished States: Historical Perspectives on the Andes," in *State and Society in Conflict: Comparative Perspectives on Andean Crises,* ed. Paul W. Drake and Eric Hershberg [Pittsburgh, Pa.: University of Pittsburgh Press, 2006], 42).

17. See Nancy P. Appelbaum, *Muddied Waters: Race, Region, and Local History in Colombia, 1846–1948* (Durham, N.C.: Duke University Press, 2003); James E. Sanders, *Contentious Republican: Popular Politics, Race, and Class in Nineteenth-Century Colombia* (Durham, N.C.: Duke University Press, 2004); Aline Helg, *Liberty and Equality in Caribbean Colombia, 1770–1835* (Chapel Hill: University of North Carolina Press, 2004); Marixa Lasso, *Myths of Harmony: Race and Republicanism during the Age of Revolution, Colombia 1795–1831* (Pittsburgh, Pa.: University of Pittsburgh Press, 2007). See also Brooke Larson, *Trials of Nation Making: Liberalism, Race, and Ethnicity in the Andes, 1810–1910* (New York: Cambridge University Press, 2004).

18. For an overview on these topics and politics, see Fernán E. González, "A Propósito de 'Las palabras de la guerra': los comienzos conflictivos de la construcción del estado nación y las guerras civiles de la primera mitad del siglo XIX," *Estudios Políticos* 24 (July–Dec., 2004): 37–70.

19. For a fuller discussion of this underdevelopment, see Botero Herrera, *Estado, nación y provincia de Antioquia*.

20. On customs houses as isolated manifestations of the state, see Jane Rausch, *Colombia, Territorial Rule and the Llanos Frontier* (Gainesville: University Press of Florida, 1999), 133.

21. Marco Palacios sums up the process whereby "nominal contractual liberty, celebrated by liberalism, ended up underwriting feudal obligations (Palacios, *Between Legitimacy and Violence: A History of Colombia, 1875–2002,* trans. Richard Stoller [Durham, N.C.: Duke University Press, 2006], 13). The process in La Salina is notable only in its intensity and the details available.

22. See Gustavo Humberto Rodríguez R., *Boyacenses en la historia de Colombia, bocetos biográficos* (Bogotá: Jurídicas Wilches, 1994); and Javier Ocampo López, *Historia del pueblo boyacense: de los origenes paleonindigenas y miticos a la culminación de la Independencia* (Tunja: Instituto de Cultura y Bellas Artes de Boyacá, 1983).

23. Depending on the criteria used, the number of regions runs from nine to fifteen. This question is not purely academic. An article in the Constitution of 1991 formed a commission to assess redrawing internal boundaries. This project was led by the sociologist Orlando Fals Borda; see his *Taller sobre sistemas de ordenamiento territorial, autonomía y descentralización, Bogotá, 1993* (Bogotá: Comisión de Ordenamiento Territorial/

Instituto Geográfica Agustín Codazzi, 1994). The concern for developing a systematic understanding of Colombia's regions is long-standing. At the beginning of the century, Francisco Javier Vergara Velaso divided the country into thirteen major regions and forty-three subregions in his work *Nueva geografía de Colombia: escrita por regiones naturales* (Bogotá, 1901). On the challenges of defining regions, see Myriam Jimeno, "Región, nación y diversidad cultural en Colombia," in *Territorios, regiones, sociedades,* ed. Renán Silva (Bogotá: CEREC, 1994), 65–78.

24. Finding a single accurate designation for Boyacá is problematic. The term *region* cannot be used without extensive qualifying descriptors. Legally, this area was the province of Tunja from 1810 to 1819, the intendencia of Boyacá from 1822 to 1828, the province of Boyacá from 1829 to 1848, the provinces of Tunja and Tundama from 1849 to 1857, and the "sovereign state" of Boyacá from 1857 to 1885. For a detailed history of these changes, see José Miguel Pinto, "División política del Departamento de Boyacá de 1821 a 1905," in *Boletín de Historia y Antigüedades* 6, no. 69 (Feb. 1911): 578–81. For clarity I will refer to it as "Boyacá" and try to use the department, province, and state designations in as accurate and clear a manner as possible. The chief executive will be the governor, though in notes the working title (intendente or presidente) is used. On internal political divisions, see Javier Guerrero Baron, *Los años del olvido: Boyacá y los orígenes de la Violencia* (Bogotá: Tercer Mundo, 1991). On politics, see Junco Velosa, *Del fraude y la violencia al clientelism: Boyacá, 1930–1990* (Tunja: UPTC, 1992). For a look at one of the regions incorporated into Boyacá during the nineteenth century, see Néstor Moyano Rodríguez, *Vélez: la gobernación, 1832–1857* (Bogotá: Ediciones Carare, 2003).

25. Anthony McFarlane uses this definition in *Colombia before Independence: Economy, Society and Politics under Bourbon Rule* (Cambridge: Cambridge University Press, 1993), 40. Marco Palacios also used this grouping in naming "four great economic regions—Antioquia, the Caribbean coast, the east-central region, and the southwest, including the marginalized Pacific coast" (*Between Legitimacy and Violence,* 5). Aquileo Parra, the finance minister in 1873, used this grouping in a discussion of economic development, speaking of the "comercio interior de los tres estados" (*Memoria de Hacienda y Fomento de 1873,* 74–83). At times Boyacá and Santander were grouped together as the Military Department of the North. See a decree reestablishing this district in *Gaceta Oficial,* September 30, 1859.

26. For example, the highlands on either side of the Boyacá-Cundinamarca border are geographically, demographically, and economically almost indistinguishable from each other. As one travels northeast from the capital, the tableland of the *sabana de Bogotá* gives way to rolling hills where, in the words of Manuel Ancízar, "all is green, gently tilted pasture" (*Peregrinación de Alpha,* 2 vols. [Bogotá: Fondo de promoción de la cultura del Banco Popular, 1984], 2:65). This easy transition forged a strong link between the history of western Boyacá and eastern Cundinamarca. William McGreevy reports that the per-kilometer cost of shipping cargo between Bogotá and Tunja was one of the cheapest in Colombia during the nineteenth century; see McGreevy, *An Economic History of Colombia 1845–1930* (Cambridge: Cambridge University Press, 1971), 43–45. Orlando Fals Borda divides the province into twelve subregions in *El hombre y la tierra en Boyacá: bases socio-históricas para una reforma agraria* (Bogotá: Ediciones Documentos Colombianos, 1957), 9–14, 148–60. For a work that uses the broader regional context, see Luis Javier Ortiz Mesa, *Fusiles y plegarias: guerra de guerrillas en Cundinamarca, Boyacá y Santander, 1876–1877* (Medellín: Universidad Nacional de Colombia, 2004).

27. Access to the Magdalena came when the province of Vélez was broken up and parts were added to Boyacá; see Pinto, "División política." On attempts to encourage settlement in western Boyacá, see Aristides Ramos Peñuela, "Colonos, empresarios y vagos: experiencias fronterizas en el Carare en la primera mitad del siglo XIX," *Fronteras* 2, no. 2 (1998): 65–92.

28. Adding to this confusion, at various points Casanare was a territory subordinate

to the government of Boyacá (in 1821–32, 1857–68, 1889–92, and 1911–75). In other eras it was administered directly by the federal government; see Jane Rausch, *The Llanos Frontier in Colombian History 1830–1930* (Albuquerque: University of New Mexico Press, 1993), 297.

29. Cantons were subprovincial administrative units that defined local histories. On cantons as a structure linking local economies and national politics, see Richard Stoller, "Liberalism and Conflict in Socorro" (PhD diss., Duke University, 1991).

30. For a twentieth-century example of the llanos' seductive powers, see Alan Weisman, *Gaviotas: A Village to Reinvent the World* (White River Junction, Vt.: Chelsea Green, 1998).

31. For example, Anthony McFarlane informs us: "At a couple of days' ride from Bogotá lay Tunja, capital of a heavily populated province of about 129,000 people. Once a close rival to Bogotá [Tunja] was in decline by the mid-eighteenth century. In the 1760s, it had only about 3,000 inhabitants within an urban precinct. . . . Most of the *population of the province* was dispersed over the countryside, living in the villages and hamlets, and the estates and farms that dotted the highland landscape" (McFarlane, *Colombia before Independence,* 50). In his essay on the colonial state, Magnus Mörner also makes little distinction between Tunja the city and Tunja the province; see Mörner, *Region and State in Latin America's Past* (Baltimore, Md.: John Hopkins University Press, 1993), 24–29.

32. Two novels by Eduardo Caballero Calderon, *Siervo sin tierra* (Madrid: Ediciones Rodas, 1973) and *El Cristo de Espaldas* (New York: Macmillan, 1967), provide portraits of this peasantry and life in Boyacá.

33. Scholars agree that the failure of the textile industry was a key element in this decline, though they differ in estimating when the decline began. Alfonso Múnera argues it began in the second half of the eighteenth century; see Múnera, *El fracaso de la nación: region, clase y raza en el Caribe colombiano (1717–1821)* (Bogotá: Banco de la República/El Áncora Editories, 1998), 35. Luis Corsi Otálora puts it after the Wars of Independence in his book *Boyacá: Atlántida andina* (Tunja: Academia Boyacense de Historia, 2005). Tovar Pinzón argues that it occurred after independence in "La lenta ruptura con el pasado colonial (1810–1850)," in *Historia Económica de Colombia,* ed. José Antonio Ocampo (Bogotá: Siglo XIX, 1987), 87–118; Maurice Philip Brungardt uses the same date in "Tithe Production and Patterns of Economic Change in Central Colombia, 1764–1883" (PhD diss., University of Texas, 1974).

34. Ancízar, *Peregrinación de Alpha,* 2:56. For a discussion on elite views and ethnicity in the eastern highlands, see Frank Safford, "Race, Integration, and Progress: Elite Attitudes and the Indian in Colombia, 1750–1870," *Hispanic American Historical Review* 71, no. 1 (Feb. 1991): 1–33; and Glenn Thomas Curry, "The Disappearance of the Resguardos Indigenas of Cundinamarca, Colombia, 1800–1863" (PhD diss., Vanderbilt University, 1981).

35. Germán Colmenares, *La Provincia de Tunja en el Nuevo Reino de Granada: ensayo de historia social (1539–1800)* (Tunja: Publicaciones de la Academia Boyacense de Historia, 1984), 9.

36. Frank Safford found that "Colombia made notably little economic progress in her first half century of Independence" and that the "people in [Boyacá] were the poorest in Colombia" (Safford, "Commerce and Enterprise in Central Colombia" [PhD diss., Columbia University, 1965], 28). For an overview, see Ines Pinto Escobar, *Progreso, industrialización y utopía en Boyacá: el caso de la Sigerúrgica de Samacá* (Tunja: Publicaciones de del Magister en Historia, UPTC, 1997).

37. Gaspard-Théodore Mollien, *Viaje por la República de Colombia en 1832,* intro. Carlos José Reyes (Bogotá: Instituto Colombiano de Cultura, 1992), 117.

38. AGN, SR, Ministerio de Hacienda, leg. 254, fols. 445–46, July 1830. Soban planned to sell shares, or "subscriptions," to purchase machinery and to bring technical advisers from the United States or Europe. He asked the Ministry of Finance to petition Congress

for protection from competitive imports. The margin note to his proposal, a common form of internal communication, cited the Constitution in explaining that the executive branch was powerless in this matter. On similar grants to protect domestic industry, see Safford, "Commerce and Enterprise," 149. On similar measures, see David Sowell, *The Early Colombian Labor Movement: Artisans and Politics in Bogotá 1832–1919* (Philadelphia: Temple University Press, 1992), 18.

39. *Memoria de Hacienda de 1831* (Márquez), 7.

40. *El Constitucional de Boyacá,* January 27, 1833. As late as 1904, Benjamín Calvo discussed promoting textiles, particularly in Boyacá and Cundinamarca (*Memoria de Hacienda de 1904,* xi).

41. Agustín Codazzi, *Geografía física y política de la Confederacíon Granadina,* ed. Camilo A. Domínguez Ossa et al., 6 vols. (Bogotá: Universidad Nacional de Colombia/ Universidad Pedagógica y Tecnológica de Colombia, 2003). Luis Ospina Vásquez also mentions ceramics in Sogamoso in *Industria y protección en Colombia, 1810–1930,* 4th ed. (Medellin: FAES, 1987), 259. Codazzi headed the Comisión Corográfica, which traveled though Colombia producing detailed reports on industry, population, travel conditions, and other matters. The impact of the commission and the support it enjoyed from the government is difficult to overstate. For a bibliography of Codazzi's writing, maps, and related scholarship, see the catalog from an exhibition of his cartography, Fiorgio Antei, *Mal de América: las obras y las días de Agustín Codazzi, 1793–1859* (Bogotá: Museo Nacional, 1993). On the subject of cartography, Colombia, and Codazzi's work, consult the website Razón Cartográfica at www.razoncartografica.wordpress.com.

42. Codazzi, *Geografía física y política,* 3:176.

43. David Bushnell puts the rate as starting at 1 percent and decreasing to 0.033 percent as an individual's wealth increased; see Bushnell, *The Santander Regime in Gran Colombia* (Westport, Conn.: Greenwood, 1970), 81–83. This data shows a range that started at 0.5 percent for those taxed with the smallest fortunes and only a slight decrease as wealth increased. For a list detailing individual contributions by municipality, see AGN, SR, Intendencia de Boyacá, leg. 2, fols. 454–632, June–August 1823.

44. Codazzi, *Geografía física y política,* 3:259.

45. Ibid., 224.

46. Anicízar, *Peregrinación de Alpha,* 2:49; see his comments on racial mixing in 1:228 and 2:105.

47. Ibid., 1:248.

Chapter 2: Change and Community in La Salina

1. See Eliécer Silva Celis, "Contribución al conocimiento de la civilización de los Laches," *Boletín de Arqueología* (Bogotá) 1, no. 5 (1945): 369–424; and Luis Duque Gómez, *Tribus indigenas y sitios arqueológicos* (Bogotá: Editorial Lerner, 1967); see also José Ignacio Avellaneda, *The Conquerors of the New Kingdom of Granada* (Albuquerque: University of New Mexico Press, 1995), 113–14.

2. Ramón C. Correa, *Monografías de los pueblos de Boyacá,* 2 vols. (Tunja: Publicaciones de la Academia Boyacense de Historia, 1987–89), 1:395–96.

3. Martin Javier Plazas Pérez, "La Salina: el pueblo de la sal, una puerta de la Casa del Sol," *Caribarare* 4, no. 4 (Feb. 1992): 50–51; Martín Amaya Roldán, *Historia de Chita* (Tunja: Imprenta Oficial, 1930).

4. Juan Friede, "La conquista del territorio y el poblamiento," in *Manual de historia de Colombia,* ed. Jaramillo Uribe, 4th ed., 3 vols. (Bogotá: Tercer Mundo, 1992), 1:144; Juan Rodríguez Freile, *El carnero (segun el manuscrito de yerbabuena),* ed. Mario Germán Romero (Bogotá: Instituto Caro y Cuervo, 1984), 9–19; and Avellaneda, *Conquerors of the New Kingdom,* 34–35.

5. On the institutionalization of La Salina in this subordinate role at the beginning

of the seventeenth century, see Alba Luz Bonilla de Pico, "Espacio poblamiento en el resguardo de Chita en la segunda mitad del siglo xvii," *Historia Critica* 26 (July–Dec. 2003): 77–92.

6. "The encomienda included the town of Chita with seven captains, and another twelve towns (Támara, Pisba, Chipa, etc.) laches, chiscas y tunebos . . . one of the largest encomiendas in the province of Tunja" (Colmenares, *La Provincia de Tunja,* 17, 43–44). On La Salina's legal elevation to the status of pueblo under the Spanish, see Belisario Arenas Paz, *Guía geográfica de Colombia (diccionario breve de geografía física, económica, política, histórica y matemática),* 3d ed. (Bogotá: Editorial Santafe, 1940), 170, 253. On the tribute being paid in salt, see Plazas Pérez, "La Salina," 52. The Castilian Crown had claimed rights over salt production in its kingdom as far back as 1348; see Robert P. Multhauf, *Neptune's Gift: A History of Common Salt* (Baltimore, Md.: Johns Hopkins University Press, 1978), 15.

7. Hector Publio Pérez Angel, "Economía y demografía de La Salina: fuentes para un estudio," *Caribarare* 4, no. 4 (Feb. 1992): 53.

8. David Cook, *Born to Die: Disease and New World Conquests, 1492–1650* (Cambridge: Cambridge University Press, 1998). Juan Friede's article on Muzo, in western Boyacá, documents a substantial decline in the indigenous population during the first half of the seventeenth century but notes that Muzo and other similar mining areas registered population increases by attracting migrants (Friede, "Demographic Changes in the Mining Community of Muzo after the Plague of 1629," *Hispanic American Historical Review* 47, no. 3 [Aug. 1967]: 338–43). Peter Marzahl found dramatic decreases in the indigenous population in southwestern Colombia; see his *Town in the Empire: Government, Politics, and Society in Seventeenth-Century Popayán,* Latin American Monographs no. 45 (Austin: Institute of Latin American Studies, University of Texas, 1978), 36.

9. "Documentos ineditos del Archivo de Indias," *Reportorio Boyacense* 4:40. The salt monopoly was established in the sixteenth century, but it was centered on the works of Cundinamarca rather than peripheral works such as La Salina (Peña Díaz, *Sal, sudor y fisco,* 55–56).

10. One was cited for forcing the population to work without pay, and another was accused of mistreating residents in a dispute regarding salt production; see AGN, SC, Caciques y Indios, leg. 22, fols. 777–79, 1625; and AGN, SC, Encomiendas, 32, fols. 1–155, 1654.

11. On the first viceroyalty, see Múnera, *Fracaso de la nación,* 32; and Gabriel Giraldo Jaramillo, introduction, *Relaciones de mando,* 9. See also John R. Fisher, Allan J. Kuethe, and Anthony McFarlane, eds., *Reform and Insurrection in Bourbon New Granada and Peru* (Baton Rouge: Louisiana State University Press, 1990), especially the articles by Kuethe, Jacques A. Barbier, and J. R. Fisher. On the resettlement of *pueblos de indios,* see the writing of Francisco Antonio Moreno y Escandón collected as *Indios y mestizos de la Nueva Granada a finales del siglo XVIII,* ed. Germán Colmenares and Alonso Valencia (Bogotá: Fondo de Promoción de la Cultura del Banco Popular, 1985); and Margarita González, *El resguardo en el Nuevo Reino de Granada,* 3d ed. (Bogotá: El Áncora Editores, 1992). For regional perspectives, see Allan Kuethe, *Military Reform and Society in New Granada, 1773–1808* (Gainesville: University Press of Florida, 1978); and his "More on 'The Culmination of the Bourbon Reforms': A Perspective from New Granada," *Hispanic American Historical Review* 58 (Aug. 1978): 477–80. On the economic impact of these reforms, see Jaime Jaramillo Uribe, "La economía colonial del virreinato (1740–1810)," in *Historia económica de Colombia,* ed. José Antonio Ocampo (Bogotá: Siglo Veintiuno, 1987), 81.

12. Anthony McFarlane explains that the "reglamento de comercio libre," the reform that loosened Cadiz's monopoly on American trade, was "the lynchpin of Caroline economic reform," for it remade vast sections of the economy when the empire was not at war. However, Spain was at war with France in 1773–75 and with England in 1779–83, 1796–1802, and 1804–8 (McFarlane, *Colombia before Independence,* 126–29, 154–61, 297–321).

On the reforms across Spanish America, see Kenneth J. Andrien and Lyman L. Johnson, eds., *The Political Economy of Spanish America in the Age of Revolution, 1750–1850* (Albuquerque: University of New Mexico Press, 1994).

13. For example, in 1804 the Real Hacienda collected 2,453,096 pesos. The monopoly on tobacco accounted for 470,000 pesos; aguardiente, for 295,000; the alcabala, for 184,880; and the salt monopoly (mainly the works at Zipaquirá), for 100,000. See Jaramillo Uribe, "La economía del Virreinato," 79. On the Bourbon monopolies in New Granada, see Eduardo Arcila Farias, *Historia de un monopolio: el estanco del tabaco en Venezuela (1779–1833)* (Caracas: Facultad de Humanidades y Educacion, Instituto de Estudio Hispanoamericanos, Universidad Central de Venezuela, 1977). For overviews on state monopolies in the colonial era, see Rafael Dobado González, "El monopolio estatal del mercurio en Nueva España durante el siglo XVIII," *Hispanic American Historical Review* 82, no. 4 (Nov. 2002): 685–718; Susan Deans-Smith, "State Enterprise, Work and Workers in Mexico: The Case of the Tobacco Monopoly, 1765–1850," in *Political Economy of Spanish America*, ed. Andrien and Johnson, 63–94; Deans-Smith, *Bureaucrats, Planters, and Workers: The Making of the Tobacco Monopoly in Bourbon Mexico* (Austin: University of Texas Press, 1992); and Myriam Ellis, *O monopólio do sal no estado de Brasil (1631–1801): contribução ao estudo do monopólio comercial português no Brasil durante o periodo colonial* (São Paulo: Universidad do São Paulo, 1955).

14. La Salina's presence in the record increased notably in this era, with dozens of indexed documents in the AGN's Sección Colonial, particularly in the collections Visitas, Caciques e Indios, Encomiendas, Resguardo-Boyacá, Juicios Civiles-Boyacá, and Poblaciones-Boyacá; see Gilberto Suarez Casteñeda, *Fondo Salinas: Sección Colonial, catálogo e indice* (Bogotá: AGN, 1992). For details on life in La Salina, see Pérez Angel, "Economía y demografía," 59n6.

15. AGN, SC, Juicios Criminales, 101, fols. 320–48, 1774; Moreno y Escandón, *Indios y mestizos*, 159.

16. Curry, "Disappearance of the Resguardos," 34–36. See also John Leddy Phelan, *The People and the King: The Comunero Revolution in Colombia, 1781* (Madison: University of Wisconsin Press), 92. On Moreno y Escandón's career, see McFarlane, *Colombia before Independence*, 207; on the removals, see Jorge Orlando Melo's introduction to Moreno y Escandón, *Indios y mestizos*, 24–36. Six pueblos in Boyacá had been targeted in 1744: Soatá, Motavita, Pare, Tensa, Garagoa, and Somondoco. Attempts were made at removals in Ramiriquí in 1763 and Sogamoso in 1767. See also "Relación del estado del Nuevo Reino de Granada que hace el Excmo. Sr. D. Francisco Gil y Lemos a su successor Excmo. Sr. D. José de Espeleta, 1789," in *Relaciones de mando*, ed. Giraldo Jaramillo, 143–59; and González, *El resguardo*, 91–98.

17. Moreno y Escandón, *Indios y mestizos*, 28–30. There were eighty-two vecino heads of household in the town. Campuzano y Lanza reported La Salina had twenty-four tribute-paying heads of household, just below the twenty-five cited as the minimum to avoid resettlement, though this seems to have been more a benchmark than a rule. Another official counted twenty-eight in La Salina. A visitor to Chita in 1635 reported resident vecinos, so it is reasonable to assume that there were also vecinos in La Salina during the seventeenth century; see Bonilla de Pico, "Espacio poblamiento," 81. The saltworks at Recetor were also developed as a Crown rent at this point (Peña Díaz, *Sal, sudor y fisco*, 58–59).

18. Moreno y Escandón, *Indios y mestizos*, 165.

19. Ibid., 164.

20. Amaya Roldán, *Historia de Chita*, 76–77. For a reference to the return of the community, see AGN, SR, Gobernación de Tundama, leg. 4, fols. 306–7, October 1852.

21. Bonilla de Pico, "Espacio y poblamiento."

22. Phelan, *The People and the King*, chap. 3, "The Crowd Riots"; see also Mario Agu-

ilera Peña, *Los Comuneros: guerra social y lucha anticolonial* (Bogotá: Universidad Nacional de Colombia, 1985). On other rebellions prompted by the reforms in New Granada, see Anthony McFarlane, "The 'Rebellion of the Barrios': Urban Insurrection in Bourbon Quito," *Hispanic American Historical Review* 69, no. 2 (May 1989): 283–330.

23. McFarlane, *Colombia before Independence,* 261.

24. Ibid., 216–20.

25. For an example of a regional history of the Comunero Rebellion that offers significant variation from the central narrative, see Jane M. Loy, "Forgotten Comuneros: The 1781 Revolt in the Llanos of Casanare," *Hispanic American Historical Review* 61, no. 2 (1981): 235–57.

26. José de Jesús Fuentes Combita, "Las salinas de Zipaquirá y Nemocón en la historia social y cultural de Colombia," Master's thesis, Universidad Externado de Colombia, Bogotá, 1990, 60. On the captains, see Gabriel Camargo Pérez, *Pueblos y jornadas Boyacenses* (Tunja: Academia Boyacense de Historia y Gobernación de Boyacá, 1994), 108.

27. Climaco Calderón, *Elementos de la hacienda pública* (Bogotá: La Luz, 1911), 394. In Zipaquirá prices went from two to three and one-half reales for an arroba, and payment had to be in currency rather than barter (Phelan, *The People and the King,* 159, 229–30). It was also in the 1770s that the royal government claimed and began to develop previously unexploited works in what would become the territory of San Martin during republican times; see Nohora Beatriz Guzmán Ramírez, "Economía extractiva: la explotación de la sal mina en el territorio de San Martín, Colombia, siglo XIX," paper presented at the Segundo Congreso Latinoamericano de Historia y Económica, February 2010.

28. Laso de la Vega "respected the free benefit that the vecinos had enjoyed" to make salt (Calderón, *Elementos de la Hacienda,* 394).

29. Moreno y Escandón, *Indios y Mestizos,* 159. On the Llanos, see Amaya Roldán, *Historia de Chita,* 76–77. On applying for parish status, see Plazas Pérez, "La Salina," 55; "Documentos," *Caribarare* 4, no. 4 (Feb. 1992): 11; on a similar attempt in 1811, see Pérez Angel, "Economia y demografia," 59.

30. AGN, SC, Fondo Salinas, leg. 1, fols. 444–53, 1804. For Caicedo's requests, see ibid., leg. 5, fols. 858–61, 1806, and leg. 7, fols. 297–313, 1806. In another incident showing salt's role, in 1801 Captain Joseph Miguel Vasquez, who was attempting to "pacify" the Tunebo Indians of eastern Boyacá and Casanare, petitioned for permission to use the proceeds from eight "hornos" (batches) of salt, free of any taxes, for the good of the Tunebos he sought to settle (AGN, SC, Caciques y Indios, leg. 12, fol. 808, August 1801).

31. The *Memoria de Hacienda de 1826,* by Jose del Castillo y Rada, offers an overview of the abolition and restoration of these taxes. On Nariño, see Luis Eduardo Nieto Arteta, *Economía y cultura en la historia de Colombia* (Bogotá: Banco de la República/El Áncora Editores, 1996 [1941]).

32. Carlos Hernádez Vargas, "La Salina: una riqueza en el olvido," *Caribarare* 4, no. 4 (Feb. 1992): 5–6; Pérez Angel, "Economia y demografía," 62; Juan Friede, *La Batalla de Boyacá—7 de Agosto de 1819—a través de los Archivos Españoles* (Bogotá: Banco de la República, 1969).

33. On Spanish forces in the area in 1816, see Humberto Enciso Diaz, *Historia del Municipio de Chita* (Tunja: n.p., 1992), 12.

34. Alejandro Osorio, "Memoria del Ministro de Guerra y Hacienda de 1819," in *De Boyaca a Cucuta: memoria administrativa, 1819–1821,* ed. Luis Horacio Lopez Dominguez (Bogotá: Biblioteca de la Presidencia de la Republica/ Administracion Virgilio Barco, 1990), 99. Bolívar's decree from September 13, 1819, "Circular para que las rentas se mantengan en el mismo estado que antes," stated that this should continue until "observation over time" revealed the reforms that could be implemented (ibid., 33).

35. Ibid., 102. Future *Memorias* were addressed to Congress rather than the executive.

36. *Gaceta de Santafé de Bogotá,* qtd. in ibid., 155. This report also noted that production

had been reorganized along the lines recommended by Alexander von Humboldt prior to independence. For his impressions, see Humboldt, *Memoria raciocinado de las Salina de Zipaquirá* (Bogotá: Banco de la República, 1952). Early laws established state control. One, passed on July 10, 1824, prohibited salt imports to protect domestic production; another, from July 28, 1824, formally declared the works to be property of the state. See Luis Horacio López Domínguez, ed., *Administraciones de Santander,* 6 vols. (Bogotá: Fundación para la Conmemoración del Bicentenario del Natalcio y el Sesquicentenario de la Muerte General Francisco de Paula Santander), 2:310.

37. For the text of the Ley fundamental de la República de Colombia, see Luis Fernando López Garavito, *Historia de la Hacienda,* 18–20. For an overview on taxes at this time in Gran Colombia, particularly on republican control over saltworks, see Bushnell, *The Santander Regime,* 83–86. The Congress at Angostora had produced an earlier "Ley fundamental de Colombia"; see http://www.alianzabolivariana.org/pdf/Ley_fundamental_Colombia.pdf.

38. Reading the writings of José María Eusebio Carlos del Rosario del Castillo y Rada is one of the pleasures of examining Colombia's early fiscal policy. For a brief and admiring biography, see Javier Ocampo López, "José María del Castillo y Rada," in *Gran enciclopedia de Colombia del Círculo de Lectores, tomo de biografías,* in the Biblioteca Virtual, Biblioteca Luis Ángel Arango (Dec. 6, 2004), http://banrepcultural.org/blaavirtual/biografias/castjose.thm.

39. *Memoria de Hacienda de 1823* (José Castillo y Rada), 12–13.

40. There was a fair amount of fluctuation in the manner of dealing with the rent. In 1821 Congress authorized the vice president of Cundinamarca to issue 200,000 pesos worth of currency, redeemable at the works at Zipaquirá, Nemocón, and Tausa, to pay debts and cover salaries (López Garavito, *Historia de la Hacienda,* 21). In contrast, in 1825 Bolívar "inflicted on Ecuador his own unique version of the salt monopoly," producing dramatically inflated prices (Bushnell, *Santander Regime,* 87, 311–13).

41. *Memoria de Hacienda de 1823* (Castillo y Rada), 36. On residents delivering salt, see AGN, SR, Intendencia de Boyacá, leg. 3, fols. 331–32, April 1825, and fol. 554, September 1825.

42. On the suggestion, see AGN, SR, Ministerio de Hacienda, leg. 213, fol. 641, 1821.

43. AGN, SR, Intendencia de Boyacá, leg. 3, fol. 758, December 1825. On Valderrama running the local liquor monopoly, see AGN, SR, Gobernación de Tunja, leg. 4, fol. 41, January 1834.

44. Malo was a prominent member of Tunja's "aristocratic camp"; see Victor Uribe-Uran, "Rebellion of the Young Mandarins: Lawyers, Political Parties, and the State in Colombia, 1780–1850" (PhD diss., University of Pittsburgh, 1993), 427. Malo was one of the signers of the Ley fundamental de la Nueva Granada in 1831. The only mention of other potential renters was a reference to interest on the part of a Pedro Ignacio Cela (AGN, SR, Intendencia de Boyacá, leg. 3, fol. 758, December 1825).

45. Complaints over protected markets were fairly common. For complaints from Zipaquirá about production in La Salina, see AGN, SC, Fondo Salinas, leg. 2, fols. 161–282, 1787–89, fols. 475–88, 1805–6; and AGN, SR, Gobernación de Tunja, leg. 1, fol. 882, February 1832, and leg. 4, fol. 781, November 1834. For complaints from Malo about lesser saltworks, such as Muneque, see AGN, SR, Gobernación de Tunja, leg. 3, fols. 498 and 518, April 1833, and fol. 774, October 1833.

46. Viceroy Gutiérrez de Piñeres introduced guías for merchandise during his *visita general* in 1780 (McFarlane, *Colombia before Independence,* 214). I have found only sporadic references to guías at La Salina over the course of the century.

47. See AGN, SR, Intendencia de Boyacá, leg. 1, 176, April 1822; AGN, SR, Intendencia de Boyacá, leg. 4, fol. 417, October 1826.

48. *Memoria de Hacienda de 1839* (Juan de Dios de Aranzazu), 37. Francisco Soto claimed

that the new laws saved a system that was "threatened with imminent ruin" (in López Domínguez, ed., *Administraciones de Santander,* 6:23).

49. Direct administration is an example of the process, described by Miguel Centeno, where "states penetrated their societies in increasingly complex forms in order to obtain resources" (Centeno, *Blood and Debt,* 103).

50. Peña Díaz, *Sal, sudor y fisco,* 93–95, 190.

51. AGN, SR, Ministerio de Hacienda, leg. 320, fol. 110, May 1836; López Domínguez, ed., *Administraciones de Santander,* 6:12.

52. See AGN, SR, Gobernación de Tunja, leg. 5, fol. 868, June 1836. Azenuega was also supposed to arrange for salt deliveries to the town of Soatá, where the Finance Ministry hoped to establish an almacén (AGN, SR, Ministerio de Hacienda, leg. 320, fol. 113, May 1836). On Morales, see Uribe-Uran, "Rebellion of the Young Mandarins," 302–3.

53. On the officials assuming office on June 14, see AGN, SR, Ministerio de Hacienda, leg. 320, fol. 110, May 1836; López Domínguez, ed., *Administraciones de Santander,* 6:12.

54. Legislation in 1832 set the administrator's deposit at 1,500 pesos, the almacenista's at 500 pesos, and so on (AGN, SR, Ministerio de Hacienda, leg. 320, fols. 113–17, June 1836). In one case an indebted applicant's wife supplied the deposit because the ministry would no longer accept his guarantee (ibid., fol. 157, June 1836). Regulations from 1857 stated that employees would share 7 percent of the profits from sales; the almacenista received 40 pesos per month plus 3.5 percent of the profits. The remaining profits were split 60:40 between the administrator and bookkeeper, though they were to cover the cost of renting an office, the guías, account books, and other supplies; see decree of December 15, 1857, "assigning the salaries of employees in the Administration of Salinas," article 2, *Gaceta Oficial,* January 9, 1858, 15. A similar law was passed in 1858 stating that the salaries of the administrator and bookkeeper together could not exceed 15 percent of the profits; see "Proyecto de lei organica de la renta de salinas de propiedad de la República," article 8, March 17, 1858, in *Gaceta Oficial,* April 28, 1858, 226–28. Another details the division of profits from 1867. The yearly salaries were as follows: administrator, 780 pesos; bookkeeper, 516 pesos; an almacenista in La Salina, 300 pesos; and an almacenista in Muneque, 400 pesos. The profit was divided as follows: 3 percent for the contractor, 1.5 percent for the administrator, 1.5 percent for the contador, 0.5 percent for the almacenista in La Salina, and 0.5 percent for the almacenista in Muneque; see article 7 of the "Decretos Sobre Salinas," in *Memoria de Hacienda y Fomento de 1867* (Alejo Morales).

55. For character references, see AGN, SR, Gobernación de Tunja, leg. 5, fols. 529–88, 770–88, 1836. On the tendency to give military officers government positions such as these, see Uribe-Uran, "Rebellion of the Young Mandarins," 187–256.

56. For letters on this issue, see AGN, SR, Gobernación de Tunja, leg. 6, fols. 100, 104, 230, 278, 282, April 1837, and fol. 537, May 1837.

57. Ibid., fol. 771, September 1837.

58. AGN, SR, Intendencia de Boyacá, leg. 8, fol. 137, April 1849. The legend of the plan is vague. I have taken a vara as 0.8 meters. Eduardo Arcila Farias's study of the tobacco monopoly in republican Venezuela contains plans for three buildings on the banks of the Santo Domingo River, in Barinas, designed in a similar fashion; see Farias, *Historia de un monopolio,* 124–25.

59. For the toll salt took on buildings, see AGN, SR, Gobernación de Tunja, leg. 6, fol. 100, April 1837; AGN, SR, Miscelánea de Salinas, leg. 8, fol. 713, March 1887. If the almacén was in bad shape, it was typical of government buildings of the era. On the state of jails in Boyacá, see Archivo Departmental de Boyacá (henceforth ADB), *Memoria de Gobierno de Boyacá de 1880,* 24–26. On poor conditions of schools, which were the responsibility of local communities, see AGN, SR, Gobernación de Tunja, leg. 1, fols. 116–17, March 1832, fols. 129–37, April 1832, and leg. 2, fol. 973, May 1833; Frank Safford,

The Ideal of the Practical: Colombia's Struggle to Form a Technical Elite (Austin: University of Texas Press, 1976), 49–55.

60. On the fallen almacén blocking the street, see AGN, SR, Gobernación de Tundama, leg. 4, fol. 793, January 1852; on its threatened collapse, see AGN, SR, Ministerio de Hacienda, leg. 335, fol. 716, June 1845. For an administrator's request for 100 pesos for repairs, see AGN, SR, Salinas de Boyacá, leg. 7, fol. 545, October 1854. One employee noted, "the offices are more like jails than places of public employment" (ibid., leg. 8, fol. 769, October 1877; see also AGN, SR, Miscelánea de Salinas, leg. 1, fol. 653, July 1896).

61. AGN, SR, Gobernación de Tunja, leg. 6, fol. 547, August 1837. This report mentions that though thirteen buildings could serve as almacenes, only three were considered secure.

62. AGN, SR, Ministerio de Hacienda, leg. 337, fol. 773, August 1846.

63. At first the unit had four captains and eighteen mounted guards, with seven guards stationed in La Salina; see AGN, SR, Gobernación de Tunja, leg. 11, fol. 490, August 1844. For a unit for tobacco, see AGN, SR, Tobaccos, leg. 73, fols. 388, 410, 430, 1841. On similar units in Venezuela, see Farias, *Historia de un monopolio,* chap. 4, "Rondas, contrabandistas y bandoleros." Until the reforms of midcentury, a unit of more than one hundred guards was spread out across the country; see, for example, an executive decree from June 6, 1851, *Gaceta de la Nueva Granada,* June 29, 1841. After 1850 the number increased radically. For specific decrees defining a force that fluctuated between two and three hundred in the 1850s, with close to half mounted, see Peña Díaz, *Sal, sudor y fisco,* 205–6.

64. On sealing other springs as the duty of guards from Zipaquirá, see AGN, SR, Gobernación de Tunja, leg. 5, fols. 258–66, August, November, and December 1835; for later reports of guards from La Salina doing the same, see ibid., leg. 10, fol. 656, May 1843. An "index of correspondence" contains references to similar activities by the Boyacá unit operating in the canton of Tenza, though I was unable to locate the documents; see ibid., leg. 11, fol. 122, February 1844.

65. Once established in Boyacá, the guards were divided into six units. Some patrolled the province while others were stationed at different works: seven guards in La Salina; two in Muneque; three in Chámeza; three in Mongua and Gamesa, which also included Recetor and Pajarito; and three in Sisbaca. In the sites where there were no captains, Muneque and Sisbaca, the guards were under the command of the administrator; in towns where there was a captain, the chain of command was less clear. See AGN, SR, Gobernación de Tunja, leg. 13, fol. 682, January 1845. For a request to increase the number to seven from a low of five, "as there were in the year [18]48," see AGN, SR, Gobernación de Tundama, leg. 1, fol. 221, November 1849; for a request for at least five or six guards, see ibid., leg. 6, fol. 663, July 1856; and for a reference to only six guards, see AGN, SR, Correspondencia de Boyacá, leg. 1, fol. 203, January 1858. The next mention of the number of guards reported fourteen in La Salina and two in Muneque (AGN, SR, Salinas de Boyacá, leg. 9, fol. 15, November 1864).

66. Decree of May 28, 1866, article 3, "Decretos Sobre Salinas," *Memoria de Hacienda y Fomento de 1866* (Morales).

67. On regular troops in La Salina, see AGN, SR, Salinas de Boyacá, leg. 1, fol. 204, October 1866. For other notes on increasing the number of guards, see ibid., leg. 9, fol. 538 August 1866; for further problems, see ibid., leg. 2, fol. 191, June 1867. For a decree limiting the unit to twenty men, see ibid., leg. 10, fol. 121, December 1868. For references to a guard unit of five corporals and twenty guards, see ibid., leg. 8, fols. 428–33, June 1875; to a unit numbering nineteen total, ibid., leg. 11, fol. 335, May 1880; of sixteen guards, ibid., leg. 7, fol. 778, November 1888; and, befitting calmer times, to a unit comprising one lieutenant, one corporal, and eleven guards, see AGN, SR, Miscelánea de Salinas, leg. 1, fol. 311, December 1904.

68. Among the powers of the administrator listed in a proposed law in 1858 was the ability to suspend guards when there was a "serious motive" to do so; see article 9.6, "Proyecto de lei organica," *Gaceta Oficial,* April 28, 1858, 226–28.

69. One official noted that guards from the area had "been notable for their poor conduct in most cases" (AGN, SR, Ministerio de Hacienda, leg. 61, fol. 858, September 1891). One wonders how much confidence was inspired when a guard accused of impropriety had residents attest to his good character; see AGN, SR, Gobernación de Tunja, leg. 15, fols. 663–72, July 1847.

70. *Memoria de Hacienda de 1826* (Castillo y Rada), 12.

71. For a reference to guards lacking weapons, see AGN, SR, Salinas de Boyacá, leg. 9, fol. 25, February 1865, and leg. 8, fol. 531, November 1875. For complaints about the high cost of living and low salaries, see ibid., leg. 11, fol. 491, May 1881.

72. On guards abandoning their posts, see AGN, SR, Miscelánea de Salinas, leg. 2, fol. 48, December 1878; and AGN, SR, Salinas de Boyacá, leg. 8, fol. 853, May 1878. For general incompetence, see AGN, SR, Miscelánea de Salinas, leg. 4, fol. 331, April 1883; AGN, SR, Salinas de Boyacá, leg. 12, fol. 427, May 1893; and AGN, SR, Gobernación de Tunja, leg. 15, fol. 649, June 1847; for excessive drinking, AGN, SR, Miscelánea de Salinas, leg. 1, fol. 1031, April 1899. For complaints from the administration that guards were not doing their jobs, see ibid., leg. 10, fol. 85, January 1891; for guards trading contraband, stealing, and leaving their posts, AGN, SR, Salinas de Boyacá, leg. 3, fol. 447, January 1871; for contraband, ibid., leg. 4, fol. 313, April 1872, and fol. 633, November 1872; and for refusing to weigh out salt for buyers, AGN, SR, ibid., leg. 3, fol. 435, January 1871.

73. On apparently random assaults, see AGN, SR, Salinas de Boyacá, leg. 5, fols. 880 and 909, May 1875. On the murder, see ibid., leg. 8, fols. 822–25, December 1877. For cases of sexual assault, see ibid., fol. 846, March 1878; and AGN, SR, Ministerio de Hacienda, leg. 61, fol. 852, June 1891.

74. AGN, SR, Gobernación de Tunja, leg. 9, fols. 435–37, May 1841. On the population harassing or abusing guards, see AGN, SR, Salinas de Boyacá, leg. 9, fol. 25, January 1865. On contrabandistas attacking guards, see AGN, SR, Miscelánea de Salinas, leg. 4, fol. 337, April 1887.

75. AGN, SR, Salinas de Boyacá, leg. 10, fol. 814, June 1873.

76. At one point a shelter built by guards patrolling the hills was destroyed by other members of the unit (ibid., leg. 10, fol. 766, August 1872).

77. One contract specified that if Honorato was unable to fulfill its terms because of accident or death, it would pass to his brothers Bernardo, Rafael, and Guillermo. See "Contracts," in *Memoria de Hacienda y Fomento de 1867* (Morales), 113–22, article 24. The Espinosas provided La Salina its closest analogue to the families of Zipaquirá, the entrenched interests that controlled production. For one public reference to these families, see a fanciful chart, "Estadistica Industrial," listing the families of Zipaquirá as holding the monopoly on salt, *El Loco,* Oct. 28, 1856. Though I have found little background on the family, Luis Garcia Hevia's 1841 painting *Muerte del General Santander,* which is in the collection of the Museo Nacional, shows a dying Santander surrounded by a number of men, one of them identified as Bonifacio Espinosa. I suspect this to be the same man. An image of the painting is available at the museum's website, www.museaonacional.gov.co, though the site does not include the explanatory text that identifies people by name.

78. AGN, SR, Ministerio de Hacienda, leg. 320, fol. 143, May 1836.

79. Solano's qualifications for the position were based on his experience in the *Ministerio de Guerra i Marina* between 1821 and 1827 (AGN, SR, Gobernación de Tunja, leg. 5, fol. 533, June 1836). On his short tenure, see ibid., fol. 587, July 1836, fol. 779, October 1836, and fol. 975, January 1837. On Espinosa as a subordinate employee in the first administration, see ibid., leg. 5, fol. 587, July 1836.

80. AGN, SR, Intendencia de Boyacá, leg. 6, fol. 438, October 1846. For the contract,

see AGN, SR, Gobernación de Tunja, leg. 14, fols. 284 and 451, October and December 1845.

81. AGN, SR, Gobernación de Tunja, leg. 14, fols. 437–44, August 1846, fols. 451–55, November 1846, and fol. 590, January 1847; AGN, SR, Gobernación de Tunja, leg. 15, 659–61, July 1847.

82. AGN, SR, Gobernación de Tunja, leg. 15, fols. 595–98, March 1847.

83. AGN, SR, Intendencia de Tundama, leg. 1, fol. 227, February 1849.

84. See *Gaceta Oficial,* May 2, 1850, 202, and March 30, 1851, 240. The ministry had fined him in 1847, though perhaps only for failing to forward records and reports in 1845, and he lost his position as contractor; see AGN, SR, Gobernación de Tunja, leg. 15, fols. 659–61, July 1847.

85. See *Gaceta Oficial,* September 20, 1851, 654; the following quotations are from this issue. For the invitation for new proposals, see *Gaceta Oficial,* April 20, 1852, 284.

86. See his proposal in *Gaceta Oficial,* May 24, 1856, 404. For a discussion on contract bids in 1858 from Rafael Niño and Evaristo de la Torre, as well as from Bernardo and Honorato Espinosa, see *Gaceta Oficial,* August 19, 1858, 459; Niño and de la Torre won the contract.

87. AGN, SR, Administración de Salinas, leg. 4, fols. 262–350, July 1870–April 1871. On the contract he won after Reyes y Niño failed to fulfill the conditions for theirs, see *Memoria de Hacienda y Fomento de 1867* (Morales), xi, and the section "Contratos," 113–22. On a provision to excavate and exploit a new spring, see AGN, SR, Salinas de Boyacá, leg. 10, fols. 331–32, November 1869.

88. AGN, SR, Gobernación de Tunja, leg. 28, fol. 846, February 1843.

89. Francisco Javier Vergara Velasco estimated that Boyacá's population quadrupled between 1778 and 1889; only Bolívar and Magdalena had lower rates of population increase (Vergara Velasco, *Nueva geografía de Colombia* [Bogotá, 1892]).

90. Oficina de Estadística Nacional de Colombia, *Anuario estadística de Colombia de 1875* (Bogotá, 1875), 33. Population figures fell for two towns, but this resulted from redistricting in both cases. Sativa-Norte's population went from 4,265 to 4,240, but that of its former vereda, Sativa-Sur, went from 1,048 to 2,398. A similar decrease was reported in the town of Covarachia (population 2,702 in 1851), which was renamed as Ricaurte and reformulated with new boundaries that gave it a population of only 1,914 in 1875.

91. AGN, SR, Gobernación de Tundama, leg. 1, fols. 69–70, July 1849.

92. AGN, SR, Salinas de Boyacá, leg. 5, fol. 426, July 1874. In a small measure intended to prompt development, the region was exempt from *diezmos* (tithes) before the decentralization of 1850; see *Memoria de Hacienda de 1848* (Florentino Gonzales), 11–12. Despite such efforts, few people relocated to Casanare during this era (Rausch, *Llanos Frontier,* 77, 104–13). Augusto Gómez's argues that ethnic conflict between the indigenous population and *colonos* after 1870 implies that there had been little in-migration earlier. His references to the works of Upín at Cumural suggest a link between the salt trade in Casanare and colonization. See Augusto Gómez F., *Indios, colonos, y conflictos: una historia regional de los llanos orientales, 1780–1970* (Bogotá: Siglo Veintiuno, 1992), 178–80. Like Cundinamarca, Boyacá is credited as having been a source of migrants to coffee-producing regions in the late nineteenth and early twentieth centuries, though there is little precise information on points of origin for these people. For this model, see Charles Bergquist, *Coffee and Conflict in Colombia* (Durham, N.C.: Duke University Press, 1986), 27–28; Elsy Marulanda, *Colonización y conflicto: las lecciones del Sumapaz* (Bogotá: Tercer Mundo, 1991), 41; Michael Jiménez, "Traveling Far in Grandfather's Car: The Life-Cycle of Coffee Estates in Central Colombia: The Case of Viotá, Cundinamarca, 1900–30," *Hispanic American Historical Review* 9, no. 2 (May 1989): 185–219.

93. AGN, SR, Salinas de Boyacá, leg. 4, fol. 61, September 1871. See also the administrator Vicente Vargas's comments on this subject in AGN, SR, Gobernación de Tunja,

leg. 24, fol. 725, July 1845; and AGN, SR, Ministerio de Hacienda, leg. 337, fol. 777, September 1846.

94. AGN, SR, Salinas de Boyacá, leg. 11, fol. 360, January 1881.

95. AGN, SR, Gobernación de Tunja, leg. 15, fol. 596, March 1847.

96. AGN, SR, Salinas de Boyacá, leg. 9, fol. 916, September 1868. The other mention of the cabildo from midcentury occurs in a list of criminal cases in El Cocuy that includes a note that the cabildo abierto of La Salina had been cleared of charges of "*fuga*" (flight), though there are no detials on what this charge involved; see AGN, SR, Gobernación de Tunja, leg. 15, fol. 793, January 1847.

97. Only seven towns had average payments equal to or higher than La Salina's, including the cantonal capitals Santa Rosa and Sogamoso. For town-by-town lists, see AGN, SR, Intendencia de Boyacá, leg. 2, fols. 454–632, June–August 1823. La Salina was exempt from earlier special contributions because of "the suffering of the town" under Spanish rule (López Domínguez, ed., *Administraciones de Santander,* 1:102).

98. This did not include his property in Tasco, one of the towns with an average payment higher than La Salina's, assessed at 15,472 pesos. In Sogamoso and Tirabitova two people were assessed as having property worth 40,000 pesos, the only fortunes to exceed Valderrama's in this sample. Santander himself declared that "hardly anyone would be so honest as to declare property worth over 20,000 pesos," though David Bushnell mentions fortunes of over 100,000 pesos in the 1820s (Bushnell, *The Santander Regime,* 83).

99. Hermes Tovar Pinzón, Camilo Tovar Pinzón, and Jorge Tovar Mora, *Convocatoria al poder del número: censos y estadísticas de la Nueva Granada, 1750–1830* (Bogotá: Archivo General de la Nación, 1995), 383.

100. Benjamín Novoa Zerda, *Los montes de La Salina de Chita y los intereses de la nación* (Bogotá: La Luz, 1889), 17.

101. AGN, SR, Gobernación de Tunja, leg. 5, fols. 997–98, January 1837. For references to the dispute, see AGN, SR, Gobernación de Tunja, leg. 4, fol. 996, May 1835.

102. AGN, SR, Gobernación de Tunja, leg. 2, fol. 749, January 1833. See also an editorial in *El Constitucional de Boyacá* (Dec. 9, 1832) arguing for the benefits of repartimiento.

103. AGN, SR, Gobernación de Tunja, leg. 5, fol. 972, January 1837.

104. For statements on this and on the high price of food in La Salina, see a report from Gómez included in "Informes de la administraciones de Salinas," comp. Pacifico E. Rivera, 9, in *Memoria de Hacienda de 1859*. In 1881 the administrator Santiago Izquierdo expressed his hope that a new wave of migrants might form the foundation of a permanent settlement but also complained about the cost of food in the town; see AGN, SR, Salinas de Boyacá, leg. 11, fol. 360, January 1881.

105. AGN, SR, Gobernación de Tunja, leg. 24, fols. 980–81, August 1839. In 1837 the *jefe* of the canton of El Cocuy complained that direct administration had hurt the cantonal economy as a whole (ibid., leg. 7, fol. 446, April 1837, and fol. 571, June 1837). The *Memoria de Gobierno de de Boyacá de 1878* contained a list of districts with the rents of each and the amount of money reserved for supporting education. La Salina was listed as having no rents, though 329 pesos was spent on schools; see Miryam Báez Osorio, *La educación radical en Boyacá: fundamento social y político* (Tunja: Academica Boyacense de Historia, 1996), 83–86.

106. In 1822 the curate could afford to contribute 100 pesos to the "war effort"; see AGN, SR, Intendencia de Boyacá, leg. 1, fol. 701, November 1822. Though the town was made a full parish in 1825, there were references to problems in paying the priest; see AGN, SR, Gobernación de Tunja, leg. 24, fols. 980–81, August 1839, leg. 28, fol. 846, February 1843, and leg. 12, fol. 620, July 1844. On the ministry's paying this salary, see AGN, SR, Ministerio de Hacienda, leg. 254, fol. 8, February 1830. Late in the century a visiting prefect of Gutiérrez noted that the unused parish register was in poor condition (*El Boyacense,* October 13, 1891, 2759).

107. Espino and Panqueba, used for comparison with La Salina elsewhere, returned 120 and 100 pesos, respectively (AGN, SR, Gobernación de Tunja, leg. 29, fols. 89–108, July 1843). Primicias were an offering to the church from the "first fruits" and distinct from diezmos; see Delfín Donadíu y Puignau, ed., *Novísmo diccionario enciclopédico de la lengua Castellana* (Barcelona: J. Espasa, 1890). In 1845 the administrator Vicente Vargas claimed that La Salina's poverty meant that the liquor monopoly was essentially worthless, whereas before direct administration it had been worth the substantial sum of 14,000 pesos (AGN, SR, Ministerio de Hacienda, leg. 335, fol. 705, March 1845). This was an exaggeration. In 1833 the estanco was devalued throughout Tunja. In La Salina it was valued at 1,110 pesos, more than the 691 pesos paid for the estanco in Chita, but a world away from its virtual worthlessness in the 1840s (AGN, SR, Gobernación de Tunja, leg. 5, fol. 688, November 1833).

108. Vicente Gomez Maz., in "Informes de las administraciones de salinas," comp. Pacifico E. Rivera, 9. This report is ancillary to the *Memoria de Hacienda de 1859* (Ignacio Gutiérrez).

109. Oficina de Estadística, *Anuario estadística de 1875,* 143.

110. Oficina de Estadística de Colombia, *Anuario de estadística de Colombia de 1883* (Bogotá, 1883), 11–18.

111. See Ancízar, *Peregrinación de Alpha,* 1:251; AGN, SR, Gobernación de Tunja, leg. 5, fols. 997–98, January 1837. On a conflict between the administrator and Guillermo Terán over access to timber, see AGN, SR, Salinas de Boyacá, leg. 12, fols. 291–92, 295, 299, July–August 1891.

112. AGN, SR, Gobernación de Tunja, leg. 13, fol. 870, May 1845.

113. AGN, SR, Salinas de Boyacá, leg. 9, fol. 916, September 1868. For a later report on the high cost of living as an impediment to production, see ibid., leg. 12, fol. 818, December 1894.

114. AGN, SR, Gobernación de Tundama, leg. 4, fol. 868, May 1853.

115. Fuentes Combita, "Las Salinas de Zipaquirá," 7, 55. See also Moreno y Escandón, *Indios y mestizos,* 269–82; Thomas, "Disappearance of the Resguardos," 92–93.

116. Nicolas Ortiz, *Estadística de las salinas de la República* (Bogotá, 1889), 7, 129–30.

117. Fuentes Combita, "Las Salinas de Zipaquirá," 7, 89–91. The indigenous community in Nemocón was involved in lawsuits to defend their access to the saltworks and coal deposits throughout the nineteenth century, though it is not clear whether this group included the families from Zipaquirá; see AGN, SR, Miscelánea de Salinas, leg. 2, fol. 120, October 1878, and leg. 3, fols. 1009–10, January 1881; and Thomas, "Disappearance of the Resguardo," 96.

118. On the saltworks in Guativita, see AGN, SR, Gobernación de Tunja, leg. 16, fol. 334, March 1843.

119. Moreno y Escandón, *Indios y mestizos,* 243.

120. AGN, SR, Gobernación de Tunja, leg. 10, fol. 768, September 1843. For Soto's comments, see López Domínguez, ed., *Administraciones de Santander,* 3:26. On the *repartimiento,* see Fals Borda, *Hombre y tierra,* 76. On the complaints from the community, see AGN, SR, Gobernación de Tunja, leg. 16, fols. 334–37, October 1843.

121. My view here differs from that of Peña Díaz, who emphasizes the benefits that ministry control had for Chámeza over time. But in discussing how production actually seized up in this town and had to be restarted in the 1840s, he underscores the challenges brought by the ministry's assumption of control. His description of the town also includes references to a ruined church, much like the one in La Salina. See Peña Díaz, *Sal, sudor y fisco,* 64–69.

122. At least one court decision was decided in their favor, but further appeals were filed, and I have found no record of the final outcome. On that decision, see AGN, SR, Correspondencia de Boyacá, leg. 1, 374, July 1858.

123. A provincial survey from 1849 mentioned a school, and in 1856 a teacher was hired to give daily lessons; see AGN, SR, Gobernación de Tunja, leg. 17, fol. 708, January 1849; AGN, SR, Gobernación de Tundama, leg. 7, fol. 130, July 1856. An 1876 survey noted that there was no information on schools in La Salina. Another in 1877 reported that the school had thirty-six pupils, though local officials complained that rebels had stolen all the teaching materials during the previous war; see Oficina de Estadística de Colombia, *Anuario de estadístico de Colombia de 1876* (Bogotá, 1876), 54; *El Boyacense,* November 17, 1877, 152. The *Memoria de Hacienda y Fomento* for 1878, by Luis Bernal, reported that there were fifty boys in a single school; see the *Memoria,* "Documentos sobre estadística," 73. A visitor in 1881 noted that the district had two "urban" schools and one "rural" school, observing with some surprise that the alcalde had not "left this duty incomplete" (*El Boyacense,* Apr. 21, 1881, 6657). A petition from 1887 signed by thirty-two people requested a particular director for the school (*El Boyacense,* Mar. 4, 1887, 228).

124. La Salina was an example of what James Sanders calls popular smallholder republicanism. As did people in the communities he examines, many of La Salina's residents signed a petition involving local rights. La Salina differed, however, in that women as well as padres de familia signed the document; see Sanders, *Contentious Republicanism,* 40–44.

125. Numerous documents mentioned women among Valderrama's heirs. They too were eventually pushed out by a new wave of male purchasers, but their inheritance extended the era of women landholders in La Salina for a few decades (Novoa Zerda, *Los montes de La Salina*).

126. There were a few references to women's participation in the salt market. A detailed list of purchase requests from a few weeks in 1874, discussed later, includes 416 residents of La Salina, over half of whom were women (AGN, SR, Salinas de Boyacá, leg. 5, fols. 646–54, July 1874). That same year, Marcelina Alfonso was outbid—a bid of five pesos beating hers of four pesos—in a public auction of kettle salt that had been seized from the contrabandista Medardo Cuadra (AGN, SR, Administración de Salinas, leg. 1, fol. 762, November 1871).

127. In Zipaquirá cooking salt had been a domestic industry until 1599; see Marianne Cardale de Schrimpf, *Las salinas de Zipaquirá: su explotación indígena* (Bogotá: Banco de la República, 1981), 149. Various saltworks relied on the labor of entire families, even in rationalized production systems. In Britain's Cheshire saltworks, boilers (or lumpers), comparable to cosecheros, were helped by their wives and children; see Brian Didsbury, "Cheshire Saltworkers," in *Miners, Quarryman and Saltworkers,* ed. Raphael Samuel (London: Routledge and Kegan Paul, 1977), 143–56. According to the Liverpool, New York, Salt Museum, the same was true of salt harvesters who worked the "blocks" of the Onondaga saltworks in Syracuse and Liverpool, New York.

128. On the repartimiento in Zipaquirá, see AGN, SR, Miscelánea de Salinas, leg. 4, fol. 334, February 1887.

129. AGN, SR, Gobernación de Tunja, leg. 7, fol. 684, April 1838. A number of these challenges came in April, perhaps because the lower production levels after several months of winter rains increased tension over labor and low wages.

130. Ibid., leg. 9, fol. 960, April 1842.

131. AGN, SR, Salinas de Boyacá, leg. 9, fol. 156, April 1865.

132. In 1867 residents in Zipaquirá petitioned to end the practice of distributing *mugre* (the lowest quality salt, used as a catalyst during the initial boiling) to the poor on the grounds that they did not need charity. They proposed that a private company take over the sale of mugre and donate a portion of the profits to the construction of an alms house; see AGN, SR, Miscelánea de Salinas, leg. 5, fol. 370, September 1867. Yet even there dismantling the share was difficult. Tomás Cuenca described a gleaning practice that included the distribution not only of salt water but also of the salt-laden pottery shards

produced by breaking the containers during the deshorno and even the soil around the ovens; see *Memoria de Hacienda y Fomento de 1865,* 12. For a reference to the use of these shards in La Salina, see *Memoria de Hacienda de 1861* (Gutiérrez), 28.

133. AGN, SR, Miscelánea de Salinas, leg. 5, fols. 152–54, April 1865 (quotation, fol. 154).

134. AGN, SR, Salinas de Boyacá, leg. 9, fol. 182, June 1865. Vargas also pointed out that the repartimiento was still practiced in Zipaquirá.

135. Ibid., leg. 10, fol. 375, November 1869.

136. Ibid., leg. 5, fol. 335, April 1873.

137. Ibid., fol. 336, April 1873. For other mentions of the Sunday market, see AGN, SR, Ministerio de Hacienda, leg. 335, fol. 696, February 1845; AGN, SR, Salinas de Boyacá, leg. 2, fols. 658–62, January 1869.

138. AGN, SR, Salinas de Boyacá, leg. 7, fol. 697, July 1885.

139. Ibid., leg. 3, fol. 691, August 1871; at one point officials had destroyed a garden plot.

140. Ibid., leg. 7, fols. 870 and 873, June and July 1889 (Gaitan's request), and leg. 12, fols. 490 and 507, March 1892, fols. 547–78, February 1892 (Terán).

141. Ibid., fols. 437–38, August 1892 (petition), fols. 521–22, April 1893 (rental terms), fol. 593, July 1892, and fol. 646, November 1893 (eviction), fol. 786, January 1894 (rental data). Certain people rented more than one piece, so the average amount of land rented to each individual was closer to 175 square meters. Not all the rental fees were recorded, however, so the average is a tenuous figure. For more on rentals, see AGN, SR, Miscelánea de Salinas, leg. 1, fol. 783, 786, February 1896, fols. 826 and 847, July 1896, and fol. 292, August 1897. Another report mentions forty-four pieces of land, some of them divided into smaller pieces; see *Memoria de Hacienda de 1894* (Pedro Bravo), "Documentos," 36.

Chapter 3: Making Salt in a Ministry Works

1. See, in addition to the studies cited in the introduction, Multhauf, *Neptune's Gift*; and Didsbury, "Cheshire Saltworkers." See also Paul E. Lovejoy, *Salt of the Desert Sun: A History of Salt Production and Trade in the Central Sudan* (Cambridge: Cambridge University Press, 1986); Philip Ehrensaft, "The Political Economy of Informal Empire in Pre-colonial Nigeria, 1807–1884," *Canadian Journal of African Studies* 6, no. 3 (1972): 451–90; and E. Ann McDougall, "Salts of the Western Sahara: Myths, Mysteries, and Historical Significance," *International Journal of African Historical Studies* 23, no. 2 (1990): 231–57. On salt making and colonial control, see K. V. Jeyaraj, *A History of the Salt Monopoly in Madras Presidency (1805–1878)* (Madurai, India: Ennes, 1984); and Sadananda Choudhury, *Economic History of Colonialism: A Study of British Salt Policy in Orissa* (Delhi: Inter-India, 1979).

2. Francisco L. G. González, "El páramo en paisaje deshumanizado: el caso de las provincias del Norte y Gutiérrez (Boyacá—Colombia)," 2.1.4, "Período Muisca (1.200–500 años A.P.)," in *El páramo: un ecosistema de alta montaña,* ed. Felipe Cárdenas Támara, Antoine M. Cleef, and Abdón Cortés Lombana (Bogotá: Instituto de Estudio Ambientales para el Desarrollo, 1996), available at http://www.banrepcultural.org/blaavirtual/faunayflora/paramo/para214.htm.

3. Fuentes Combita, "Las salinas de Zipaquirá," 181.

4. The most detailed list names twelve springs. La Salina had eight: Bravo, de la Virgen, Azul, Grande, Santa Barbara, Canal de Piedra, el Taro, and el Volcan. Córdoba had la Cabuya, el Paloma, Cuicas, and los Taches (AGN, SR, Ministerio de Hacienda, leg. 193, fol. 735, December 1884). A later report counted sixteen, though it did not name them. This report claimed there were twelve in La Salina and four in Córdoba (AGN, SR, Salinas de Boyacá, leg. 12, fol. 546, February 1892).

5. On earthquakes, see AGN, SR, Intendencia de Boyacá, leg. 8, fol. 211, April 1849.

One official noted that vertientes in Córdoba were "discovered in all parts" (AGN, SR, Gobernación de Tundama, leg. 1, fol. 221, December 1849).

6. AGN, SR, Miscelánea de Salinas, leg. 5, fol. 14, May 1835.

7. For the road, see AGN, SR, Gobernación de Tunja, leg. 5, fol. 971, January 1837.

8. AGN, SR, Intendencia de Boyacá, leg. 8, fol. 211, April 1849.

9. AGN, SR, Salinas de Boyacá, leg. 10, fols. 253, 255, 300, 312, 331, September–November 1869. An 1871 visit yielded reports on the flow rates of the springs. The main spring in Córdoba produced a liter of water every five seconds, but officials were unable to measure the rate of the Pozo Principal or the Pozo Azul (report from Jacinto Corredor in *Memoria de Hacienda y Fomento de 1871* (Camacho Roldán), "Documentos Sobre Salinas," 153.

10. On Santa Bárbara, see AGN, SR, Miscelánea de Salinas, leg. 7, fols. 625, 666, December 1892.

11. Malo asked to use 8,000 pesos of his rental fee to build this wall, about a quarter of the annual proceeds generated at La Salina; see AGN, SR, Ministerio de Hacienda, leg. 254, fols. 179–81, April 1830; and AGN, SR, Gobernación de Tunja, leg. 3, fol. 649, June 1833. He also petitioned to introduce the large "calderos," or kettles, used in Zipaquirá. He had extensive dealings with the ministry, borrowing money and, like Valderrama before him, collecting the *estanco de aguardiente*; see ibid., leg. 3, fol. 536, October 1832, fols. 645 and 647, August 1833; AGN, SR, Ministerio de Hacienda, leg. 254, fol. 181, April 1830. One official estimated the wall at La Salina would increase production by as much as 50 percent (AGN, SR, Miscelánea de Salinas, leg. 5, fol. 21, April 1835). Entrepreneurs from Zipaquirá complained that these changes violated their right to a protected market, which was guaranteed by ministry contract (ibid., fols. 21–35, April–May 1835, fol. 50, February 1836).

12. The wall at La Salina cost 1,320 pesos, the second-largest expense after the fábrica and ovens; see an inventory by Rafael Niño, AGN, SR, Gobernación de Tundama, leg. 36, fol. 82084, December 1854.

13. See a proposal to develop Santa Bárbara, AGN, SR, Miscelánea de Salinas, leg. 12, fol. 441, December 1892.

14. AGN, SC, Fondo Salinas, leg. 7, fol. 304, 1806.

15. The size of these wells was rarely noted. In 1892 the administrator Guillermo Terán described the Pozo Grande as having a "regular shape" measuring 5.25 by 2.28 meters. In contrast, the Pozo Azul had an "irregular" shape, measuring 7 meters by 1.25 to 2 meters (AGN, SR, Salinas de Boyacá, leg. 12, fol. 603, July 1892).

16. Samples were sent to the National Museum in Bogotá. The Pozo Azul had a higher saline content than the Pozo Grande, with la Paloma having the lowest (ibid., leg. 3, fol. 16, January 1870). A report from 1855 stated the salinity to be between 18 percent and 22 percent, but one from 1898 put it between 14 percent and 18 percent (AGN, SR, Gobernación de Tundama, leg. 6, fol. 257, June 1855; AGN, SR, Miscelánea de Salinas, leg. 1, fol. 926, October 1898). Rafael Núñez reported that the saturation was 21 percent in the main pozos and ranged from 12 to 14 percent in the others (*Memoria de Hacienda de 1857*, 33). There were numerous comments on the quality of salt from La Salina. Ancízar described the salt there as being better than the salt at Zipaquirá (*Peregrinación de Alpha*, 1:249). On crystals forming freely, see AGN, SR, Salinas de Boyacá, leg. 12, fol. 666, December 1892.

17. Ancízar, *Peregrinación de Alpha*, 1:253.

18. On oven placement affecting the water supply, see AGN, SR, Gobernación de Tunja, leg. 13, fols. 774–85, February 1845; AGN, SR, Intendencia de Boyacá, leg. 9, fol. 4, February 1865. For a poorly built well affecting supply, see AGN, SR, Ministerio de Hacienda, leg. 337, fol. 758, January 1846. The only unexplained reports on diminished

water supply are at AGN, SR, Gobernación de Tundama, leg. 6, fol. 168, March 1855; and AGN, SR, Salinas de Boyacá, leg. 10, fol. 255, September 1869.

19. AGN, SR, Ministerio de Hacienda, leg. 388, fol. 278, July 1895.

20. For comments on the abundance of the springs, see AGN, SR, Gobernación de Tundama, leg. 2, fol. 29, October 1850.

21. On moving water, see ibid., leg. 5, fol. 971, January 1837. An inventory from 1866 includes thirty-five leather *canoas* (troughs or open pipes), six of which could not be used, and seven wooden ones (AGN, SR, Salinas de Boyacá, leg. 1, fol. 198, October 1866). A report from 1875 mentioned twenty-eight canoas worth seventy-nine pesos (three pesos for those in good condition, two for those in poor condition); see ibid., leg. 8, fol. 352, April 1875. An inventory from Muneque mentioned two wood canoas worth 10 pesos each (AGN, SR, Miscelánea de Salinas, leg. 12, fol. 23, January 1895). On canoas in Upín, see Emiliano Restrepo Echavarría, *Una excursion al territorio de San Martín en Diciembre de 1869* (Bogotá: Banco de la República, 1955), 81. Inventories taken after midcentury always mention pipes and pumps. One described "pipes that take the salt water to the ovens" near the Pozo Azul (AGN, SR, Gobernación de Tundama, leg. 6, fols. 82–84, November 1854). Another inventory lists twenty-one pipes and eight pumps (AGN, SR, Salinas de Boyacá, leg. 1, fol. 198, October 1866). Another mentions *albercas,* or water tanks, "for saturation" and pipes in La Salina and Córdoba (ibid., leg. 8, fol. 352, April 1875). Albercas were built into the wall next to the river or the fábrica. One for the Pozo Azul measured 3.8 meters deep, with a surface area of 1 by 1.4 to 1.5 meters (ibid., leg. 8, fol. 77, March 1874). The most lucid description of the system appears in a proposal, written by Felix A. Soto representing the elaboradores Tomás Carrea and Francisco Calderón, for improvements in Córdoba in 1892. It included plans to "[w]all the source of salt water . . . , cover the well with a good roof, install the pump and build pipes to take the salt water to the tank in the fábrica" (AGN, SR, Miscelánea de Salinas, leg. 12, fols. 440–43, December 1892).

22. See the inventory in AGN, SR, Salinas de Boyacá, leg. 8, fol. 635, March 1876.

23. The ministry asked for "three iron cauldrons that the government need[ed] to make kettle salt" (ibid., leg. 7, fol. 140, November 1874). Cortés paid £339 10s. 6d. for three cauldrons to be delivered to Baranquilla in 314 bundles in November 1881 by the steamship *America* (AGN, SR, Miscelánea de Salinas, leg. 4, fols. 37–39, November 1881). I have found no reference to them in La Salina. On smaller iron cauldrons, see an inventory in AGN, SR, Salinas de Boyacá, leg. 8, fol. 633, March 1876.

24. On the initial boiling, see Humboldt, *Memoria raciocinado,* 23. Production in La Salina was often modeled on that in Zipaquirá—in this case, on eighteenth-century Zipaquirá. Didsbury describes the skill needed to work open pans in "Cheshire Saltworkers," 146–50, though the system at Cheshire differed from that employed in La Salina. For a system at Lake Chad that bears some parallels to that of La Salina, see Lovejoy, *Salt of the Desert Sun,* 63–74. For an illustration of a Chinese oven like those used in La Salina, see Multhauf, *Neptune's Gift,* 32.

25. In Zipaquirá an inferior salt called *migajon* was used as a catalyst. For problems with salt manufactured in this fashion, see AGN, SR, Salinas de Boyacá, leg. 10, fol. 300, October 1869. Though several substances would work, only other salts were mentioned in the official correspondence. For a reference to substances used to "dope the pan," see Didsbury, "Cheshire Saltworkers," 145–47. For a prohibition of catalysts, see AGN, SR, Gobernación de Tunja, leg. 10, fol. 106, June 1842.

26. AGN, SR, Salinas de Boyacá, leg. 10, fol. 227, August 1869.

27. Ibid., fol. 251, September 1869.

28. Two descriptions are particularly useful for understanding this process: Humboldt, *Memoria raciocinada,* 21; and Ancízar, *Peregrinación de Alpha,* 2:251.

29. For descriptions of the ovens, see AGN, SR, Salinas de Boyacá, leg. 8, fol. 38, February 1874, and leg. 10, fol. 589, May 1870.

30. AGN, SR, Gobernación de Tunja, leg. 7, fol. 571, June 1837.

31. See a note discussing production in the 1840s in AGN, SR, Salinas de Boyacá, leg. 9, fol. 4, February 1865.

32. AGN, SR, Gobernación de Tunja, leg. 10, fol. 837, October 1843. Specifically, before the repair, 1,200 bundles of wood produced only 1,300 arrobas (one arroba was 12.5 kilograms) of salt; after the repair, the same amount of wood produced 1,620 kilograms of salt.

33. On the ovens in La Salina, see AGN, SR, Salinas de Boyacá, leg. 8, fols. 38 and 77, March 1874, and fols. 632–33, March 1876; AGN, SR, Ministerio de Hacienda, leg. 388, fol. 279, July 1895. On the ovens in Córdoba, see AGN, SR, Gobernación de Tundama, leg. 6, fols. 83–84, December 1854; AGN, SR, Salinas de Boyacá, leg. 8, fol. 633, March 1876; AGN, SR, Ministerio de Hacienda, leg. 388, fol. 279, July 1895; *Memoria de Hacienda de 1896* (Ruperto Ferreira), 229–32; and AGN, SR, Miscelánea de Salinas, leg. 1, fol. 927, 1898.

34. AGN, SR, Salinas de Boyacá, leg. 10, fol. 589, May 1870.

35. Ibid., fol. 460, December 1869. The following figures are from this report.

36. For other problems with oven 4, see ibid., fols. 589–90, May 1870, and leg. 3, fols. 48 and 178, November and February 1870.

37. The only reference to a smaller efficient oven comes from late in the century, when oven 2 was measured at eighteen by four meters but reported to produce 225,000 kilograms a month. Given that La Salina's maximum production never topped 150,000 kilograms a month, this report seems suspect. This report also estimated that the oven named "Caro" in Córdoba produced 75,000 kilograms of salt per batch, though an 1895 estimate had put the batch capacity at 25,000 kilograms. See *Memoria de Hacienda de 1896* (Ferreira), 229.

38. AGN, SR, Miscelánea de Salinas, leg. 10, fols. 92–93, February 1892. In England the sheds had open sides, "leaving as much open as possible so that steam could escape" (Didsbury, "Cheshire Saltworkers," 142). The only reference to a church appeared in Caycedo's 1806 drawing, which shows a "ruined church," so the industry had the only large buildings in town.

39. AGN, SR, Miscelánea de Salinas, leg. 10, fols. 92–93, February 1892. On the size of the almacén, see AGN, SR, Salinas de Boyacá, leg. 8, fol. 77, March 1874.

40. Qtd. in AGN, SR, Ministerio de Hacienda, leg. 388, fol. 378, July 1895.

41. *Memoria de Hacienda de 1896* (Ferreira), 229–332.

42. *Memoria de Hacienda de 1894* (Bravo), 57. In a similar fashion, two decades earlier Jacinto Corredor wrote that it was possible to produce between 150,000 and 187,500 kilograms per month; see *Memoria de Hacienda y Fomento de 1871* (Camacho Roldán), "Documentos Sobre Salinas," 154.

43. For a proposal to move La Salina to cooking kettle salt, see *Memoria de Hacienda y Fomento de 1869* (Cuenca), 56 and 73.

44. *Memoria de Hacienda de 1884* (Anibal Galindo), 145. For an earlier proposal in this direction, see the proposed Law 17 of 1873. Debates on sites at which to introduce open production often questioned the continued viability of the salt monopoly; see Nicolás Esguerra's comments that Zipaquirá should not be included in the change, *Memoria de Hacienda y Fomento de 1876,* 185–86. In 1888 Felipe Paúl explained that under open production the salt monopoly still brought in money (*Memoria de Hacienda de 1888,* xliv). By 1883 there were thirty-one different businesses using steam to make caked salt in Zipaquirá (*Memoria de Hacienda de 1883,* [Galindo], 50). In 1890 there were fifty-six such operations in Zipaquirá, thirty in Nemocón, and three in Sesquilé; see *Memoria de Hacienda de 1890* (Felipe Paúl), "Documentos Sobre Salinas," 142–43.

45. Codazzi, *Comision corogáfica,* 224.

46. *Gaceta Oficial,* January 5–21, 6–32. Subsequent issues included articles on related conditions and news about vaccinations.

47. *Memoria de Hacienda y Fomento de 1868* (Jorge Gutiérrez de Lara), 30. In a related development that resonated with increasing concern over social hygiene, finance ministers began to use the metaphor of Colombian society as a body prone to illness. In 1856 Rafael Núñez used it when he began to discuss foreign debt: "Citizen Legislators, while we are unable to eradicate to the root this cancer that devours all of the National wealth . . ." (*Memoria de Hacienda de 1856,* 59). In 1871 Salvador Camacho Roldán asserted that political unrest was a "gangrene" that had eaten away at Colombia since independence, devouring the "fruit of [the nation's] labors." He continued, "Our periodic revolutions . . . feed and support a growing illness in the social body" (*Memoria de Hacienda y Fomento de 1871,* viii). Jorge Gutiérrez de Lara used similar terms when he critiqued *empleomanía,* the politicization of and excessive interest in government employment, arguing that it maintained "the hopes and continued aspiration of a multitude of citizens. From [t]here the result is the empleomanía, [a] deadly cancer of society that threatens public peace, disturbs progress and prejudices the general well-being" (*Memoria de Hacienda y Fomento de 1868,* 4).

48. *El Porvenir,* January 8, 1856. The earliest attempt at using steam to convert kettle salt into caked salt occurred in 1855 when Diego Davidson, who had served as Ignacio Morales's representative in La Salina in the 1830s, was granted a twenty-year concession to introduce the process at Zipaquirá. While Davidson did not have much success with steam, it was later used in Zipaquirá; see *Memoria de Hacienda y Fomento de 1875* (Parra), "Documentos sobre Salinas," 168–70. Until 1889 any reference to steam power implicitly referred to an imported steam engine, for that was the first year a working engine was built in Colombia; see Alberto Mayor Mora, "Las fábricas de hierro en Colombia, en el siglo XIX," *Boletín Cultural y Bibliográfico,* available at Biblioteca Virtual, Biblioteca Luis Ángel Arango (May 2005), http://www.lablaa.org/blaavirtual/revistas/credencial/julio1993/julio2.htm.

49. *Memoria de Hacienda y Fomento de 1866* (Cuenca), xiv.

50. *Memoria de Hacienda i Fomento de 1867* (Morales), xiv–xvii.

51. AGN, SR, Miscelánea de Salinas, leg. 6, fol. 289, August 1884.

52. See his report to the Finance Ministry, ibid., leg. 4, fol. 685, June 1883.

53. See his report in *Memoria de Hacienda y Fomento de 1872* (Camacho Roldán), 32. For reports on coal, see *Memoria de Hacienda de 1885* (Felipe Angulo), 44; on a contract with Enrique Luque L. to excavate the salt bank and find a coal mine, see *Memoria de Hacienda de 1892* (José Manuel Goenaga G.), xxxviii.

54. AGN, SR, Salinas de Boyacá, leg. 4, fol. 219, February 1872; AGN, SR, Miscelánea de Salinas, leg. 10, fols. 87–89, February 1891.

55. Gomez, in "Informes de las administraciones de Salinas," comp. Rivera, 9, in *Memoria de Hacienda de 1859.*

56. See Safford, *Ideal of the Practical,* 191–92.

57. See James William Park, *Rafael Núñez and the Politics of Colombian Regionalism, 1863–1886* (Baton Rouge: Louisiana State University Press, 1984), 169; and Ines Pinto Escobar, *Progreso, industrialización y utopía,* 27. The original attempt in the 1850s was headed by two English engineers, Martin Perry and Santiago Bruco. See Pablo Castillo Muñoz, "Ferrería de Samacá: Samacá–Boyacá," Monumentos Nacionales de Colombia, online exhibition, available at Biblioteca Virtual, Biblioteca Luis Ángel Arango, http://www.lablaa.org/blaavirtual/exhibiciones/monu/samaca.htm. In 1864 the government in Boyacá established a company to build an ironworks but managed only to oversee the construction of a few buildings at 3,600 pesos each (see *El Tornillo,* December 28, 1864, 12; see also *El Boyacense,* January 17 and June 25, 1870): "[The ironworks] is not yet as it needs

to be, a center of movement and activity[;] . . . it is [only] a group of five buildings" (*El Occidente,* October 26, 1872, 62–63).

58. For the 1874 law that created this ironmaking company, see *Memoria de Hacienda de 1881* (Antonio Roldán), 32. See also Park, *Rafael Núñez,* 67–70, 169–72.

59. AGN, SR, Correspondencia de Boyacá, leg. 2, fol. 786, September 1878. For an optimistic and rather technical report on the prospects of the works, see a report by an engineer, Trorimo Verany, printed in the *Diario Oficial,* Jan. 7, 1876, 3569.

60. On national funding, see AGN, SR, Miscelánea de Salinas, leg. 3, fol. 982, December 1881; and the Ley 36 de 1880, in *Memoria de Fomento de 1884,* 91–92. For a defense of these contributions, see *Memoria de Gobierno de Boyacá de 1880,* 93.

61. *El Boyacense,* July 18, 1879, 2036. An editorial the year before offered a more mundane endorsement: "In a short time the whistle of the locomotive will sound in the same place where 59 years ago the whistle of bullets worked for the liberty of Colombia, announcing a new regeneration that represents the industrial transformation of the nation" (ibid., October 3, 1878, 398). On equipment imported from the United States, see ibid., July 17, 1878, 150.

62. For claims that the ironworks would produce enough iron for railroads and bridges, see *El Boyacense,* February 11, 1879, 895 and 897. For hopeful comparisons to works in Europe and the United States, see ibid., March 19, 1880, 5359. On the federal government's providing 100,000 pesos in 1879, see *Memoria de Hacienda de 1881* (Antonio Roldán), 32. Predictably, proposals discussed the likelihood that workers from the United States would come and settle with their families (Escobar, *Progreso, industrialización y utopía,* 28). Of 100,000 pesos from the federal government, 25,000 was set aside to promote the immigration of skilled workers, who would be given land and not be taxed on tools they brought; see the Ley de 30 de Abril de 1879 and Tomás Carrasquilla, *Immigración u colonización: informe que presenta Tomás Carrasquilla al Ministro de Obras Públicas* (1906), 17, archived at the Biblioteca de la Universidad Pedagógica y Tecnológica de Colombia (UPTC), Fondo Posada, 3739/10.

63. See pictures in Escobar, *Progreso, industrialización y utopía,* 37; and descriptions in *El Boyacense,* January 28, 1881, 6385; for measurements, see ibid., November 5, 1881, 7187.

64. Escobar, *Progreso, industrialización y utopía,* 31.

65. *El Boyacense,* September 28, 1881, 7182, 7187; November 5, 1881, 7313, 7315; December 10, 1881, 7378; April 12, 1887, 307–8.

66. *Memoria de Fomento de 1881* (Paúl), 32. Attacks on Otálera only increased. He served briefly as interim president after Núñez's first term, but by the end of his tenure, he was a discredited figure; see Park, *Rafael Núñez,* 241.

67. See Escobar, *Progreso, industrialización y utopía,* 36–38.

68. *Memoria de Hacienda de 1882* (Antonio Roldán), 45–48. Ironically, Mora cites a report by Nepomuceno Rodríguez from 1883 that works in Samacá were capable of producing several tons of iron a day (Mayor Mora, "Las fábricas de hierro en Colombia"). This article contains a number of images illustrating the scope and size of various projects.

69. Park, *Rafael Núñez,* 252–53. In 1884 Galindo maintained that with a further investment of 200,000 pesos, the ironworks would produce workable iron (*Memoria de Fomento de 1884,* 93).

70. Escobar, *Progreso, industrialización y utopía,* 38–40.

71. *El Boyacense,* October 26, 1877, 610–11. On the textile company, see AGN, SR, Gobernación de Boyacá, leg. 1, fols. 107, 208, January 1888. This effort took another decade, and its modest success was overshadowed by the earlier failure: "After the disasters suffered by the various businesses in the ironworks of Samacá, destroying almost completely the hopes that existed for its construction and the production, and when the valuable resources in that place are threatened with complete ruin, the idea to dedicate

that establishment to another type of industry has been accepted" (*Memoria de Gobierno de Boyacá de 1896,* 58).

72. The literature on sugar has produced useful works for considering forest use in reduction processes; see Stuart B. Schwartz, *Sugar Plantations in the Formation of Brazilian Society: Bahia, 1550–1835* (Cambridge: Cambridge University Press, 1985), 118, 141, 219; Franklin W. Knight, *Slave Society in Cuba during the Nineteenth Century* (Madison: University of Wisconsin Press, 1970), 28–29; and Shawn W. Miller, "Fuelwood in Colonial Brazil: The Economic and Social Consequences of Fuel Depletion for the Bahian Recôncavo, 1549–1820," *Forest and Conservation History* 38, no. 4 (Oct. 1994): 181–92.

73. There are reports that production in La Salina continued in this fashion late into the twentieth century. Filmmakers Sara Bright and Peter Goodhew wrote that they felt as if they were "filming something five hundred years old" when they recorded production there in the 1980s; see a citation in Fuentes Combita, "Las Salinas de Zipaquirá," 14–16. On pre-Columbian cone-shaped ceramics and salt, see Anthony P. Andrews, *Maya Salt Production and Trade* (Tucson: University of Arizona Press, 1983), 68–69, 82–111.

74. Ancízar, *Peregrinación de Alpha,* 1:252. For other references to different-sized ceramics, see Cardale de Schrimpf, *Las Salinas de Zipaquirá,* 30; and Fuentes Combita, "Las Salinas de Zipaquirá," 16–17. Lovejoy provides documentation of a similar practice in the Lake Chad region that includes illustrations (*Salt of the Desert Sun,* 67–69).

75. Clímaco Silva reported on different sizes of containers, though he did not mention prices; see AGN, SR, Ministerio de Hacienda, leg. 388, fol. 279, July 1895. On ceramics workshops in Muneque, see AGN, SR, Salinas de Boyacá, leg. 8, fol. 633, March 1876; and AGN, SR, Miscelánea de Salinas, leg. 12, fol. 23, January 1895. On breaking open the ceramics to extract the salt, see Fuentes Combita, "Las salinas de Zipaquirá," 16.

76. Codazzi, *Comisión corográfica,* 264, 320.

77. The name Rodrigoque was synonymous with ceramics production in La Salina. A kiln in Muneque was named Rodrigoque (AGN, SR, Salinas de Boyacá, leg. 8, fol. 634, March 1876). One kiln in Rodrigoque was named Valderrama (ibid., leg. 1, fol. 199, October 1866). Reports on ceramics production included impressive lists of paraphernalia. Inventories listed four kilns, one in the fábrica principal measuring 10 by 20 meters, while another measured 6.2 by 3.5 meters (ibid., leg. 8, fol. 120, May 1874, and fols. 633–35, March 1876).

78. Ibid., leg. 8, fol. 453, June 1875.

79. See, for example, ibid., fol. 120, May 1874.

80. Another even smaller industry in La Salina was brick making. One of the few references I have found is a contract for the delivery of 8,000 bricks measuring twenty-three by twelve by six centimeters. This effort's similarity to ceramics production is unclear, though the lack of complaints seems a sign that local needs, even accounting for new construction projects, were not hard to meet; see AGN, SR, Miscelánea de Salinas, leg. 10, fol. 340, July 1892.

81. See Humboldt, *Memoria raciocinada,* 16. The University of Chicago Spanish-English dictionary describes a *tercio* as a "bundle"; McFarlane gives it as 100 pounds, roughly the equivalent of a *quintal* in La Salina (McFarlane, *Colombia before Independence,* xiii). McGreevy mentions "[h]uman carriers, called tercios because they carried approximately one-third of the 200–[2]50 lbs. carried by a mule" (*Economic History of Colombia,* 45).

82. AGN, SR, Salinas de Boyacá, leg. 4, fol. 336, February 8, 1872. For less dramatic descriptions of tercios, see ibid., leg. 2, fol. 205, October 1867, and leg. 8, fol. 261, December 1874; see also AGN, SR, Ministerio de Hacienda, leg. 338, fol. 279, July 1895.

83. Humboldt, *Memoria raciocinada,* 16. In the 1840s the rate was given at between twelve and eighteen kilograms of salt per bundle in La Salina; see the discussion on oven efficiency in note 32, and AGN, SR, Gobernación de Tunja, leg. 10, fol. 873, October 1843. In 1845 the administrator reported that 700 bundles produced 12,500 kilograms

of salt, almost 18 kilograms per bundle (AGN, SR, Ministerio de Hacienda, leg. 335, fol. 635, January 1845). From the 1870s a report on the *bosques de la Iglesia,* or the church's woods, estimated the woods would yield between 15,000 and 22,000 bundles, enough for three months' production. Projecting backward from monthly averages in 1879 and 1880 (108,111 kilograms of compact salt and 97,954 kilograms of compact salt, respectively) suggests an average of 103,032 kilograms for a month and 309,906 kilograms for three months, meaning each bundle produced between 14 and 20 kilograms of caked salt; see AGN, SR, Salinas de Boyacá, leg. 11, fol. 41, August 1878.

84. A clause specified that the bundles of twenty sticks delivered by Várgas Santos would be worth twenty-five centavos, while those of twenty sticks of "very good quality" would fetch twenty-seven and a half centavos; see "Contrato sobre suministro de leñas para la elaboracion de sal en La Salina de Chita," in *Memoria de Hacienda y Fomento de 1879* (Luis Carlos Rico), "Documentos sobre Salinas," 50–52. A detailed list of payments from 1876 includes sums of ten, fifteen, and seventeen and a half centavos per bundle, with the bulk spent on the higher-priced bundles; see AGN, SR, Salinas de Boyacá, leg. 8, fols. 711–18, July 1876.

85. The only monthly report on bundles delivered is from July 1876, when 12,141 were brought to the fábrica (AGN, SR, Salinas de Boyacá, leg. 12, fol. 304, July 1876). These totals do not include the timber used to fire ceramics.

86. AGN, SR, Gobernación de Tundama, leg. 4, fol. 693, February 1854; AGN, SR, Salinas de Boyacá, leg. 10, fol. 291, October 1869, leg. 8, fol. 182, August 1874, leg. 11, fol. 48, August 1878, and leg. 12, fol. 488, March 1893. A report predicted that within a few years the price of wood would be so high it would make profits impossible (ibid., leg. 12, fol. 296, August 1891).

87. "The fuelwood on the mountain is a league and a half from where it is used, and laborers must carry them on their backs or, in many cases, to the system of cables" (ibid., leg. 11, fol. 359, January 1881). On the cables, see ibid., leg. 12, fol. 577, June 1893, and fol. 815, December 1894; and *Memoria de Hacienda de 1896* (Ruperto Ferreira), 331. There was a proposal to use carts to carry wood, but it was never implemented (AGN, SR, Salinas de Boyacá, leg. 12, fol. 304, August 1891).

88. On controlling where people cut timber, see Vicente Gomez Maz., "Chita," in "Informes de las administraciones de salinas," comp. Rivera, 9. In 1869 an employee of the contractor complained that a timber supplier was exploiting a stand of young timber near the fábrica, resulting in inefficient production; see AGN, SR, Salinas de Boyacá, leg. 31, fols. 692, 725–26, October–November 1869. An administrator might bring charges against those who cut young trees; see ibid., leg. 12, fol. 136, April 1890; and AGN, SR, Miscelánea de Salinas, leg. 1, fol. 909, September 1897.

89. One employee estimated that it would take the forest estancia eight to ten years to regrow (AGN, SR, Salinas de Boyacá, leg. fol. 51, August 1871).

90. This strategy was suggested as early as 1840; see *Memoria de Hacienda de 1840,* 16.

91. Novoa Zerda, *Los montes de La Salina,* 17.

92. AGN, SR, Gobernación de Tunja, leg. 5, fols. 997–99, January 1837.

93. Novoa Zerda, *Los montes de La Salina,* 19.

94. AGN, SR, Gobernación de Tunja, leg. 10, fol. 925, November 1843, and leg. 13, fol. 716, October 1844. The purchase was made under Article 16 of the 1844 Law on Salinas, passed to ensure access to fuel; see *Memoria de Hacienda de 1845* (Ordoñez), 5. This land could not support production on its own: "The church woods, which belong to the nation, are completely cleared. It will be at least six years until there is wood of medium quality to cut" (AGN, SR, Salinas de Boyacá, leg. 1, fol. 199, October 1866). In 1857 Rafael Núñez's advised Congress to allot 25,000 pesos so the Finance Ministry could buy land in La Salina (*Memoria de Hacienda de 1857,* 33).

95. AGN, SR, Gobernación de Tunja, leg. 6, fol. 419, November 1855; at one point

they offered to sell various pieces of land, but the administrator estimated that the deforested area was not worth the 30,000 peso asking price. At times these heirs were accused of limiting access to timber on their land. They were sued to allow access in 1861 (AGN, SR, Correspondencia de Boyacá, leg. 1, fol. 827, November 1861).

96. The Reyes y Niño Company's activities were not limited to La Salina, as is revealed by a contract between the ministry and Gabriel Reyes Patria for supplying coal to Zipaquirá; see *Memoria de Hacienda y Fomento de 1878* (Luis Bernal), "Documentos sobre salinas," 5. The working relations between the company and the Espinosas remains unclear.

97. For notarized documents recording land transfers, see AGN, SR, Intendencia de Boyacá, leg. 7, fol. 385, June 1848. On rentals, see AGN, SR, Gobernación de Tundama, leg. 1, fols. 231–22, February 1849, leg. 3, fol. 686, June 1851, and leg. 6, fol. 419, November 1855; AGN, SR, Intendencia de Boyacá y Antioquia, leg. 1, fol. 99, July 1860; AGN, SR, Correspondencia de Boyacá, leg. 1, fol. 829, November 1861; AGN, SR, Salinas de Boyacá, leg. 9, fol. 344, September 1866, and leg. 4, fol. 51, August 1871; AGN, SR, Miscelánea de Salinas, leg. 2, fol. 416, July 1878; AGN, SR, Salinas de Boyacá, leg. 11, fol. 19, May 1878.

98. Estevan Varon, *Alegato de conclusion defendiendo al Juez del distrito de La Salinas,* pamphlet, archived at BNC, Fondo Quijano, vol. 201, entry 4, May 1860.

99. Novoa Zerda, *Los montes de La Salina,* 17.

100. Ibid., 5. According to this pamphlet, the only land with clearly documented sales records was Quilambria, which had not been part of the original communal lands. In fact, Quilambria supplied firewood for the fábrica in Córdoba raher than the fábrica in the center of La Salina (AGN, SR, Salinas de Boyacá, leg. 4, fol. 51, August 1871).

101. AGN, SR, Gobernación de Tundama, leg. 6, fol. 430, December 1855. For example, Rafael Niño used control over these forests to his advantage in negotiations with the ministry during the Melista rebellion of 1854.

102. The contract included favorable terms for the company, though it did mandate that the company sell its forested lands and a ceramics workshop in La Salina to the ministry; see *Memoria de Hacienda y Fomento de 1866* (Cuenca), xxviii; for a copy of the contract, see the section "Contratos," 87–89. Cuenca's successor, Alejo Morales, explained that the Reyes y Niño Company had been unable to fulfill that contract and that another had been made with Honorato Espinosa (*Memoria de Hacienda y Fomento de 1867,* xi).

103. Novoa Zerda, *Los montes de La Salina,* 17. The ministry had acquired the land according to the terms of a contract from 1865 (*Memoria de Hacienda y Fomento de 1867,* "Contratos," 87–89).

104. *Diario Oficial,* Feb. 2, 1876, 3660.

105. Novoa Zerda, *Los montes de La Salina,* 8–9.

106. At that time a fiscal commissioner in Chita ordered the seizure of these forests because the owners were "hostile to the government," though it is not clear whether this owner was Cantini (AGN, SR, Salinas de Boyacá, leg. 7, fol. 557, May 1885).

107. On the government ownership at the end of the century, see AGN, SR, Miscelánea de Salinas, leg. 1, fol. 925, December 1898. For a critique of the dealings between Cantini and the ministry, see S. A. L. Ustio (pseud.), *Las minas de la sal (artículos publicado en El Relator y El Diario de Cundinamarca)* (Bogotá: Imprenta de Llera, 1893), archived at BNC, Miscelánea, vol. 850, item 3. On the case heard by the Supreme Court, see *Memoria de Hacienda de 1890* (Paúl), 31–32. The lands and real properties that were eventually transferred included Resguardo, Rodrigoque, Rionegro, La Estancia de Quilambria, a house in the point called La Plazuelita, a house in Muneque, an almacén in Muneque, the shelter of the fábrica in Muneque, and three "well-built bridges"—one over the Casanare, one over the Río Negro, and the other over the Ucubí Gorge—for a total value of

108,450 pesos ("Informe del administrador de las salinas de Chita y Muneque," in *Memoria de Hacienda de 1890,* "Documentos sobre salinas," 165–69).

108. See Catherine LeGrand, *Frontier Expansion and Peasant Protest in Colombia, 1850–1936* (Albuquerque: University of New Mexico Press, 1986).

109. For Vargas's report, see AGN, SR, Ministerio de Hacienda, leg. 335, fol. 725, July 1845. A few weeks after Gutiérrez made his admission, he wrote complaining about a labor shortage. It is impossible to know whether this complaint was prompted by his sudden awareness of the number of workers necessary; see AGN, SR, Gobernación de Tunja, leg. 9, fol. 813, December 1841.

110. AGN, SR, Gobernación de Tunja, leg. 9, fol. 810–12, repeated on fol. 895, December 1841. For another report that La Salina employed more than one thousand workers, see ibid., fol. 717, December 1841.

111. Ibid., leg. 15, fol. 596, March 1847.

112. *Gaceta Oficial,* May 2, 1850.

113. See AGN, SR, Salinas de Boyacá, leg. 8, fols. 711–18, July 1876. The sheet lists fifty-three workers in the fábrica of La Salina, sixteen of them working either days or nights; eighteen workers making ceramics in Rodrigoque; and fifteen delivering timber at the site Alta del Costal. Assuming that the people whom subcontractors hired to deliver wood made about the same amount as did workers in Alta del Costal, whose monthly wage averaged 6.59 peso, then the twenty-four contractors, who delivered an average of 505.9 bundles and received an average of 81.01 pesos, employed as many as ten workers each.

114. *El Boyacense,* November 11–December 27, 1871; the data came from responses to the newspaper's questionnaire asking about conditions in the Canton del Centro and the Valle de Tenza. In Miraflores wages were twelve pesos a month—when work could be found; in Tunja, eighteen pesos a month. Safford puts the daily wage in Boyacá at two reales in the 1870s, perhaps twenty-five centavos, but he does not specify the part of the province to which he is referring ("Commerce and Enterprise," 28). In describing the poor state of roads in Boyacá, Aquileo Parra offered this assessment: "The salaries of workers in the State of Boyacá and in the department of Vélez, which constitutes the most populated part of the Union, does not pass 20 centavos a day, so that to improve conditions they have to travel to points distant from their homes in search of work" (*Memoria de Hacienda y Fomento de 1873,* 73). The only other direct reference to wages in La Salina comes from 1865 when Medardo Cuadra, a manager working for Reyes y Niño, claimed that those in his employ earned three pesos a week, more than the wages for six days of paid work under the shift system of 1876 (AGN, SR, Salinas de Boyacá, leg. 9, fol. 182, June 1865).

115. AGN, SR, Salinas de Boyacá 10, fol. 538, March 1870. Espinosa used the word *enganchar,* though probably not in the sense that he was entrapping workers. When another contractor admitted that wages were low, he was criticizing the administration for late payment of salt deliveries, which he claimed was pushing his employees to work for profiteers; see ibid., leg. 1, fol. 980, January 1867, and leg. 8, fol. 179, August 1874. One administrator suggested using convict labor as in Zipaquirá; see AGN, SR, Ministerio de Hacienda, leg. 337, fol. 777, September 1846. For more complaints about worker scarcity, see *Memoria de Hacienda de 1896* (Ferreira), "Documento F," 336–39.

116. AGN, SR, Salinas de Boyacá, leg. 8, fol. 220, September 1874; see also ibid., leg. 10, fols. 453, 463, July 1875.

117. AGN, SR, Miscelánea de Salinas, leg. 6, fol. 306, May 1884.

118. AGN, SR, Salinas de Boyacá, leg. 2, fol. 197, October 1867.

119. Ibid., leg. 10, fol. 226, August 1869, and fol. 589, May 1870.

120. Rufino Cuervo, "Noticia Biográfica," in *Cómo se evapora un ejército,* by Angel Cuervo (Bogotá: Editorial incunables, 1984), x.

121. Ibid., xii–xiii.

122. AGN, SR, Miscelánea de Salinas, leg. 12, fol. 31, March 1895.

123. Restrepo Echavarría, *Una excursión,* 80.

Chapter 4: The Ministry Monopoly and the Market Monopoly

1. In López Domínguez, ed., *Administraciones de Santander,* 2:26.

2. Ibid., 2:296 and 327.

3. Ibid., 305. On the government finances in the 1820s, see Nieto Arteta, *Economía y cultura,* 82–83; Bushnell, *The Santander Regime,* 340; and López Garavito, *Historia de la Hacienda,* 32–34 and 39.

4. Congress passed legislation to rent the saltworks on April 24, 1826, and, influenced by Castillo y Rada's desire for a competitive market, a law permitting the import of salt on March 13, 1826; see López Domínguez, ed., *Administraciones de Santander,* 2:310. On the survival of neo-Bourbon taxes, see Margarita González, "Las rentas del estado," in *Era republicana,* ed. Jaime Jaramillo Uribe, vol. 2 of *Nueva historia de Colombia,* 8 vols. ed. Alvaro Tirado Mejía (Bogotá: Planeta, 1889), 185–96.

5. In López Dominguez, ed., *Administraciones de Santander,* 5:270–71.

6. Ibid., 275–76.

7. *Memoria de Hacienda de 1839* (Juan de Dios de Aranzazu), 37. The year before he had pointed out, "It will never be prudent to base the state's income on the uncertain yield of customs, which could be diminished from a number of causes, or completely destroyed by a maritime war. For this reason it is important not only to conserve internal rents, but to prevent them from the destruction with which they are threatened, and among them the salt rent is one of the most productive" (in López Domínguez, ed., *Administraciones de Santander,* 6:277).

8. Petitions had been made for these reforms frequently between 1809 and 1840 (Sowell, *Early Colombian Labor Movement,* 57). The *Memoria* of 1845 included a separate report from the director of the tobacco monopoly, Ignacio Gutiérrez, "Esposición del Director Jeneral de la Renta del Tabaco en 1844." His report of the following year ("Esposición del Director Jeneral de la Renta del Tabaco en 1845") discussed the Ley Organica del Tabaco, passed in June of 1844.

9. González, "Las rentas del estado," 194–95.

10. *Memoria de Hacienda de 1848* (Florentino González), 9.

11. See Oscar Rodríguez Salazar and Desci Arévalo Hernandez, "La historiografía económica colombiana del siglo XIX," in *La historia al final del milenio,* 2 vols., ed. Bernardo Tovar Zambrano and Mauricio Archila (Bogotá: Departamento de Historia, Universidad Nacional de Colombia y Editorial Universidad Nacional, 1994), 1:192–205; and David Johnson, *Santander: siglo XIX—cambios socioeconómicos* (Bogotá: Carlos Valencia Editores, 1984), 119–66. On this topic, consult McGreevy, *Economic History of Colombia,* 100–116; and Jorge Orlando Melo, "La evolución económica de Colombia, 1830–1900," in *Manual de historia de Colombia,* ed. Jaime Jaramillo Uribe, 4th ed., 3 vols. (Bogotá: Tercer Mundo, 1994), 2:135–210. This boom was followed by brief cycles of market success in quinine, cotton, and indigo that also acted as temporary tonics for the economy. For the debates on the benefits of these booms, see McGreevy, *Economic History of Colombia,* chaps. 4–7; David Bushnell, *The Making of Modern Colombia: A Nation in Spite of Itself* (Berkeley: University of California Press, 1993), 101–40; Nola Reinhardt, "The Consolidation of the Import-Export Economy in Nineteenth-Century Colombia," *Latin American Perspectives* 13, no. 1 (Winter 1986): 75–98. For a general overview of this period that includes statistics, see Jorge Orland Melo, "Las vicisitudes del modelo liberal (1850–1899)," in *Historia económica de Colombia,* ed. José Antonio Ocampo (Bogotá: Siglo Veintiuno Editores, 1988), 119–72. In addition, among other things, the remaining slaves were to be set free on January 1, 1852.

12. *Memoria de Hacienda de 1850* (Manuel Murillo Toro), 8. For a discussion of the reform, see Nieto Arteta, *Economic y cultura,* 136, 142–43. Murillo Toro's liberal change was still linked to Castillo y Rada's sense of the Finance Ministry's moral duties. Murillo Toro favored lower import taxes on utilitarian items but not on liquor or playing cards (*Memoria de Hacienda de 1851*).

13. Murillo Toro's report from 1851 included a proposal, never adopted, to end the monopoly on January 1, 1859; see *Memoria de Hacienda de 1851,* 54.

14. Peña Díaz, *Sal, sudor y fisco,* 159.

15. A look at the averages in the eleven years for which complete figures for this sporadic data exist gives similar results. On average, total intake was 2,409,969 pesos a year, with 332,449 pesos coming from salt; 622,665 pesos, from customs; and 639,918 pesos, from tobacco.

16. Nieto Arteta, *Economía y cultura,* 143. National revenue fell 47 percent from 1849 to 1851 (McGreevy, *Economic History of Colombia,* 86). The paper *El Neo-Granadino* published an article entitled "Abajo los Monopolios" that was reprinted in the *Gaceta Oficial,* April 20, 1852, 284–86.

17. *Memoria de Hacienda de 1856* (Núñez), 21.

18. Ibid., 10, 59. The absolute importance of customs and saltworks was stressed numerous times in the following decades. Froilán Largacha began his report with a discussion of "the two most important sources for the life of the union, customs and saltworks" (*Memoria de Hacienda de 1864,* 4). For similar comments, see *Memoria de Hacienda y Fomento de 1878* (Bernal), 3.

19. Nieto Arteta, *Economía y cultura,* 201–2.

20. Rafael Pombo, *Renta de salinas* (Bogotá, 1863), archived at BNC, Fondo Quijano vol. 201, item 7, p. 6. Tomás Cuenca cited monopolies with "special rules" and their unequal tax burdens (*Memoria de Hacienda de 1865,* 14). On this era and monopolies, see Hernán Clavijo Ocampo, "Monopolio fiscal y guerras civiles en el Tolima, 1865–1899," *Boletín Cultural y Bibliográfico* 30, no. 32 (1993), available at Biblioteca Virtual, Biblioteca Luis Ángel Arango, http://www.banrep.gov.co/blaavirtual/publicacionesbanrep/boletin/boleti5/bol32/monopo1.htm.

21. *Memoria de Hacienda y Fomento de 1866* (Cuenca), xviii. This was followed by a discussion about static versus flexible demand and the assertion that lowering prices would increase demand and therefore bring in more revenue (ibid., xx). The issue was far from settled. Cuenca's successor argued that events had shown "precisely the opposite," meaning demand was static and lower prices would automatically mean lower revenue; see *Memoria de Hacienda y Fomento de 1867* (Morales), v.

22. López Garavito, *Historia de la Hacienda,* 152–53. By 1865 customs once again brought in over a million pesos; see Nieto Arteta, *Economia y cultura,* 131–32.

23. Summing up years of frustration with a flair for understatement, Nicolás Esguerra wrote, "[T]he results have not equaled our hopes" (*Memoria de Hacienda y Fomento de 1876,* 50).

24. For the contracts with traders to deliver quantities of salt ranging from 50 to 100 cargas (625 to 1,250 kg), see AGN, SR, Administración de Salinas, leg. 1, fols. 5–222, August–September 1871. Traders from ten different towns in central and eastern Boyacá were paid 30 centavos for each arroba delivered to Sogamoso, where kettle salt sold for 1.10 pesos an arroba. This rate of 3 pesos per carga was less than an earlier estimate that set the cost for bringing a carga from Zipaquirá to Sogamoso at 4.20 to 6.40 pesos (AGN, SR, Intendencia de Boyacá and Antioquia, leg. 1, fol. 185, December 1860). See also Corredor, *Informe del administrador principal,* AGN, SR, Salinas de Boyacá, leg. 4, fols. 146, 162, December 1871. For earlier attempts to wean the region from La Salina in the 1850s, see AGN, SR, Correspondencia de Boyacá, leg. 1, fols. 12, 93 and 325, October 1857, and fol. 623, March 1859.

25. AGN, SR, Administración de Salinas, leg. 1, fol. 799, January 1871.
26. *Memoria de Hacienda y Fomento de 1871* (Camacho Roldán), lxvi.
27. Ibid., xi.
28. See "Proposición hecha a los acreedores estranjeros de darles las salinas de Zipaquirá en pago de la deuda esterior," in ibid., lxiv–lxxiii.
29. Salvador Camacho Roldán, "Negociación con los acrededores extranjeros para la dación, en pago de la deuda exterior, de las salinas de Zipaquriá, Nemocón y Tausa," in *Escritos varios,* 2d ed., 3 vols. (Bogotá: Editorial Incunables, 1983), 3:91–116. Colombia would also give the company two million hectares of *tierras baldias* with the "the obligation of populating and cultivating the land by means of foreign immigration, European or American, of no less than ten thousand families or fifty thousand people over the course of ten years." Roldán assured potential investors reading his report that salt from Boyacá was not marketed near Bogotá, dwelling instead on the absence of saltworks in Tolima and Cauca's dependence on salt imported from Peru (ibid., 139).
30. *Memoria de Hacienda y Fomento de 1872,* qtd. in Salvador Camacho Roldán, "Negociación con los acreedores," 3:174.
31. *Memoria de Hacienda y Fomento de 1875* (Parra), 15.
32. *Memoria de Hacienda y Fomento de 1876* (Nicolás Esguerra), 33. Two different pieces of legislation promoted free production: article 2 of Law 33 of May 19, 1875, and decree 22 of the Law of February 20, 1876.
33. *Memoria de Hacienda y Fomento de 1877* (Januario Salgar), 37–38.
34. *Memoria de Hacienda y Fomento de 1879* (Rico), 31.
35. *Memoria de Hacienda de 1881* (Antonio Roldán), 46; the following quotation is from 46–47.
36. Ibid., 49. Roldán's *Memoria de Hacienda de 1882* followed in the same revisionist manner.
37. Ibid., 4.
38. *Memoria de Hacienda de 1884* (Galindo), 145.
39. A contractual clause from 1865 exemplifies the degree of state control attempted: "The almacenes where salt is deposited will have two sets of keys[;] one will be managed by the agents of the government and the others by Reyes y Niño. Salt will not be taken from the almacenes unless both almacenistas are there" (*Memoria de Hacienda y Fomento de 1866* [Cuenca], "Contratos," 87).
40. On monopoly, see Raymond Williams, *Keywords: A Vocabulary of Culture and Society* (New York: Oxford University Press, 1983), 209–10.
41. For estimates of 150,000, see *Memoria de Hacienda y Fomento de 1866* (Cuenca), xxix; *Memoria de Hacienda de 1881* (Camacho Roldán), 112. For one of the high estimates of 350,000 people, see *Memoria de Hacienda de 1898,* 59.
42. This figure is from Multhauf, who cites estimates of annual consumption that range from 0.3 to 7.5 kilograms and concludes that 4.5 is an acceptable compromise (*Neptune's Gift,* 3–4).
43. Zipaquirá's market was immense. It sent salt to the cities of Bucaramanga, Manizales, and Popayan (AGN, SR, Miscelánea de Salinas, leg. 5, fol. 454, September 1870). For a canton-by-canton breakdown of consumption, see AGN, SR, Gobernación de Tunja, leg. 11, fols. 233–40, December 1843. On salt from Zipaquirá sold around Tunja, see ibid., leg. 10, fol. 419, October 1842, and fol. 789, September 1843; Corredor, *Informe del administrador principal,* AGN, SR, Salinas de Boyacá, leg. 4, fol. 154, 1871; and Camacho Roldán, "Negociación con los acreedores." Contracts with traders to deliver salt to an almacén established in Sogamoso in 1874 show that they came from ten towns in Boyacá; regarding those towns, detailed records from sales in 1843 and 1874 listed only Belén as a town of origin for customers in La Salina, implying that consumers might use salt from either place but that the trade networks were distinct.

44. AGN, SR, Gobernación de Tunja, leg. 7, fol. 571, June 1837, and leg. 11, fols. 233–40, December 1843; *Memoria de Hacienda de 1876,* 50.

45. AGN, SR, Intendencia de Boyacá, leg. 6, fol. 780, July 1847; AGN, SR, Salinas de Boyacá, leg. 3, fol. 704, April 1871, and leg. 5, fol. 642, July 1874; AGN, SR, Ministerio de Hacienda, leg. 193, fol. 456, January 1895. Tomás Cuenca also described this consumption radius (*Memoria de Hacienda y Fomento de 1866,* xxix).

46. AGN, SR, Gobernación de Tunja, leg. 6, fol. 767, October 1837; AGN, SR, Salinas de Boyacá, leg. 8, fol. 96, April 1874; AGN, SR, Miscelánea de Salinas, leg. 4, fol. 311, October 1882.

47. AGN, SR, Salinas de Boyacá, leg. 5, fol. 608, December 1872. For example, one official reported that he had ensured that "there was no one left from the llanos who did not take salt."

48. For example, the Finance Ministry made an unfavorable contract with Rafael Niño for production in La Salina during the civil war of 1854 because, it said, "Zipaquirá [had] been cut off from communication [and] demand had increased" (AGN, SR, Gobernación de Tundama, leg. 6, fol. 281, June 1854). Four years later officials who were concerned that political turmoil in Venezuela might threaten the salt supply in the eastern plains saw to it that additional salt from Zipaquirá was marketed in central Boyacá. They also made arrangements to transport more salt from La Salina to Casanare and Arauca if needed; see AGN, SR, Correspondencia de Boyacá, leg. 1, fol. 324, May 1858.

49. See AGN, SR, Miscelánea de Salinas, leg. 5, fols. 14–15, May 1835; and AGN, SR, Gobernación de Tunja, leg. 7, fol. 571, June 1837. In another example of the impact of direct administration, the political head of the canton of El Cocuy included the following in an 1843 report on the Tunebo, a relatively unacculturated indigenous population in the districts of Chiscas and Guicán: "To settle them in contemporary life . . . they need to have access to a road that would run between the provinces of Casanare and Tunja. Primarily this would satisfy them with regard to their need for salt, which is important to their subsistence. To do this it would be necessary to return La Salina de Chita to its old system of administration, so that working there they would obtain a share. The opportunity of employment would give them the chance to leave their homes as they used to" (AGN, SR, Gobernación de Tunja, leg. 29, fols. 747–51, December 1843). Ancízar reported that in his travels he had met some Tunebo Indians who maintained their isolation from the "Spanish population" by intentionally making access to their territory difficult (*Peregrinación de Alpha,* 1:242–43).

50. Ancízar, *Peregrinación de Alpha,* 1:249.

51. Codazzi, *Comisión corográfica,* 321.

52. AGN, SR, Miscelánea de Salinas, leg. 10, fol. 832, January 1892.

53. Departmental officials also weighed in on these issues; see, for example, "Memoria de Tundama," 20, in *Memoria de Presidente de Boyacá de 1869.*

54. These improvements included work on a road linking La Salina and Muneque and building another connecting the town center of La Salina to the vereda of Córdoba (AGN, SR, Gobernación de Tunja, leg. 5, fol. 566, July 1836, and fol. 971, January 1837).

55. In the 1840s the contractors were expected to make an annual contribution of 100 pesos for road upkeep (AGN, SR, Gobernación de Tundama, leg. 3, fol. 725, November 1851).

56. Gomez, in "Informes de las administraciones de salinas," comp. Rivera, 10.

57. On problems with corvée labor, see AGN, SR, Ministerio de Hacienda, leg. 388, fols. 392–93, October 1896; and *El Boyacense,* November 20, 1889, 1787. For shortages of government funds, see AGN, SR, Miscelánea de Salinas, leg. 4, fol. 808, July 1883; and AGN, SR, Salinas de Boyacá, leg. 9, fol. 916, September 1868. For winter rains and floods, see ibid., leg. 10, fol. 158, June 1869. For landslides, see AGN, SR, Intendencia de Boyacá,

leg. 6, fol. 360, October 1846; and AGN, SR, Gobernación de Tundama, leg. 6, fol. 649, August 1857.

58. AGN, SR, Gobernación de Tundama, leg. 3, fols. 727–78, October 1851.

59. For example, plans from 1865 for a new Zapo bridge on the road to Chita called for bricks and a cement foundation, though nothing indicates that anything like this was ever achieved; see AGN, SR, Salinas de Boyacá, leg. 9, fol. 255, January 1865. For another attempt, see AGN, SR, Miscelánea de Salinas, leg. 3, fol. 164, April 1880. On repairs, see AGN, SR, Gobernación de Tunja, leg. 30, fol. 386, July 1846, and fol. 710, May 1847; and AGN, SR, Correspondencia de Boyacá, leg. 1, fol. 892, June 1862. For other complaints about the bridge, see AGN, SR, Intendencia de Boyacá, leg. 7, fol. 455, July 1848. For the comment about the lack of local supplies, see AGN, SR, Gobernación de Tunja, leg. 11, fol. 758, November 1844. On 600 pesos for repairs, see ibid., leg. 14, fol. 622, March 1846. One report claimed that a mule carrying 7,000 pesos of ministry receipts fell off a poorly repaired bridge (AGN, SR, Salinas de Boyacá, leg. 3, fol. 679, July 1871).

60. AGN, SR, Intendencia de Boyacá, leg. 7, fol. 466, July 1848.

61. *El Boyacense,* June 20, 1889, 1566. In some parts of Boyacá people felt poor roads benefited them. One official wrote, "Additionally it must be pointed out that the populace of the canton of Leyva are dead set against the opening of new roads. It could ruin the trade in honey that the locals carry on with the town of Moniquirá" (AGN, SR, Gobernación de Tunja, leg. 28, fol. 983, April 1843). See similar comments in in ibid., leg. 29, fols. 335–37, August 1843.

62. On bad roads between La Salina and El Cocuy, see AGN, SR, Gobernación de Tundama, leg. 6, fol. 678, August 1855; on Chita as an alternate route for this circuit, see AGN, SR, Intendencia de Boyacá, leg. 6, fol. 360, October 1846. One letter specified that road conditions had declined, making it difficult for regional residents to reach La Salina and leading to an increase in profiteering (AGN, SR, Gobernación de Tunja, leg. 30, fol. 733, June 1847). Contracts for road work between Chita and La Salina at times relied on labor from Chita or involved the cabildo of Chita (AGN, SR, Gobernación de Tundama, leg. 1, fol. 717, March 1850). In 1870 Jacinto Corredor stressed the "deplorable condition" of the roads, predicting that the roads would be "totally destroyed" (AGN, SR, Salinas de Boyacá, leg. 3, fols. 16–25 [quotation, fols. 22–23]). Residents of Chita had options for accessing regional salt markets. One official reported that they were bypassing La Salina to sell goods in Sácama and buy salt in Muneque (AGN, SR, ibid., leg. 12, fol. 286, July 1891).

63. AGN, SR, Intendencia de Boyacá, leg. 6, fol. 360, October 1846.

64. AGN, SR, Salinas de Boyacá, leg. 9, fol. 916, September 1867.

65. This excludes several hundred thousand kilograms of sea salt that, beginning in the 1880s, was reported in a separate category.

66. Oficina de Estadística Nacional, *Anuario estadística de 1875,* 211. The tobacco rent had also been expensive, with almost half its revenue going to administrative costs (Melo, "Las vicisitudes del modelo liberal," 149). As an example, in 1863–64, when salt returned 766,625 pesos in gross profits and customs returned 636,960 pesos, salt earned net profits of 564,402 pesos, and customs, of 551,142 pesos (*Memoria de Hacienda y Fomento de 1865* [Cuenca], 10). Salvador Camacho Roldán calulated the profit per 100 kilograms of salt for the fiscal year 1869–70 as follows: Zipaquirá, 4.35 pesos; Nemocón, 4.70 pesos; Tausa, 5.09 pesos; Sesquilé, 4.75 pesos (Nemocón, Tausa, and Sesquilé were satellites of Zipaquirá and undoubtedly benefited from its infrastructure); La Salina de Chita, 2.77 pesos; and Chámeza, 1.90 pesos, with some smaller works in the eastern highlands and Casanare having even lower rates (*Memoria de Hacienda y Fomento de 1871,* 40). On the percentage of revenue spent in management at various customs offices in 1864, see López Garavito, *Historia de la Hacienda,* fig. 9, p. 154.

67. Not counting small amounts of low-grade salt, records show the almacén as

having a salt suplus at month's end for 150 months, less than a quarter of the months reported. The first period of excess occurred in 1837, from June to December, following the implementation of direct administration and year-round production. There was simply more salt available than had been the case in previous years, and perhaps distribution bottlenecks kept the salt at the point of production, for later in the century, managers commented that such periods were produced by flaws in distribution networks rather than excess production (AGN, SR, Salinas de Boyacá, leg. 9, fol. 559, May 1867, and fol. 906, September 1868). The second period, from April 1846 to January 1848, happened after Bonifacio Espinosa began his tenure as contractor by delivering more than 400,000 kilograms of salt, though how he did this is unclear. Bonuses for producing amounts above a predetermined threshold elevated the contractor's per-arroba remuneration, so it was to his benefit to concentrate surplus deliveries in a single month. The following three periods of excess, March 1867 to August 1869, August 1887 to February 1888, and June to December of 1897, seem to have resulted from increased production capacity for one reason or another. Over time the regional market adjusted, and the surpluses disappeared.

68. AGN, SR, Gobernación de Tunja, leg. 5, fol. 971, January 1837.

69. Ibid., leg. 7, fols. 277, 281, and 312, February 1838.

70. Ibid., leg. 9, fols. 617 and 626, November 1841.

71. *Memoria de Hacienda y Fomento de 1868,* 33–34.

72. On other small purchases, see AGN, SR, Gobernación de Tunja, leg. 8, fol. 951, December 1839; for a receipt for a purchase of two arrobas by a buyer from Soatá, see ibid., fol. 959, December 1839. On sales on Sundays, see AGN, SR, Ministerio de Hacienda, leg. 335, fol. 696, February 1845.

73. Fifteen of the towns listed were between 30 and 78 kilometers from La Salina, while one was 115 kilometers distant (Codazzi, *Comisión corográfica,* 261–71). The towns in García Rovira were from 50 to 130 kilometers as the crow flies, though much farther on existing roads.

74. AGN, SR, Administración de Salinas, leg. 3, fols. 948–99, October 1842–July 1843. The monthly sales average between June 1843 and May 1844 was 119,121 kilograms, more or less what was sold during this month. Records from Muneque show a similar pattern. A monthly record lists sales on twelve different days (ibid.).

75. AGN, SR, Ministerio de Hacienda, leg. 335, fol. 725, July 1845.

76. *Gaceta Oficial,* September 20, 1851.

77. Ibid., August 17 and 19, 1852.

78. Ibid., May 18, 1852.

79. AGN, SR, Gobernación de Tundama, leg. 4, fol. 349, July 1852. Similar letters were sent by the cabildos of Busbansa, Socha, and Beitiva (ibid., fols. 263–66, July 1852).

80. The details were supplied by the alcalde, who sent them to the governor of Tunja; see *Gaceta Oficial,* August 11, 1853. Complaints from the two preceding years cited the high prices for salt and focused on the term *monopoly.* One letter mentioned that a carga of salt sold for eight pesos in the almacén but was resold for ten pesos in Santa Rosa, a markup that was apparently beyond the norm (AGN, SR, Gobernación de Tundama, leg. 4, fol. 208, March 1852). For other complaints about resales, though without the term *monopoly,* see ibid., leg. 3, fol. 905 and 933, October 1851.

81. AGN, SR, Gobernación de Tundama, leg. 4, fol. 207, March 1852. I have seen no other reference to Dr. Bernal.

82. Ibid., fol. 863, May 1853. Peña Díaz discusses the existence of a "private monopoly" in Chámeza. He refers to the general control of the industry by an individual, often Antonio Roldán, who later served as the finance minister. While Peña Diaz does not discuss specific market practices, his description confirms the widespread application of the term (*Sal, sudor y fisco,* 107, 222).

83. Most accounts were written by ministry employees, so this underreporting is

unexceptional, though it makes it difficult to engage the extent of their participation in these practices. A few reports implicated employees. In 1862 the inspector Angel Galan wrote that one informant reported seeing "around 25 cargas" (more than 3,000 kilograms) stored separately from the salt dedicated to public sale, though he could find no one to corroborate this statement (AGN, SR, Correspondencia de Boyacá, leg. 1, fols. 887–93, June 1862). In 1897 an employee was accused of withholding ten cargas (1,250 kilograms) from public sales every week (AGN, SR, Miscelánea de Salinas, leg. 1, fols. 519–20, September 1897, and fols. 905–6, December 1897).

84. AGN, SR, Gobernación de Tundama, leg. 4, fol. 272, August 1852.

85. Ibid., leg. 6, fol. 345, September 1855.

86. *El Loco,* October 28, 1856.

87. See Sowell, *Early Colombian Labor Movement,* 54–80.

88. Gustavo Arboleda, *Historia contemporánea de Colombia,* 12 vols. (Bogotá: Banco Central, 1990), 7:37–59.

89. *Gaceta Oficial,* April 24, 1854, 351.

90. The governor of Tundama to the secretary of war, March 10, 1855, AGN, Gobernación de Tundama, leg. 5, fol. 849. For another report of residents' loyalties, see Gobernación de Tundama, leg. 5, fol. 762, December 1854.

91. On rising prices, see ibid., leg. 4, fols. 863–64, May 1853; and AGN, SR, Salinas de Boyacá, leg. 11, fol. 48, August 1878.

92. AGN, SR, Gobernación de Tundama, leg. 6, fol. 76, January 1855. In an awkwardly worded section, the administrator notes that residents earned twelve reales each time they listed an inappropriate buyer, such as a child, presumably referencing the expected profits from the illegal resale.

93. *Memoria de Hacienda de 1857* (Núñez), 27–29. Ministry officials were aware of the political situation. Writing about high sales in March 1856, Rafael Núñez reported, "The explanation for this phenomenon lies in the exceptional nature of La Salina de Chita and in the short duration of the contracts that ran last year. It is likely that the fear of salt shortages in the months following August were the real cause of the increased sales" (ibid., 29).

94. AGN, SR, Gobernación de Tundama, leg. 7, fol. 652, October 1857.

95. For Espinosa's letter, see AGN, SR, Salinas de Boyacá, leg. 2, fols. 509–10 and 512, July and August 1868; his source was *Memoria de Hacienda y Fomento de 1866* (Cuenca), xxvii–xxx.

96. Rivera, comp. "Informes de las administraciones de salinas," 10, in *Memoria de Hacienda de 1859.*

97. AGN, SR, Correspondencia de Boyacá, leg. 1, fols. 887–93, June 1862.

98. Ibid., fol. 887, June 1862. Galán also saw residents making purchases larger than those of outside buyers and noted that some simply sold their guías to traders for twice the set price, making money without handling the salt.

99. Ibid., fols. 888 and 890, June 1862. For a claim that the administrator had never been to La Salina, see AGN, SR, Salinas de Boyacá, leg. 1, fol. 988, February 1867.

100. AGN, SR, Salinas de Boyacá, leg. 9, fol. 202, August 1866. In a request to increase the number of guards to twenty-five, the administrator mentioned that the guards had been attacked by a party of fifty men (ibid., fol. 539, May 1867).

101. Ibid., leg. 1, fols. 228–31, December 1866. Guillermo Terán made a similar complaint about the road a few decades later; see AGN, SR, Miscelánea de Salinas, leg. 9, fol. 981, November 1890.

102. AGN, SR, Salinas de Boyacá, leg. 1, fol. 986, May 1867, and leg. 9, fol. 535, May 1867. A few months later Espinosa complained that it was common knowledge that people stole both timber and ceramics from the fábrica on a nightly basis (ibid., leg. 2, fol. 197, October 1867).

103. In response to Espinosa's petition, a ministry correspondent explained that "the special circumstances of the saltworks and the politics of the country" prevented the ministry from making the necessary repairs (ibid., leg. 2, fol. 511, September 1868). Espinosa claimed he had lost 3,000 pesos in production the year before. He also complained that local officials were imprisoning his workers on false charges (ibid., fol. 515, December 1868). In a following note, he detailed serial abuses by local officials who used demands for "special contributions" to harass and imprison workers (ibid., fol. 520, December 1868).

104. Ibid., leg. 2, fols. 658–62, January 1869.

105. Ibid., fol. 680, January 1869.

106. An employee reported that Barón had instructed his worker to collect timber in a forest that was supposed to remain unharvested. Espinosa claimed Barón was particularly suspect in his dealings with timber (ibid., fols. 725–26, December 1869).

107. Ibid., fol. 680, January 1869.

108. Ibid., fol. 520, December 1869, 4, fols. 61, 104 and 530, September and October 1871, and leg. 12, fols. 408, 413, May 1893; AGN, SR, Ministerio de Hacienda, leg. 193, fol. 449, March 1896; and AGN, SR, Miscelánea de Salinas, leg. 1, fol. 796, May 1896. Those employed by the ministry were supposed to be excused from various types of service. Article 63 of the Ley Organica of March 21, 1832, stated that employees were excused from being named as defendants in trials; see a reference in AGN, SR, Asuntos Criminals, leg. 82, fol. 86, June 1842.

109. AGN, SR, Salinas de Boyacá, leg. 2, fols. 677–78, April–July 1869.

110. Ibid., leg. 4, fol. 61, September 1871.

111. Ibid., leg. 2, fol. 691, October 1869 (quotation), and leg. 3, fols. 16–25, January 1870.

112. Ibid., leg. 4, fols. 335–39, February 1872. All quotations in this and the next paragraph are drawn from this source.

113. A less eloquent note signed by twenty-six people contained numerous complaints, including one that the administration allowed only the original purchaser to collect salt: "the son cannot represent his father, nor the father his son, nor the husband his wife or vice versa, . . . and if after buying some salt someone, god forbid, is sick or has died, nobody can take their purchase from the almacén." The note also claimed that people coming for their shares of salt water were being abused. The administrator responded, in part based on information from his employees, that people who had come to collect water on Sundays were not treated in this fashion, though he did not deny the claim of abuse on other days; see ibid., leg. 5, fol. 335, April 1873. For complaints about workers "hanging around the almacén waiting" for sales, see ibid., leg. 3, fol. 64, November 1870; for complaints that residents were not interested in honest labor, see ibid., leg. 5, fol. 426, July 1874, and leg. 8, fols. 179 and 182, August 1874.

114. For accounts of salt being resold at 1.20 pesos, see ibid., leg. 8, fol. 122, May 1874; for salt bought at 0.60 pesos and resold for either 1.30 pesos or 1.60 to 1.80 pesos, ibid., fol. 177, August 1874, and fol. 242, October 1874. The report of prices between 3 and 4 pesos also noted that in La Salina, amounts of less than a kilogram were sold at a rate equivalent to 2.50 pesos an arroba (ibid., leg. 5, fol. 642, June 1874).

115. For other reports of such fires, see ibid., leg. 4, fol. 776, March 1872; and for one in Munequé the following month, ibid., fol. 290, April 1872. For a fire in La Salina the following year, see ibid., leg. 5, fol. 196, February 1873, and fol. 202, April 1873. For one in Muneque at almost the same time, see ibid., fol. 290, March 1873. For later reports, see ibid., leg. 8, fol. 63, March 1874, and fol. 220, September 1874. While these few years brought the greatest concentration of reports of arson, others followed; see AGN, SR, Miscelánea de Salinas, leg. 2, fol. 163, February 1878; AGN, SR, Ministerio de Hacienda,

leg. 193, fol. 302, March 1897; and, for an incident when the almacén in Córdoba was burned, ibid., leg. 196, fols. 213, 218, March–April 1899.

116. AGN, SR, Salinas de Boyacá, leg. 5, fols. 640–41, July 1874.

117. One official claimed that residents needed about 500 kilograms for personal consumption, though he did not specify how long this quantity would last them (ibid., leg. 3, fol. 36, October 1870).

118. Vicente Gomez Maz. commented on the tension this practice caused, adding that the high price of food encouraged the arrival of more outsiders who came with goods to sell and then sought to participate in resales (in "Informes de las administraciones de salinas," comp. Rivera, 9–10).

119. A report from July 14 specified that residents were to be given only a tenth of the salt sold. Outsiders asking for less than an arroba were to be served first (AGN, SR, Salinas de Boyacá, leg. 5, fol. 644, July 1874). This approach had precedent. In 1853 the administrator reported that he sold two-thirds of the available salt to outsiders and only one-third to people from La Salina (AGN, SR, Gobernación de Tundama, leg. 4, fol. 936, July 1853). Officials stopped noting the quantity requested, for that figure had no relation to how much salt was available (AGN, SR, Salinas de Boyacá, leg. 5, fol. 672, September 1874).

120. See AGN, SR, Salinas de Boyacá, leg. 8, fol. 133, May 1874; see also similar comments a decade later, ibid., leg. 4, fol. 792, November 1884. One employee suggested that outsiders be required to leave town immediately after buying salt (ibid., leg. 12, fol. 643, October 1893).

121. In 1876 an official reported that some traders were entering La Salina repeatedly, using a new name and receiving a new guía from the guards on the road into town each time, thereby frustrating efforts to control sales (ibid., leg. 8, fol. 661, May 1876). Two years later a note from the administration claimed that a dozen or so monopolists worked actively with the alcalde and that people went as far as breaking ceramics to limit the supply of salt (AGN, Miscelánea de Salinas 2, fol. 163, February 1878).

122. AGN, SR, Salinas de Boyacá, leg. 8, fol. 179, August 1874.

123. The ministry saw fit to publish this message; see *Diario Oficial* October 25, 1877, 5195.

124. *Memoria de Hacienda y Fomento de 1877* (Esguerra), 206–7.

125. *Memoria de Hacienda y Fomento de 1878* (Salgar), 38.

126. *Memoria de Hacienda y Fomento de 1879* (Rico), 31–32.

127. *Memoria de Hacienda de 1881* (Camacho Roldán), 46–47.

128. AGN, SR, Salinas de Boyacá, leg. 11, fol. 524, August 1882. A report from two years later claimed that traders simply sold their salt permits to "monopolists." In one example, a resident arrived in the almacén with sixty guías, presumably obtained from traders, technically entitling him to purchase sixty cargas (ibid., fol. 792, November 1884).

129. Ibid., leg. 12, fol. 469, February 1892.

130. Ibid., leg. 12, fol. 147, July 1890; and AGN, SR, Miscelánea de Salinas, leg. 1, fol. 856, May 1896.

131. AGN, SR, Miscelánea de Salinas, leg. 11, fol. 311, September 1893. Nor was the issue of monopoly ended elsewhere. In 1894 Bogotá police agents uncovered what they deemed a workers' conspiracy against the government, finding thousands of seditious pamphlets in the shop of Bernadino Ranjel, including one titled, "Down with monopolies" (Sowell, *Early Colombian Labor Movement,* 113).

132. AGN, SR, Miscelánea de Salinas, leg. 1, fol. 614, March 1896. For accusations that officials were involved, see AGN, SR, Salinas de Boyacá, leg. 12, fol. 255, April 1891; AGN, SR, Miscelánea de Salinas, leg. 12, fols. 519–20 and 633, September 1897 and November 1898, and leg. 1, fols. 612–26, April 1896.

133. AGN, SR, Salinas de Boyacá, leg. 12, fol. 545, February 1892.

134. Ustio, *Las minas de la sal,* 111.

135. AGN, SR, Salinas de Boyacá, leg. 12, fols. 469–71, February 1893.

136. For this note and correspondence describing his other actions, see ibid., fols. 477–83 and 488–89, March 1893.

137. Ibid., fol. 408, May 1893.

138. Ibid., fols. 433–37, August 1893. For Sánchez's self-defense against accusations of monopoly, see AGN, SR, Ministerio de Hacienda, leg. 388, fol. 243, April 1895. On comments that he obstructed improvements, see AGN, SR, Miscelánea de Salinas, leg. 12, fols. 353, 356, and 363, February and April 1896.

139. Ustio, *Las minas de la sal,* 101. Of course, employees defended their records. Alejo Luque wrote about his attempts to form a junta when he arrived with the goal of controlling costs and regulating these practices; see *Memoria de Hacienda de 1896* (Ferreira), "Documento F," 336–39. Two years later he reported on poor conditions, indicating his lack of success; see "Informe del administrador de La Salina de Chita," in *Memoria de Hacienda de 1898* (Esguerra), "Documentos sobre salinas," 175–77.

140. "Informe del administrador de La Salina de Chita," in *Memoria de Hacienda de 1898* (Esguerra), "Documentos sobre salinas," 58.

141. Although they presented another potential point of conflict, weights and measures never provoked more than a passing comment, perhaps because caked salt was cooked in easily identifiable sizes. Weights and measures were standard equipment in the almacén. For a list of weights of one to five arrobas (12.5, 25, 37.5, 50, and 62.5 kilograms), see AGN, SR, Miscelánea de Salinas, leg. 5, fol. 839, February 1875. The balance consisted of two plates and leather cords or, later, iron chains and itself weighed as much as fifty pounds; how well it worked remains unclear. On this issue, see AGN, SR, Intendencia de Boyacá, leg. 7, fol. 668, October 1848; AGN, SR, Salinas de Boyacá, leg. 8, fol. 1024, December 1879, and leg. 2, fol. 926, August 1873. For complaints about the weights, see AGN, SR, Gobernación de Tunja, leg. 7, fol. 708, March 1838, and leg. 14, fols. 857–59, September 1846; AGN, SR, Gobernación de Tundama, leg. 6, fols. 629–30, May 1856; AGN, SR, Salinas de Boyacá, leg. 2, fol. 688, September 1869; and AGN, SR, Miscelánea de Salinas, leg. 1, fol. 654, July 1896.

142. By law, a house where contraband liquor was made or tobacco had been found could be seized, while the discovery of illegal salt resulted only in the seizure of the contraband itself (McGreevy, *Economic History of Colombia,* 66).

143. For early complaints, see AGN, SR, Intendencia de Boyacá, leg. 1, fols. 65 and 243, March and May 1822. On fraud stemming from avoiding the tax on butchering, see ibid., leg. 4, fol. 417, October 1826.

144. In his *Memoria* of 1865, Tomás Cuenca, considering Zipaquirá, estimated that the 148,325 kilograms apprehended in the previous year represented about one-quarter of the illegal trade, meaning that contraband production supplied about one-tenth of the salt made in government works in Cundimarca (*Memoria de Hacienda y Fomento de 1865,* 13). On smuggling and contraband, see McGreevy, *Economic History of Colombia,* 28–32; and Lance Graham, *The Political Economy of Smuggling: Regional Informal Economies in Early Bourbon New Granada* (Boulder, Colo.: Harper Collins, 1997).

145. AGN, SR, Gobernación de Tunja, leg. 10, fol. 67, May 1842.

146. For an account of the guards finding five kettles, see AGN, SR, Salinas de Boyacá, leg. 5, fol. 375, July 1873. At one point, three "broken" cauldrons were found with 112 kilograms of *sal de caldero* (ibid., leg. 12, fol. 445, January 1893); on another occasion, sixteen cauldrons were found (ibid., fol. 853, August 1895). Contraband salt and equipment were auctioned; see AGN, SR, Administración de Salinas, leg. 1, fols. 761–62, November 1871. On confiscated contraband so low grade that no one would buy it, see AGN, SR, Salinas de Boyacá, leg. 9, fol. 690, March 1868.

147. AGN, SR, Gobernación de Tunja, leg. 10, fol. 656, May 1843. In 1834 the administrator of La Salina estimated that a contraband operation in the town of Sisguara produced 2,500 kilograms a week (AGN, SR, Gobernación de Tunja, leg. 4, fol. 266, October 1834). In 1861 the political head of Tundama reported that salt traders from Cundimarca claimed contraband trade amounted to at least 37,500 kilograms a month (AGN, SR, Correspondencia de Boyacá, leg. 1, fol. 807, 1861).

148. For complaints that people spied on guards for contrabandistas, see AGN, SR, Salinas de Boyacá, leg. 12, fol. 444, January 1893.

149. AGN, SR, Gobernación de Tunja, leg. 4, fol. 349, November 1835; and AGN, SR, Salinas de Boyacá, leg. 1, fol. 986, May 1867.

150. Early in the century, contractors from Zipaquirá wrote to the ministry complaining that "far from prosecuting contraband, the lesser authorities [in Boyacá] protect[ed] it" (AGN, SR, Gobernación de Tunja, leg. 4, fol. 367, February 1834).

151. For such "squatting" in Sirgagua, see ibid., leg. 10, fol. 824, July 1843; in Muneque, see AGN, SR, Ministerio de Hacienda, leg. 196, fol. 296, September 1899.

152. Oszlak, "Historical Formation," 27.

Chapter 5: La Salina and Colombian History to 1857

1. For a list, see Alvarado Tirado Mejía, "El estado y la política en el siglo XIX," in *Manual de historia de Colombia,* ed. Jaramillo Uribe, 2:328. For works that discuss party competition as a factor in these wars, see James Park's analysis in *Rafael Núñez,* 268–69; and Reinhardt, "Consolidation of the Import-Export Economy." For studies that examine other factors, see Germán Colmenares, *Partidos politicos y clases sociales* (Bogotá: Ediciones Universidad de los Andes, 1968); Gustavo Vargas Martínez, *Colombia 1854: Melo, los artesanos y el socialismo* (Bogotá, 1972); and Marulanda, *Colonización y conflicto.*

2. For a discussion of campaigns and the role of taxes, see Méndez, *Plebeian Republic,* chap. 6.

3. On the valleys from the highlands to the llanos, see Rausch, *Llanos Frontier,* 6–8.

4. This observation is based on regional lists of prosecutions and crimes. The only two homicides cited in an 1839 list of criminal cases in cantonal courts occurred in Chita (AGN, SR, Gobernación de Tunja, leg. 11, fol. 946, January 1839). A list of eighteen murder cases in the Superior Court of Tundama between January 1890 and May 1893 ("homicidio, asesinato, y infanticidio") includes one from Chita (AGN, SR, Ministerio de Gobierno, sección cuarta, leg. 4, fol. 121, May 1893; for similar case lists, see the following, all from AGN, SR, Gobernación de Tunja: leg. 13, fol. 271, June 1845, and fol. 522, October 1845; leg. 14, fols. 983 and 1008, November 1845, fols. 46 and 171, February 1846, fol. 450, September 1846, fol. 792, January 1847, and fol. 985, April 1847; leg. 16, fol. 98, May 1847, and fol. 610, January 1848; leg. 14, fol. 423, September 1849; leg. 19, fol. 613, April 1851. For a list of suspects accused of crimes who were still at large, with some having committed crimes in Chita and thought to still be there, see ibid., leg. 15, fols. 1 and 46, February 1846, and leg. 16, fol. 55, May 1847, and fols. 768 and 771, January 1848.

5. *El Boyacense,* April 5, 1887, 302.

6. Hermes Tovar Pinzón has called the decline in per-capita public spending in the early republic "an unequivocal sign of the absolute poverty" of the new country ("La lenta ruptura," 114).

7. For a brief biography, see Ocampo López, "José María del Castillo y Rada." On Castillo y Rada's appointment, see López Garavito, *Historia de la Hacienda,* 21–22. In a letter to Bolívar in December 1826, Santander reported that he had left the work of creating the administration to Castillo y Rada (ibid., 29).

8. See Nieto Arteta, *Economía y cultura,* 63.

9. The most explicit data on this point comes from the second half of the century. A chart comparing state contributions to federal taxes in 1864 and 1870 shows that Boyacá

had the lowest per-capita contribution. The amounts were as follows: Antioquiá, 0.94 pesos; Bolívar, 1.03 pesos; Boyacá, 0.22 pesos; Cauca, 0.59 pesos; Cundinamarca, 0.81 pesos; Madgalena, 0.81 pesos; Panamá, 1.35 pesos; Santander, 0.43 pesos; and Tolima, 0.50 pesos. The national average was 0.67 pesos; see *Memoria de Hacienda y Fomento de 1871* (Camacho Roldán), xxii. In the following *Memoria,* Camacho Roldán listed per-capita wealth for states based on a cadastral survey: Cundinamarca, 146 pesos; Boyacá, 31 pesos; and Santander, 37 pesos (*Memoria de Hacienda y Fomento de 1872,* 12–13).

10. In six of the seven months reported during these two years, federal rents in Boyacá produced a total of 34,287 pesos. Salt returned 10,684 pesos; the contribución directa, 9,335.5 pesos; the alcabala, 1,512.8 pesos; aguardiente, 3,531 pesos; and tobacco, only 950 pesos (AGN, SR, Intendencia de Boyacá, leg. 5, fols. 270–869, 1823–24).

11. AGN, SR, Gobernación de Boyacá, legs. 3–14, 1832–46. La Salina accounted for 95 percent of the salt rent in Boyacá. There are no figures reported for tobacco for 1832–33, so that number is roughly 20,000 pesos low. I found records for 147 months of the 162 months from January 1832 through June 1846. Salt returned more than any other rent ninety-five times and was the second highest twenty-four times.

12. López Garavito, *Historia de la Hacienda,* 66. On salt receipts paying government salaries, see AGN, SR, Gobernación de Tunja, leg. 3, fol. 538, May 1833; on using these receipts for military salaries, see ibid., leg. 5, fol. 496, August 1835.

13. Bushnell, *Making of Modern Colombia,* 82–85; Antonio Manuel Pombo and José Joaquín Guerra, eds., *Constituciones de Colombia: recopiladas y precedidas de una breve reseña histórica,* 4 vols. (Bogotá: Ministerio de Educación Nacional, 1951), 3:230–49.

14. Uribe-Uran, "Rebellion of the Young Mandarins," 257–58.

15. See Helen Delpar, *Red against Blue: The Liberal Party in Colombian Politics, 1863–1899* (Tuscaloosa: University of Alabama Press, 1981), 1–5. The division between "aristocratic" and "provincial" factors comes from Victor Uribe-Uran, *Honorable Lives: Lawyers, Families, and Politics in Colombia 1780–1850* (Pittsburgh, Pa.: University of Pittsburgh Press, 2000).

16. Uribe-Uran has documented the spread of *empleomanía* and protests against the practice as one of the causes of the war. He explains that elites "tried to reshape the state bureaucracy by favoring their peers and excluding upwardly mobile individuals" (*Honorable Lives,* 133–37 [quotation, 136]). In 1840 Aranzazu objected to the system as difficult and inappropriate for positions that were "poco lucrativos" (*Memoria de Hacienda de 1840,* 17).

17. *Memoria de Hacienda de 1841* (Mariano Calvo), 9–10.

18. See Fernán González, *Para leer la política: ensayos de historia política colombiana,* 2 vols. (Bogotá: CINEP, 1997), 2:83–162; and Fernán González, "A propósito de 'Las palabras de la guerra.'" Uribe-Uran mentions that "provincials" were upset in part over the cancellation of a ten-year contract for production in Muneque: "[T]he contract had been granted in 1835 to an individual associated with provincial General Juan N. Moreno of Casanare. It was prematurely terminated in 1838 under alleged pressure from aristocratic Tunja lawyer and Senator Antonio Malo . . . [and] Ignacio Morales, who managed other salt deposits in the neighboring region of [La Salina de] Chita" (*Honorable Lives,* 119).

19. González, *Para leer la política,* 128–29; see also Rausch, *Llanos Frontier,* 44–46; and *Memoria del Ministerio de Interior* (1841), 9. On officials anticipating problems in Sogamoso, see AGN, SR, Gobernación de Tunja, leg. 25, fol. 890, September 1840.

20. Uribe-Uran, "Rebellion of the Young Mandarins," 415.

21. They particularly stressed the need to avoid "the hydra of discord" ("Nota official que el JENERAL PATRIA DIRIJIO al presidente J. Y. Marques," September 20, 1840, BNC, Fondo Pineda, vol. 888, item 64).

22. See accounts in the short-lived paper published by the rebels *El Rejenerador del Pueblo,* December 10, 1840, archived at BNC, Fondo Pineda, vol. 888, item 71. The

government was particularly concerned about maintaining order in the canton of El Cocuy (AGN, SR, Gobernación de Tunja, leg. 25, fol. 952, September 1840). For a discussion of elite relations and popular support that indicates how political leaders handled such dynamics and of bloc voting in Tunja, see David Bushnell, "Voter Participation in the Colombian Election of 1856," *Hispanic American Historical Review* 51, no. 2 (May 1971): 242–46.

23. AGN, SR, Gobernación de Tunja, leg. 26, fols. 36–37, January 1841. The following description comes completely from Melgarejo's account cited here. For his activities in Casanare, see Rausch, *Llanos Frontier,* 44–46.

24. Uribe-Uran lists Hoyos, a native of Cauca, as one of the lawyers who supported the sitting government along with other "plutocratic elites" (*Honorable Lives,* 138–39).

25. *Triunfo de las armas del gobierno contra los rebeldes de las provincias del Norte* (January 1841), pamphlet, archived at BNC, Fonda Pineda, vol. 888, item 75.

26. See AGN, SR, Gobernación de Tunja, leg. 26, 1841; for lists of people to be arrested, see ibid., fol. 38, January 1841; for a divisional commander seeking people who had signed documents supporting the rebellion, ibid., fol. 68, February 1841; and for appeals for clemency from people convicted for taking part in the rebellion, ibid., fols. 334–37, May 1841.

27. "CUADRO: Que manifiestas los individuos que tomaron parte en la rebellion del año anterior en esta provincia i que se hallan," AGN, SR, Gobernación de Tunja, leg. 26, fol. 560, June 1841. On the subject of postwar punishment, see María Teresa Uribe de Hincapié and Liliana María López Lopera, *Las palabras de la guerra; metáforas, narraciones y lenguajes políticos: un estudio sobre las memorias de las guerras civiles en Colombia* (Medellín: Instituto de Estudio Políticos, Universidad de Antioquia, 2006).

28. On death sentences given to participants in the War of the Supremes and refusals on the part of acting president Aranzazu to stay the sentence, see *Gaceta de la Nueva Granada,* August 8, 1841; for another death penalty, see ibid., August 22, 1841, and September 26, 1841.

29."CUADRO."

30. Valderrama offered his hacienda, Santa Barbara, located in Tasco, as an assurance for Lozano while he served in the administration. (AGN, SR, Gobernación de Tunja, leg. 5, fol. 866, September 1836). In the end Lozano's wife provided the deposit (ibid., fol. 783, October 1836).

31. Numerous reports deal with these charges and the follow-up; see AGN, SR, Gobernación de Tunja, leg. 6, fols. 295–390, May–November 1837. Azuenaga claimed they attempted this theft while he was carrying out his inspection.

32. See ibid., leg. 7, fols. 20 and 25, December 1837. For his excuses about the almacén, see ibid., fols. 107–216, February–May 1837. Lozano and Solano were temporarily removed from their posts and tried for corruption, though the disorganized records produced by their negligence ensured a lack of evidence. This was enough vindication for them; see *El Patriota,* May 1, 1838; *Auto del Tribunal de Boyacá declarando con lugar la formación de causa a los empleados de Salinas de Chita* and *Vindicación de los ex-empleados de Chita,* both pamphlets archived BNC, Fondo Piñeda, items 466 and 466a, 1838. The issue of requiring regular hours in the almacén had been addressed as early as 1826; see López Garavito, *Historia de la Hacienda,* 33.

33. Charges brought against Manuel Salvador Cárdenas explained that he had absconded with funds and brought them to Solano (AGN, SR, Gobernación de Tunja, leg. 26, fol. 560, June 1841; see another table of suspects to prosecute, AGN, SR, Gobernación de Tunja, leg. 27, fol. 239, November 1841).

34. Solano is mentioned as the patriotic director of a school in Santa Rosa in the next decade (Ancízar, *Peregrinación de Alpha,* 2:14).

35. AGN, SR, Gobernación de Tunja, leg. 27, fols. 318–19, 331–32, November 1841.

Urrea later held official posts, serving as the political head of the canton of El Cocuy; see a petition from residents of Chita, AGN, SR, Gobernación de Tundama, leg. 1, fol. 220, November 1849.

36. AGN, SR, Gobernación de Tunja, leg. 9, fols. 323–24, May 1841.

37. Ibid., fol. 794, January 1842. Legal proceedings about restoring guards' salaries and their behavior during the rebellion continued into 1843; see ibid., leg. 10, fol. 597, May 1843.

38. *Memoria de Hacienda de 1842* (Jorge Hoyos), 14–15.

39. See AGN, SR, Gobernación de Tunja, leg. 13, fol. 870, May 1845, leg. 7, fol. 684, April 1838, leg. 19, fol. 873, October 1843, leg. 9, fol. 960, April 1842, and leg. 10, fol. 925, November 1843.

40. Ibid., leg. 13, fol. 682, January 1845, and leg. 9, fols. 717 and 810–12, December 1841.

41. Centeno, *Blood and Debt,* 103.

42. *Memoria de Hacienda de 1842,* 17.

43. Uribe-Uran, "Rebellion of the Young Mandarins," 418.

44. Several studies use a periodization that emphasizes the importance of the mid-nineteenth century as the point of rupture with the colonial order. See, e.g., McGreevy, *Economic History of Colombia*; the volume of papers inspired by his book, Instituto de Estudios Colombianos, *Historia económica de Colombia: un debate en marcha* (Bogotá: Biblioteca Banco Popular, 1979); Tovar Pinzón, "La lenta ruptura"; and José Escorcia, *Desarrollo politico, social y económica 1800–1854,* vol. 3 of *Sociedad y economía en el Valle del Cauca,* ed. Universidad del Valle, 5 vols. (Bogotá: Biblioteca Banco Popular/Universidad del Valle, 1983).

45. Bushnell has pinpointed the third quarter of the nineteenth century in Colombia as the period that most "clearly exemplified" developments in Latin America (*Making of Modern Colombia,* 101). The reforms were crucial in ushering in that era. Among the other changes, land owned by indigenous communities, already divided during the *repartimiento de resguardos* in the 1830s, could now be legally sold. Records in the Archivo Regional de Boyacá show a marked increase in land sales in indigenous towns that had been divided into individual plots in the 1830s; see Notario Primero 244, *Protocolos de Chivata, 1850–1852;* ibid., 245, *Protocolos de Sotaquirá, 1850–1855*; and Notario Segundo 276, *Protocolos de Siachoque y Oicata, 1850–1856.* In contrast, the abolition of slavery was almost meaningless in Boyacá, for few slaves remained in the region; see David Rueda Mendez, *Introducción a la historia de la esclavitud negra en la provincia de Tunja: siglo XVIII* (Tunja: UPTC, 1989); see also a report on funds for manumission that put the number of slaves in Boyacá as fewer than thirty as early as 1842 (BNC, Fondo Pineda 733/500, *Memoria de Gobernador de Tunja de 1843,* 3).

46. See Rodríguez Salazar and Arévalo Hernandez, "La historiografía económica colombiana"; Johnson, *Santander,* 119–66; McGreevy, *Economic History of Colombia,* 100–116; and Orlando Melo, "La evolución económica de Colombia." This boom was followed by brief cycles in the markets for quinine, cotton, and indigo that also acted as temporary tonics for the economy. For the debates on the benefits of these booms, see McGreevy, *Economic History of Colombia,* chaps. 4–7; Bushnell, *Making of Modern Colombia,* 101–40; Reinhardt, "Consolidation of the Import-Export Economy." For a general overview of this period that includes statistics, see Orlando Melo, "Las vicisitudes."

47. On this conflict and the nature of the changing political culture, see Sanders, *Contentious Republicans.*

48. *Memoria de Hacienda de 1853* (Nepomuceno Gómez), 2. On politics and fiscal policy between 1850 and 1853, see Nieto Arteta, *Economía y cultura,* 117, 125–26; and López Garavito, *Historia de la Hacienda,* 122–23.

49. See Bushnell, "Voter Participation"; and Sanders, *Contentious Republicans,* 100–124.

50. Popular societies were formed in Bogotá as early as 1838, often under elite direction. The right-leaning society was La Sociedad Católica, while a more progressive one was the Sociedad Democrática-Republicana de Artesanos i Laboradores Progresistas de la Provincia de Bogotá (Sowell, *Early Colombian Labor Movement,* 32–40). Sowell cites 1847 as the first year during which popular societies were organized and run by artisans, as does Mauricio Archila Neira in *Cultura e identidad obrera: Colombia 1910–1945* (Bogotá: CINEP, 1991), 45–46. Societies also emerged in Boyacá and the then-independent province of Tundama. In Sogamoso there was the Sociedad Santander (AGN, SR, Gobernación de Tundama, leg. 3, fol. 26, January 1850) and the Sociedad de Instrucción Popular (ibid., 4, fol. 727, November 1852); in Tota, the Sociedad de Hombres Libres (ibid., leg. 5, fol. 333, April 1853); in Tunja, the Sociedad de Instrucción Popular, "identical to that of artisans in Bogotá" (ibid., leg. 19, fols. 15 and 18, January 1850); in Villa de Leyva, the Sociedad Democrática y Instrucción Popular, with forty-two members (ibid., fol. 209, April 1850); in Samacá, the Sociedad Democrática (ibid., fol. 210, April 1850); and in Tibirita, an otherwise unspecified "society" (ibid., fols. 518 and 585, April 1850). For the literature on this topic, see Fabrio Zambrano P., "Historiografía sobre los movimiento sociales en Colombia, siglo XIX," in *La historia al final del milenio,* ed. Bernardo Tovar Zambrano and Mauricio Archila, 2 vols. (Bogotá: Editorial Universidad Nacional), 1:147–82.

51. Bushnell, *Making of Modern Colombia,* 112–13. Melo financed his government with funds in the treasury, forced loans, and salt sales in Zipaquirá; see Antonio Pérez Aguirre, *25 años de historia colombiana 1853 a 1878: del centralismo a federación* (Bogotá: Editorial Sucre, 1959), 20.

52. Park states that "Melo received little support outside of the Bogotá savanna except from the populist governor of Cartagena, Juan José Nieto" (*Rafael Núñez,* 18). Mario Aguilera Peña and Renan Vega have considered the influence of the artisans' rebellion, emphasizing the role of French revolutionary ideology as popular groups not only inserted themselves into national politics but also wove the "symbolic ideas" of the French Revolution into their rituals and charters in order to "realize the Republic" (Aguilera Peña and Vega Cantor, *Ideal democrático y revuelta popular: bosquejo histórico de la mentalidad política popular en Colombia 1781–1948* [Bogotá: Ismac, 1991], 119). See also Christina Rojas, *Civilization and Violence: Regimes of Representation in Nineteenth-Century Colombia* (Minneapolis: University of Minnesota Press, 2001). For a regional study, see Victor Negrete B., *Origen de los luchas agrarias en Córdoba* (Monteria, Colombia: Fundación del Caribe, 1981). On artisans, see Carmen Escobar, *La revolución liberal y la protesta del artesanado* (Bogotá: Fundación Universitaria Autónoma de Colombia, 1990); Sergio Guerra Vilaboy, *La República artesana en Colombia* (La Habana, Colombia: Editorial de Ciencias Sociales, 1980); Eugenio Gutiérrez Cely, "Las luchas populares in Bogotá en el siglo XIX: El Motín del Pan," *Revista Universidad Distrital Francisco José de Caldas* 3–4 (1988): 33–48.

53. Pinto, "Division política"; AGN, SR, Gobernación de Tundama, leg. 4, fol. 301, October 1852. This province was confirmed by the Constitution of 1853, which created thirty-six provinces (Sowell, *Early Colombian Labor Movement,* 58). On the excitement in the region over the establishment of Tundama, see articles in *La Fraternidad,* May 13, 1849, a newspaper published in Santa Rosa.

54. Concerns about public order and rebellion surfaced in 1850 with fears that a rebellion in Pasto would provoke sympathetic uprisings elsewhere. In the end, the only incidents involved suspicious cattle dealers trafficking between Tundama and Casanare. For patriotic manifestations from Sogamoso during this rebellion, see AGN, SR, Gobernación de Tundama, leg. 2, fol. 725, May 1851; from Socotá, ibid., fol. 736, May 1851; and for further support in the region, ibid., fol. 743, June 1851. On comments about political tension, see ibid., leg. 5, fol. 194, January 1853.

55. Ibid., leg. 2, fol. 754, July 1851.

56. Ibid., fol. 651, November 1854.

57. In explaining to the finance minister how difficult it was to make a production contract, the governor of Tundama noted that the rebels had "dispersed and almost annihilated the Constitutional forces of Zipaquirá and Fiquisa and this province [was] threatened immediately by the rebels" (ibid., leg. 6, fols. 257–60, June 1855). On the occupation of Zipaquirá, see Arboleda, *Historia contemporánea,* 4:60, 93.

58. On the Northern Army, see the governor's report, AGN, SR, Gobernación de Tundama, leg. 5, fol. 849, March 1855. After the war, the Finance Ministry submitted reports on the terms of repaying the 120,000 pesos raised through forced loans, noting that one-quarter of this sum was raised in Tundama (Arboleda, *Historia contemporánea,* 4:419).

59. Arboleda, *Historia contemporánea,* 4:117–18.

60. Governor's report to the minister of war, AGN, SR, Gobernación de Tundama, leg. 5, fol. 844, February 1855.

61. AGN, SR, Administración de Salinas, leg. 4, fol. 188, June 1854.

62. AGN, SR, Gobernación de Tundama, leg. 6, fol. 145, December 1854.

63. For the inventory mentioning the town's four ovens, see ibid., fols. 81–84, January 1855.

64. See letters from the governor of Tundama to the minister of war, ibid., leg. 5, fols. 642–45 and 763, October 1854.

65. See letter from the governor of Tundama to the minister of war, ibid., leg. 6, fol. 145, December 1854.

66. Rico qtd. in governor's letter, ibid., leg. 5, fols. 642–43, October 1854.

67. Governor of Tundama to secretary of war, ibid., fol. 849, March 1855. For comments on the way residents in La Salina were treated during the war, see ibid., fol. 762, December 1854.

68. See a note to the Finance Ministry, ibid., leg. 6, fol. 145, December 1845.

69. Report from Juan Nepomuceno Rico, ibid., leg. 5, fol. 763, October 1854.

70. Ibid., leg. 6, fols. 129–30, August 1854.

71. Uribe de Hincapié and López, *La palabras de la guerra,* 466–74.

72. Governor's report to the Ministry of Finance, AGN, SR, Gobernación de Tundama, leg. 6, fols. 257–60; Decree of June 14, 1854, ibid., 281–82, June 1854.

73. Ibid., leg. 6, fol. 399, October 1855.

74. Bushnell, *Making of Modern Colombia,* 118.

75. In the election of 1848 Boyacá had 298 electoral votes; 101 votes went to the Liberals, 184 to the Conservatives, and 13 to other parties. During the first election with universal male suffrage, in 1852, there were reports that Obando, leader of the Draconians, collected 239 out of the 263 electoral votes in Boyacá (AGN, SR, Gobernación de Tundama, leg. 4, fol. 622, June 1852; AGN, SR, Gobernación de Tunja, leg. 20, fol. 439, June 1852). According to Delpar, presidential elections of 1856 produced 12,813 Liberal votes, 27,866 Conservative votes, and 1,821 votes for the National Party (*Red against Blue,* 34). Bushnell provides incomplete yet intriguing district voting for this same election: in the province of Tunja, the Conservative Ospina received 18,310 votes; the Liberal Murillo received 3,303; and another Liberal, Mosquera, received 1,078, with four votes cast for other candidates. In Tundama, these numbers were 8,924, 5,381, 358, and 7, respectively. Bushnell points to the 71 percent participation in Tunja and the high number of precincts with extraordinarily high turnouts as evidence of clientalism and bloc voting in *The Making of Modern Colombia* (118) and provides a more complete discussion in "Voter Participation."

76. In his correspondence, the governor of Tundama quoted a letter from an official in Sativa Norte who argued that during an election, "the most lawful and honorable [course of action is] arms"; see AGN, SR, Gobernación de Tundama, leg. 7, fol. 199, October 1856.

77. There were also discussions about establishing a customs office to monitor the trade corridor that ran through La Salina. The governor of Tundama pointed out the advantages of considering La Salina in this effort. See ibid., leg. 6, fols. 686 and 717, August and October 1856.

78. Ibid., leg. 5, fol. 725, 729, July 1854. López Garavito describes the nature of provincial power under decentralization: "The governors were authorized to administer the personal possessions of the State in their jurisdiction, working as true agents of the Secretary of Finance in each territory" (*Historia de Hacienda,* 111).

79. AGN, SR, Gobernación de Tundama, leg. 4, fols. 712–13, November 1852.

80. Boyacá itself was not wealthy. From 1857 to 1867 its tax intake averaged 83,742 pesos, while its expenses averaged 108,449 pesos. In only two of those ten years did the government spend less than it took in, and then by less than 1,000 pesos. In the eight years with deficits, only twice were they less than 1,000 pesos; see Báez Osorio, *La educación radical,* 29.

81. Qtd. in Arboleda, *Historia contemporánea,* 4:257.

82. *Memoria de Hacienda de 1858,* 47. Ironically, just one year earlier, in defending the salt rent against popular hostility, Rafael Núñez had stated unequivocally that the presidency would never see the salt rent abolished because it was too valuable. Apparently this did not include La Salina.

83. The numbers for the three years given in total revenue intake and profits were, respectively, 49,569 and 28,163 pesos, 75,230 and 47,054 pesos, and 95,912 and 56,813 pesos.

84. Peña Díaz, *Sal, sudor y fisco,* 204–5.

85. *Memoria de Hacienda de 1858,* 48.

86. AGN, SR, Gobernación de Tundama, leg. 7, fol. 656, September 1857. On the reintegration of Tundama into Boyacá, see Pinto, "División política," 580.

87. AGN, SR, Gobernación de Tundama, leg. 7, fol. 652, September 1857.

88. Ibid., leg. 5, fols. 642–45, October 1854.

89. AGN, SR, Gobernación de Bogotá, 35, fols. 43–53, January 1855 (quotation, fol. 53).

90. For another reference to Dovieski as the commander, see Peña Díaz, *Sal, sudo y fisco,* 123.

91. Literally thousands of accused rebels were granted indultos in 1854 and 1855. For lists, see *Gaceta Oficial,* December 21, 1854, January 1, 5, 15, and March 1, 13, 1855.

92. Peña Díaz, *Sal, sudo y fisco,* 123.

93. AGN, SR, Gobernación de Tundama, leg. 6, fol. 341, September 1855; see also AGN, SR, Ministerio de Hacienda, leg. 357, fols. 1126–27, November 1855. This contract was "fairly advantageous" for the ministry (*Memoria de Hacienda de 1856* [Núñez], 15–16). For more on his contracts, loss of contracts, and fines, see *Gaceta Oficial,* August 22, November 13, 1855, and March 29, 1857.

94. AGN, SR, Gobernación de Tundama, leg. 6, fol. 496, January 1856.

95. A list of congressional decrees from February 1858 mentions a resolution in August 1857 returning 10,000 pesos to Dovieski and canceling other debts; see *Gaceta Oficial,* February 26, 1858.

96. On his contracts to repair an oven and the Zapo bridge, see AGN, SR, Gobernación de Tundama, leg. 6, fol. 710, January 1856.

97. Ibid., fol. 717, October 1856.

98. The only record of Dovieski producing salt under this contract before the uprising in September is an undated receipt for the delivery of 32,250 kilograms of salt at 2.90 pesos per arroba; see ibid., leg. 7, fol. 448.

99. Ibid., fol. 656, September 1857.

100. Ibid., fol. 655, September 1857.

101. *Gaceta Oficial,* October 6, 1857. The following accounts are from Pinzon's report

published in this issue and a letter to the governor of Boyacá from Ignacio Gutiérrez. For corroboration, see AGN, SR, Correspondencia de Boyacá, leg. 1, fol. 92, October 1857.

102. On officials' automatic inclination to denigrate the population's ability to act and the "prose of counterinsurgency," see Mark Thurner, *From Two Republics to One Divided: Contradiction of Postcolonial Nationmaking in Andean Peru* (Durham, N.C.: Duke University Press, 1997), 84; and Ranajit Guha, *Elementary Aspects of Peasant Insurgency in Colonial India* (Oxford: Oxford University Press, 1983).

103. AGN, SR, Correspondencia de Boyacá, leg. 1, fols. 23 and 48, October 1857.

104. Ibid., fol. 53, October 1857.

105. Ibid., fol. 54, October 1857. The following quotations are from this letter.

106. Ibid., fol. 48, October 1857 (quotation), and fol. 40, October 1857.

107. AGN, SR, Gobernación de Tequendama, Santander, y Boyacá, leg. 1, fol. 613, October 1857.

108. *Gaceta Oficial,* October 6, 1857. Gutiérrez later described the rebellion as "the mutiny and theft of the saltworks for a month and a half" (*Memoria de Hacienda de 1858,* 47–48). He had an unsympathetic understanding of Colombian behavior, citing "the peculiarly impatience of the American character" as an obstacle to reforms and economic development (ibid., 5). Later he wrote on the link to the country's Spanish legacy with reference to "the peculiar impatience of our race" (*Memoria de Hacienda de 1860,* 5).

109. AGN, SR, Correspondencia de Boyacá, leg. 1, fol. 103, October 1857. One official suggested returning to a form of open production; see ibid., fol. 76, October 1857.

110. Ibid., fols. 40–42, October 1857. The legal reasoning behind the request for national intervention was that though saltworks belonged to the government, they were a form of private property protected by article 5 of the federal constitution, therefore placing the matter in the realm of federal responsibility. For the letter from the prefect of Tundama about his health, see ibid., fol. 36, October 1857.

111. See ibid., fols. 48 and 67, October 1857 (written on the tenth and fifteenth, respectively).

112. The rebellion was usually described as lasting six weeks; see *Memoria de Hacienda de 1858* (Gutiérrez), 48.

113. AGN, SR, Correspondencia de Boyacá, leg. 1, fol. 125, November 1857; see also *Gaceta Oficial,* October 31, 1857. De la Torre had held provisional contracts before, but he found success as a contractor only after he joined with Niño.

114. AGN, SR, Correspondencia de Boyacá, leg. 1, fol. 148, November 1857.

115. Ibid., fols. 162–64, November–December.

116. Ibid., fol. 240, February 1858.

Chapter 6: La Salina, Boyacá, and Colombia after 1857

1. The writings of Javier Ocampo López provide careful documention of Boyacense contributions to liberalism in the nineteenth century. Taking a different tack, Guillermo Dueñas Vargas has argued against the notion that the population of the Boyacense highlands was passive; see Dueñas Vargas, "Algunas hipótesis para el estudio de la resistencia campesina en la región central de Colombia, siglo XIX," *Anuario Colombiano de Historia Social y de la Cultura* 2 (1992): 90–106.

2. On links between church dominance and economic power, see Juan Manuel Robaya Avendano, "La propiedad rural como elemento del crédito en la colonia: la provincia de Tunja en la segunda mitad del siglo XVIII" (master's thesis, Universidad Pedagógica y Tecnológica de Colombia, 1977), 255.

3. Ibid.

4. Arboleda, *Historia contemporánea,* 5:378–79.

5. Ibid., 6:12.

6. Julio Mondragon Castañeda, *Las ideas de paz y de constitucionalidad de los adalides*

Boyacenses en el radicalismo Colombiano, 1850–1885: con el texto de la Constitución Política para el Estado de Boyacá (Septiembre 3 de 1869) (Tunja: Publicaciones del Magister en Historia, UPTC, 1991), 41.

7. See Rausch, *Llanos Frontier,* 83; Jesus C. Torres Almeida, *Manuel Murillo Toro: caudillo radical y reformador social* (Bogotá: Ediciones El Tiempo, 1984), 221–22.

8. Francisco de Paula Borda, *Conversaciones con mis hijos,* intro. Juan Lozano y Lozano, ed. Jose M. de Mier, 3 vols. (Bogotá: Banco Popular, 1974), 1:213.

9. AGN, SR, Intendencia de Boyacá and Antioquia, leg. 1, fols. 43–46, August 1860. For related comments about the military crisis forwarded by the intendente in El Cocuy, see ibid., fol. 85, September 1860. Rico had been serving as the contractor in partnership with Gabriel Reyes Patria in Muneque (ibid., fols. 119–21, October 1860).

10. Arboleda, *Historia contemporánea,* 6:552.

11. Officials reported, with satisfaction, "because of Colonel Guerrero's maneuvers, and the determined opposition in Jerico, where all of the citizens armed themselves to prevent [the rebels] from passing, they returned to Sogamoso where they are now trying to reorganize" (AGN, Intendencia de Boyacá and Antioquia, leg. 1, fol. 187, December 1860). On successfully maintaining control of La Salina, see ibid., fols. 140 and 167, October and November 1860. On this campaign and Reyes Patria, see Rodríguez R., *Boyacenses en la historia de Colombia,* 212–15.

12. Torres Almeida, *Manuel Murillo Toro,* 224. Froilán Largacha discussed how to use monthly returns from Zipaquirá to purchase rations for troops stationed in Bogotá; see *Memoria de Hacienda de 1864,* 9.

13. The Law of May 19, 1863, offered 15 percent of gross returns from future salt receipts as security on a loan of a million pesos (*Memoria de Hacienda de 1864,* 8). Alejo Morales discussed using returns from Zipaquirá as a guarantee on government notes in 1866 (*Memoria de Hacienda de 1867,* vii). In an observation reversing normal complaints about profiteering, Jorge Gutiérrez de Lara observed that the increasing number of notes that were redeemed for salt meant that in some areas, individuals had so much salt they sold it below the official price, thereby undercutting the market (*Memoria de Hacienda y Fomento de 1868,* 29).

14. For a note about problems with salt harvesters, see AGN, SR, Correspondencia de Boyacá, leg. 1, fols. 887–93, June 1862. Also see a letter from the administrator to the jefe politico of the Departamento del Norte about similar problems, including Rafael Niño's management of a lozaria owned by the Valderáma family (ibid., fols. 997–1000, September 1865).

15. See "Decreto de 6 de mayo de 1861, concediendo amnistía a los comprometidos en la revolución contra la soberanía de los estados," *Rejistro Oficial,* September 6, 1861, 54. The desire for a break with existing politics touched even the name of the government's official publication. The *Diario Oficial* was renamed and its volume number set back to year one, a hint of the profundity some saw in this Liberal triumph.

16. See letters from commanders, AGN, SR, Administración de Salinas, leg. 4, fol. 185, June 1854, and fols. 239–41, April 1871.

17. *Memoria de Hacienda de 1865,* 20.

18. Under the federal system, each "state" had one vote in presidential elections. With these three states voting together, Radical Liberal candidates needed only two more state votes to ensure a majority in a national election. Most candidates could count on only the vote of their home state. See González's comments on the importance of equilibrium and Boyacá's role in *Para leer la política,* 194 and 297, as well as Delpar's discussion of elections in *Red against Blue,* 96–100.

19. R. Perez, "Informe del administrador de salinas de Zipaquirá," 4, in *Memoria de Hacienda y Fomento de 1864.* Perez suggested, somewhat impossibly, that the sites of production be moved. Finance Minister Friolán Largacha agreed that the tax was trouble-

some but advocated technical and organization improvements rather than such dramatic and improbable changes (*Memoria de Hacienda y Fomento de 1864*, 10).

20. To comply with this law, each month the administration in La Salina reported the amount taken in and the amount due each state according to the formula. See "Relación de las cantidades que debe la administración de Salinas de Chita i Muneque al Estado de Boyacá por su participación en la renta, de conformidad con el artículo 440 del Codigo Fiscal," AGN, SR, Miscelánea de Salinas, leg. 2, fols. 174–77, February 1878. Reports are filed under "correspondiente á los estados participas en la renta de salinas" in AGN, SR, Salinas de Boyacá, leg. 8, 1874–79. For adjustments to the plan, see "Lei de 15 de abril de 1872 por la cual se ordena el pago de las cantidades ques el Tesoro nacional debe a ciertos estados por la participación que tienen en la renta de salinas," in *Memoria de Hacienda y Fomento de 1873* (Aquileo Parra), 41; article 8 of Law 89 of 1873; and Law 31 of 1874. Salvador Camacho Roldán's "Presupuestos de rentas y gastos del estado de Cundinamarca en el año 1873 y 1874" listed the receipt of 20,000 pesos, about 5 percent of the budget, for participating in the revenue-sharing plan (*Escritos varios,* 3:7).

21. A survey of the correspondence in the AGN, SR, Salinas de Boyacá, leg. 8, shows that the almacén in La Salina should have sent around 1,000 pesos a month to the fund. In theory, the fund would in turn have disbursed 1,500 pesos to Boyacá, a fair sum considering that the state revenues rarely surpassed 100,000 pesos. In the fiscal year 1874–75 the fund took in 194,058 pesos, giving Boyacá 24,935 pesos; Cauca, 2,137 pesos; Cundinamarca, 21,373 pesos; Santander 15,673 pesos; and Tolima, 7,124 pesos; see *Memoria de Hacienda y Fomento de 1876* (Esguerra) 37. In 1885 Felipe Angulo reported that funds were shared as follows: Boyacá, 27,595 pesos; Cundinamarca, 23,636 pesos; Santander, 17,332 pesos; Tolima, 7,878 pesos; and Cauca, 2,362 pesos (*Memoria de Hacienda de 1885,* 35).

22. Letter from La Salina, B. Gonazález, AGN, SR, Salinas de Boyacá, leg. 8, fol. 199, September 1874. The government of Cundinamarca kept trying to access the wealth of Zipaquirá, but its 1875 attempt to tax each load of coal brought in was defeated by a federal lawsuit. Ley 84 of 1875 passed on November 17; see *Memoria de Hacienda y Fomento de 1876* (Esguerra), 42.

23. On changes involving the church with regard to the state after 1862, see Fernando Díaz Díaz, *La desamortización de bienes eclesiasticos en Boyacá* (Tunja: Ediciones la Rana y el Aguila, 1977), 45. On the impact of the 1870 reform in Boyacá, see Jane Meyer Loy, "Primary Education during the Colombian Federation: The School Reform of 1870," *Hispanic American Historical Review* 51, no. 2 (May 1871): 275–94; and Gonzalo Cataño, "Los radicales y la educacion," *Revista Credencial Historia* 66 (June 1995), available at http://www.lablaa.org/blaavirtual/revistas/credencial/junio1995/junio2.htm.

24. Ocampo López, *Los hombres y las ideas en Boyacá* (Tunja: Publicaciones de UPTC, 1989), 11. For another example of Boyacense radicalism, see the *Memoria de Gobernador de Boyacá de 1869*.

25. AGN, SR, Ministerio de Interior, 75, fol. 842, February 1868. That same year the state government passed a law dismantling religious communities; see Báez Osorio, *La educación radical,* 25.

26. *El Tiempo,* December 5, 1871.

27. "Revista," *El Tiempo,* May 9, 1871. On the military oligarchy of Boyacá, see "Orden publico," *El Tiempo,* April 18, 1871. In a biography, Enrique Pérez blamed the Boyacense legislature for the atmosphere; see Pérez, *Vida de Felipe Pérez* (Bogotá: La Luz, 1911), 126. The *Diario de Cundinamarca* leveled similar charges about the elections of José Eusebio Otálora almost a decade later; see Escobar, *Progreso, industrialización y utopía,* 56.

28. Felipe Pérez, qtd. in Mondragon Castañeda, *Las ideas de paz,* 16.

29. The only reference to the rebellion I found in a secondary source is Delpar, *Red against Blue,* 86. Nonetheless, the rebellion did follow the general pattern Bushnell outlines: "When a state government was overturned, . . . it usually meant that one Liberal

faction had ousted another" (Bushnell, *Making of Modern Colombia,* 125). On Salgar's manuevering against the governor, see Pérez, *Vida de Felipe Pérez,* 25–26, 126.

30. Aquileo Parra, *Memorias de Aquileo Parra* (Bogotá, 1912), 576. He did not specify whether the dispute was over one of the plans authored by Salvador Camacho Roldán already discussed or a separate scheme.

31. On reports of conflicts in the region and concern that it would spread, see *El Boyacense,* December 31, 1870, 753. On the outbreak of the rebellion, see *Diario de Cundinamarca,* January 1871, 291.

32. *Diario de Cundinamarca,* January 18, 1871, 310–11, and January 25, 1871, 332; "Sección Boyacá," *El Tiempo,* February 28 and March 7, 1871.

33. "Revista," *El Tiempo,* March 28, 1871.

34. Several peace commissions from Bogotá failed; see *El Tiempo,* April 11 and 25, 1871.

35. "Revista," *El Tiempo,* April 4, 1871; *Diario de Cundinamarca,* April 13, 1871.

36. *El Tiempo,* April 18, 1871. Pérez had also argued that that the reliance on Boyacense soldiers in other wars served as an impediment to the regional population becoming full citizens (ADB, *Memoria de Gobernador de Boyacá de 1870,* 10).

37. Parra, *Memorias,* 578.

38. "Revista," *El Tiempo,* April 4, 1871.

39. *Diario de Cundinamarca,* April 13, 1871.

40. *El Tiempo,* April 11, 1871.

41. *Diario de Cundinamarca,* April 8, 1871.

42. Ibid., April 17, 1871.

43. On April 25 *El Tiempo* still had no word of the battle. For an estimate that Pérez's army consisted of 400 men, see *Diario de Cundinamarca,* April 20, 1871; for one that put the number at 1,200, see "Revista," *El Tiempo,* April 25, 1871. For reports of a minor victory in Soatá, see *El Tiempo,* April 18, 1871; and for another in eastern Boyacá, see *Diario de Cundinamarca,* April 14, 1871.

44. *El Boyacense,* May 31, 1871, 765–67.

45. On the resignation, see *El Tiempo,* May 30, 1871.

46. ADB, *Memoria de Gobernador de Boyacá de 1872,* 5.

47. Nineteen accounts were published; see *El Boyacense,* May 13, 1871, 767–76, and June 17, 1871, 805.

48. *El Boyacense,* May 13, 1871, 767. Other witnesses said the order was also given by the rebel commanders Sarmiento and Cardoso. One soldier admitted that he set fire to a number of houses "out of fear" of his officers (ibid., 774).

49. Other testimonies highlighted the role of the priest Miguel Puerto, who was "short of stature, fat and dark" (ibid., 776).

50. Only one witness, a rebel soldier, testified that he saw houses burning (*El Boyacense,* June 17, 1871, 805).

51. For the orders of the jefe politico de Canton del Centro, see *El Boyacense,* May 13, 1871, 767.

52. "Revista," *El Tiempo,* May 23, June 6, and June 29, 1871.

53. To provide evidence of this horror, the article explained that there was a religious school in Chinquinquira that espoused the death penalty and was against press freedom ("Revista," *El Boyacense,* December 5, 1871).

54. "Revista," *El Tiempo,* August 22, 1871. On forced loans, the role of Boyacá, and peace commissions, see Eduardo Posada-Carbo, "Elections and Civil Wars in Nineteenth-Century Colombia: The 1875 Presidential Campaign," *Journal of Latin American Studies* 26, no. 3 (Oct., 1994): 621–49.

55. *El Boyacense,* May 13, 1871, 770.

56. AGN, SR, Ministerio de Gobierno, 81, fols. 487–89, September 1871. It is not clear

whether this was the same Enrique Cortés involved in the purchase of iron cauldrons for La Salina.

57. See Helen Delpar, "Aspects of Liberal Factionalism in Colombia, 1875–1885," *Hispanic American Historical Review* 51, no. 2 (May 1971): 250–74.

58. AGN, SR, Salinas de Boyacá, leg. 4, fol. 153. For the "decree on the establishment of a principal administration of Salinas in the State of Boyacá," see *Memoria de Hacienda y Fomento de 1871* (Camacho Roldán), "Documentos sobre salinas," 146–48. Corredor had links to the salt monopoly dating back to at least 1857, when he was listed as a guarantor for the contract bid of Juan Nepomuceno Rico in Chámeza (Peña Díaz, *Sal, sudor y fisco,* 84–5).

59. AGN, SR, Salinas de Boyacá, leg. 4, fol. 162.

60. On these almacenes, see ibid., leg. 3, fol. 704, April 1871, and leg. 5, fol. 642, July 1872. By the middle of the decade, there were reports from Soatá that crude salt from Zipaquirá was selling at 1.70 pesos per arroba and that La Salina's consumption radius had not suffered (ibid., leg. 8, fol. 96, April 1874). Compact salt sold for only 1.50 to 1.60 pesos per arroba (AGN, SR, Miscelánea de Salinas, leg. 5, fol. 824, December 1875). On consumer preference for salt from La Salina, see ibid., leg. 2, fol. 107, October 1878, fol. 324, October 1878, fol. 598, March 1879, and fol. 608, June 1879. On the poor condition of the salt in the almacén, see ibid., leg. 3, fol. 45, February 1880.

61. During the fiscal year 1872–73, the almacén of Sogamoso sold 998,343 kilograms for gross profits of 80,607 pesos and net profits of only 33,398 pesos, a rate of return below that of La Salina (*Memoria de Hacienda y Fomento de 1874* [Parra], 46).

62. *Memoria de Hacienda y Fomento de 1871* (Camacho Roldán), 131–32.

63. Hernán Horna, "Transportation Modernization and Entrepreneurship in Nineteenth Century Colombia," *Journal of Latin American Studies* 14, no. 1 (May 1982): 38n19.

64. *Memoria de Hacienda y Fomento de 1873* (Parra).

65. In addition, several parties, including both governmental entities and private individuals, loaned funds in support of this project. A list of the shareholders receiving a portion of the 7 percent interest due (a total of 8,039 pesos) includes the following: the state of Cundinamarca (C$1,733), the state of Santander (C$1,764), the state of Boyacá (C$1,680), the district of Bogotá (C$288), the district of Vélez (C$171), the district of Chiquinquirá (C$29), Aquileo Parra (C$42), Jacinto Corredor (the administrator of the saltworks of Boyacá at the beginning of the decade, C$8), Felipe Pérez (C$2.97), Salvador Camacho Roldán (C$2.97), Januario Salgar (C$2.90), and Nicolás Esguerra (C$8.40) (*Memoria de Hacienda y Fomento de 1878* [Bernal], "Documentos sobre fomento," 71–73). Solon Wilches, the governor of Santander and one of the most prominent Liberals of the era, worked for the project; see Gustavo Otero Muñoz, *Wilches y su época* (Bucaramanga, Colombia: Gobernación de Santander, 1990), 140–42. On the issue of paying for this work, see uncomplimentary articles by Salvador Camacho Roldán in the *Diario de Cundimarca,* collected as "Ferrocarril del Norte," in *Escritos varios,* 3:31–90. The proposed route was to run through Bogotá, Usaquén, Chía, Cájica, Zipaquirá, Cogua, Nemocón, Ubatá, Cucunubá, Guacheta, Chinquinquirá, and Saboya, terminating in Bolívar. For the political context, see Park, *Rafael Núñez,* 66–70.

66. *Memoria de Hacienda y Fomento de 1874* (Parra), 44–45. These were not empty words. Article 211 of the Ley de 1874, empowering the president to complete the railroad, increased the price of salt in ministry works and the fee for bringing sea salt into the interior to aid the project; see *Memoria de Hacienda y Fomento de 1875* (Parra), 48. The railroad interfered with the plan to share revenue with the states. The administrator B. González wrote, "since the publication of the contract for the construction of the Northern Railroad, the States have stopped participating" (AGN, SR, Salinas de Boyacá, leg. 8, fol. 395, May 1875).

67. *El Occidente,* July–September 1873, 1–9; *Memoria de Gobierno de Boyacá de 1874,* 26.

Other regional projects were also praised. See Luis Bernal on the Southern Railroad intended to link Cauca to the Pacific coast, which would have "produce[d] the immediate effect of letting [the nation] take advantage of the privileged position that Colombia occupies on the map of the New World" (*Memoria de Hacienda y Fomento de 1878,* 5); and Rafael Uribe Uribe, *La Regeneración conservadora de Núñez y Caro,* ed. and intro. Otto Morales Benítez (Bogotá: Instituto Para el Desarrollo de la Demoracia Luis Carlos Galan, 1995), 237–58.

68. *Memoria de Hacienda y Fomento de 1874* (Parra), 71–72.

69. Bushnell, *Making of Modern Colombia,* 135. For the list of fourteen working lines in 1897, none of which were in Boyacá, see *Memoria de Hacienda de 1896* (Ferreira), "Informe del Jefe de la Seccion 3a," 30. For the context of this effort, see Ospina Vásquez, *Industria y protección,* 267–68; Safford, *Ideal of the Practical,* 5–6, 185–208; and Pamela S. Murray, *Dreams of Development: Colombia's National School of Mines and Its Engineers, 1887–1970* (Tuscaloosa: University of Alabama Press, 1987).

70. Bergquist, *Coffee and Conflicts,* 67.

71. *Memoria de Hacienda de 1898* (Esguerra), 179. In 1898 the project was referred to as the "ferrocarril de Zipaquirá ó del Norte," and only sixty kilometers of track had been laid (ibid., 160).

72. On the attitude of these insurgents, see "La jornada de 'la Calleja' episodios de la Revolución de Agosto de 1876," in José Manuel Marroquín, *Escritos historicos* (Bogotá: Banco Popular, 1982), 191–99.

73. *Memoria de Hacienda y Fomento de 1877* (Salgar), 3.

74. Park, *Rafael Núñez,* 168.

75. *Memoria de Hacienda y Fomento de 1877,* 9–10.

76. On action in western Boyacá, see a letter from the president of the state to the minister of the interior, AGN, SR, Ministerio de Interior, leg. 15, fols. 617–30, September–October 1876. For the "Guerrilla de Guasca," who operated on the border with Cundinamarca in November 1877, nominally after the war had ended, see ibid., fol. 675, November 1877.

77. In 1874 problems in Chita were calmed only when the Departmento del Norte transferred its administration there (Báez Osorio, *La educación radical,* 31).

78. AGN, SR, Ministerio de Interior, 15, fol. 592, May 1877.

79. See a letter from the administrator, AGN, SR, Miscelánea de Salinas, leg. 2, fol. 233, October 1878. Luis Bernal's report of 1878 discussed the degree to which rebels had occupied ministry saltworks: "In the state of Cundinamarca it was necessary to end work and the sales in the saltworks of Sesquilé, Nemocón, Tausa and Gachetá, because the guerrillas often occupied these works at the time of the deshorno. . . . Further, despite the measures taken by the Government of Boyacá, La Salina de Chita was very poorly served, and because of the entrance of a guerrilla, there has been no rendering of accounts of its products for the entire fiscal year, nor is it possible to obtain any information about the quantity of salt made or sold" (*Memoria de Hacienda y Fomento de 1878,* 42). The war was followed by restrictive measures, including forced loans, and decrees intended to limit the influence of the clergy (Delpar, *Red against Blue,* 118–24). In central Boyacá there was continued concern over the influence of the clergy, as can be seen in a reference to "the exaggerated intentions of a considerable part of the catholic clergy . . . all supported by the fanaticism of the credulous and superstitious pueblo" ("Informe del Jefe del Departamento del Centro," at ADB, *Informe de Gobierno de Boyacá de 1877*). For lists of local priests classified by their support of the government, see AGN, SR, Ministerio de Interior, leg. 15, fols. 519–90, September 1877.

80. See a letter to the ministry from the almacén, AGN, SR, Ministerio de Hacienda, leg. 4, fol. 772, October 1876.

81. While I have found no prosecutions after the occupation, the administrator

admitted that they did not have a clear understanding of who had done what during the occupation; see his letter, AGN, SR, Miscelánea de Salinas, leg. 2, fol. 233, October 1878.

82. *Memoria de Hacienda y Fomento de 1878* (Bernal), 40; and López Garavito, *Historia de la Hacienda,* 261, 272–73.

83. *Memoria de Hacienda y Fomento de 1877* (Salgar), 1.

84. Ibid., 50.

85. Trujillo received seven of the nine presidential votes, enjoying the support of Costeño Liberals, Mosqueristas from Cauca, and even some Conservatives; see Park, *Rafael Núñez,* 160. On Núñistas in Boyacá, see Delpar, "Aspects of Liberal Factionalism," 261–74. The president of Boyacá was José Eusebio Otálora, a longtime Radical Liberal who would back Núñez after 1875 (Park, *Rafael Núñez,* 167). A letter from "the municipality of La Salina" showed its "true republican sentiment" by congratulating the Senate on the appointment of Pablo Arosemena as the secretary of the interior (AGN, SR, Miscelánea de Salinas, leg. 2, fols. 720 and 732, March 1879).

86. Park, *Rafael Núñez,* 239.

87. On the Regeneration, see Lenin Flores G. and Adolfo Atehortúa C., *Estudios sobre la Regeneración* (Cali: Imprenta Departmental del Valle, 1987); and Fernando Guillén Martinez, *La Regeneración: primer frente nacional* (Bogotá: Carlos Valencia, 1986).

88. For a lucid, if brief, explanation, see Frank Safford and Marcos Palacios, *Colombia: Fragmented Land, Divided Society* (New York: Oxford University Press, 2002), 245. For a more detailed account, see Antonio Pérez Aguirre, *Las Radicales y la Regneración* (Bogotá: Editorial Cromos, 1941), 215–38.

89. Foción Soto, *Memorias sobre el movimiento de resistencia a la dictadura de Rafael Núñez, 1884–85,* 2 vols. (Bogotá: Editorial Incunables, 1986), 1:151.

90. On this campaign, see Gonzalo España, *La guerra civil de 1885: Núñez y la derrota del radicalismo* (Bogotá: El Áncora Editores, 1985), 102–12.

91. Soto, *Memorias,* 175.

92. On his tenure in La Salina, see Rausch, *Llanos Frontier,* 184–95. See also Novoa Zerda, *Los montes de La Salina,* 8.

93. See Soto's comments quoted in España, *La guerra civil,* 112.

94. Soto, *Memorias,* 245–46, 253.

95. España, *La Guerra civil,* 179.

96. *Diario Oficial,* March 5, 1885, 4365.

97. General Valderáma, who was stationed in La Salina, wrote about the challenges in providing for his troops there during the war, noting that it was difficult to pay for purchases with paper money; see AGN, SR, Miscelánea de Salinas, leg. 7, fol. 47, June 1885.

98. Letter from the administrator, ibid., leg. 6, fols. 981–82, July 1885. The administrator suggested permanently stationing a battalion of infantry in La Salina (ibid., fol. 994, November 1885). In 1882 the administrator wrote to the president requesting "at least fifty men," adding that they would be "absolutely essential" in those regions where there was "a special situation" if officials were "to maintain the security of National interests" (AGN, SR, Salinas de Boyacá, leg. 11, fol. 566, May 1882). In April 1887 the Ministry of War informed the administrator that it was sending an escort of twenty-five men with an officer to aid the guard unit (AGN, SR, Miscelánea de Salinas, leg. 8, fol. 837, April 1887).

99. Park, *Rafael Núñez,* 197 and 204.

100. *Memoria de Hacienda de 1884* (Galindo), 145.

101. Ibid., 147.

102. Bergquist, *Coffee and Conflict,* 35–39.

103. Telegrams from the commander of the fuerza nacional in Santa Rosa, AGN, SR, Gobernación de Boyacá, leg. 1, fols. 255 and 257, January 1889.

104. Ibid., fols. 270–71, February 1889.

105. Ibid., fol. 282, May 1889.

106. On these events, see a letter from Governor Próspero Pinzón to the Ministry of Government, ibid., fol. 332, September 1889.

107. Ibid., fol. 506, October 1891. On the other disturbances, including confiscating "weapons of war," see ibid., fols. 381 and 385, March 1890.

108. Correspondence from Próspero Pinzón to the Ministry of Government, ibid., fols. 649 and 652, January 1893. On the relationship between artisans and the government, see Sowell, *Early Colombian Labor Movement,* 119–29.

109. Novoa Zerda, *Los montes de La Salina,* 29. The ministry worked with Cantini, though there were claims he did not own the land he offered as security. Hipolito Morales and Mercedes Morales de Quintero complained to the ministry that the production company was harvesting timber on their land; see AGN, SR, Miscelánea de Salinas, leg. 11, fol. 555, May 1894. For critiques of the Regeneration regime, see Uribe Uribe, *La Regeneración conservadora.*

110. Ustio, *Las minas de la sal,* 1.

111. Ibid., 10, 86, 96.

112. Ibid., 101, 106. The commander of the guard unit, José Puentes, supported this assertion; see his report to the ministry, AGN, SR, Ministerio de Hacienda, leg. 388, fol. 301, September 1895. Instead of leading to reform, these articles engendered harassment. The later articles of the series describe an incident when the prefect of police in Bogotá visited the editors of the *Diario de Cundinamarca,* whom they questioned about the author and his sources of information, despite the fact he was mostly quoting from the ministry's own *Memorias,* which were published as individual volumes and serialized in the *Diario Oficial* (ibid., 130).

113. For example, the *Memoria de Hacienda* of 1892, by Goenaga G., opens with a brief discussion about declining salt revenues and a request to create a statistics office (iii–iv).

114. Bravo added that they often worked longer hours than required, a far cry from reports from earlier in the century (*Memoria de Hacienda de 1894,* 7).

115. Miguel Antonio Caro, *Mensaje que el vice-presidente de la república, encargado del poder ejecutivo, dirije al Congreso Nacional* (Bogota: Zalamea Hermanos, 1894), 39, archived at UPTC, Fondo Posada, vol. 3640, item 4.

116. Bergquist, *Coffee and Conflict,* 48–49.

117. *Informe del administrador de La Salina de Chita,* qtd. in *Memoria de Hacienda de 1896* (Ferreira), "Documento E," 333–35.

118. For information on retaking La Salina, see the administrator's letter, AGN, SR, Ministerio de Hacienda, leg. 388, fol. 243, April 1895; on rebel activity in Casanare, ibid., 193, July 1895.

119. AGN, SR, Miscelánea de Salinas, leg. 11, fol. 716, April 1895.

120. Ibid., fol. 255, May 1895 (Sánchez quotations), and fol. 300, September 1895 (Puentes quotations).

121. Ibid., leg. 1, fol. 741, March 1896. Such clashes also had precedent; see notes from an administrator upset over soldiers addressing him disrespectfully (AGN, SR, Salinas de Boyacá, leg. 8, fols. 1013–14, May 1879).

122. See various notes, AGN, SR, Miscelánea de Salinas, leg. 12, fols. 353, 356, and 363, February and April 1896.

123. On these events, see AGN, SR, Salinas de Boyacá, leg. 12, fols. 408–88, February–August 1893.

124. AGN, SR, Miscelánea de Salinas, leg. 1, fol. 612, April 1896.

125. Ibid., fol. 627, June 1896 (Muñoz), fol. 628, June 1896 (Sánchez), fols. 636–42, June 1896, and fol. 646, June 1896 (both Luque).

126. He recommended that if tenants could not present the original rental agreement, the land in question should be confiscated; see AGN, SR, Ministerio de Hacienda, leg. 388, fol. 380, July 1896.

127. See a general report on La Salina, AGN, SR, Miscelánea de Salinas, leg. 12, fol. 375, July 1896.

128. Letters from the commander were forwarded to the ministry by the prefect of Gutiérrez, AGN, SR, Miscelánea de Salinas, leg. 1, fols. 653–56, 674–78 (quotation on 678), August 1896, and fol. 737, April 1896. He also reported that Luque had abused his position as the head of the elections junta, refusing to include soldiers in the electoral lists. For context, see reports on a judge voiding a local election because of fraud in La Salina and Chita, AGN, SR, Gobernación de Boyacá, leg. 1, fols. 581–90, June 1892.

129. See letters to the ministry accusing Luque of involvement in resales through his subordinates Pantaleon Guarin and Eliser Niera and his rebuttal, AGN, SR, Miscelánea de Salinas, leg. 12, fols. 519–20, September 1897, and fol. 633, November 1898.

130. AGN, SR, Ministerio de Hacienda, leg. 196, fol. 40, March 1898.

131. Bergquist, *Coffee and Conflict,* 133–94.

132. Bergquist puts the meeting in Tame (ibid., 90). Guilllermo Plazas Olarte names the site of the meeting as La Salina (*La Guerra civil de los Mil Dias* [Tunja: Publicaciones de la Academia Boyacense de Historia, 1985], 25). He later states, "In 1901, groups of guerrillas left the Llanos by way of La Salina. Between Chita and El Cocuy they punished the forces of the Government General Eloy Caycedo, threatening Soatá" (ibid., 152). For a reference to fighting around La Salina in May 1901, see *Memoria del Gobernador de Boyacá de 1903,* 5.

133. *Memoria de Hacienda de 1904* (Calvo), vii.

134. Calvo's comments are in "Informe del administrador de las salinas de Chita y Muneque," in *Memoria de Hacienda de 1904,* sect. 3, ii and vi.

135. *Memoria de Hacienda de 1910.* For problems with fulewood in La Salina, see ibid., 28 and 29.

136. See *Memoria de Hacienda de 1919,* lxvii; and *Memoria de Hacienda de 1928,* 53–55.

137. For example, the Colombian government put a great deal of energy into developing a national telegraph system during the 1870s. This included constant calls for construction and maintenance contracts. Material relevant to these attempts appears in most issues of the *Gaceta Oficial* from 1876 and 1877. For an example showing how studying the telegraph informs these themes, see Todd A. Diacon, *Stringing Together a Nation: Cândido Mariano da Silva Rondon and the Construction of Modern Brazil, 1906–1930* (Durham, N.C.: Duke University Press, 2004).

138. Raymond B. Craib, *Cartographic Mexico: A History of State Fixations and Fugitive Landscapes* (Durham, N.C.: Duke University Press, 2004), 9–10.

139. I have referred to some of their works throughout this study. To gain a sense of this collective effort, see the website of Razón Cartográfica at http://razoncartografica.wordpress.com/. The site offers a number of tools for those interested in this topic and will likely remain a resource and clearinghouse for scholarship on Colombian cartography.

140. Elizabeth Dore, "One Step Forward, Two Steps Back: Gender and the State in the Long Nineteenth Century," in *Hidden Histories of Gender and the State in Latin America,* ed. Elizabeth Dore and Maxine Molyneux, 3–32 (Durham, N.C.: Duke University Press, 2000), 9.

141. On this topic, see James C. Scott, *Seeing Like a State: How Certain Schemes to Improve the Human Condition Have Failed* (New Haven, Conn.: Yale University Press, 1999), 342.

142. On the aerial bombing during la Violencia, see Hernández Vargas, "La Salina," 5. Municipal elections were canceled in La Salina and Chita in October 1994. This made the

front pages of the Bogotá papers; see "Ante denuncias de violencia contra candidatos, 'No somos indiferentes': Serpa," *El Tiempo,* October 27, 1994. Elections postponed during October were held later in a few municipalities, namely, Santo Domingo (Antioquia), Achi, Barranco de Loba, Pinillos, Rioviejo (Bolívar), Chita (Boyacá), Guavatá (Santander), La Salina (Casanare), and Sácama (Santander); see "Consejo electoral comienza hoy el escrutinio final: alcaldes cívicos en zonas rojas," *El Tiempo,* December 20, 1994. When I asked a Universidad de Javeriana agronomist who worked in that area about these events, he made it clear that such violence was relatively common but not usually deemed worthy of national coverage. In 1995 Bogotá papers reported on battles between the army and the Fuerzas Armadas Revolucionarias de Colombia (FARC) patrols on the road from La Salina to Sácama and a simultaneous attack by 150 members of FARC on the government in Chita (see *El Tiempo,* April 15, 1996).

143. When I accessed the website on November 29, 2010, the page had a note saying that it had last been updated on November 24, 2010. This inflammatory statement had been taken down by the time I revisited the site the following week. I have attempted to access the version of the page I saw by using "The Internet Archive Wayback Machine" (web.archive.org), but the La Salina site (and presumably other municipal sites in Colombia) block the program that this site uses.

144. See http://lasalina-casanare.gov.co/nuestromunicipio.shtml?apc=M-xx-1–and m=fands=mhistoria.

Bibliography

Archives and Collections

Archivo Departmental de Boyacá (ADB)
Archivo General de la Nación de Colombia (AGN):
  Sección Colonial (SC)
  Sección Republica (SR)
Archivo Regional de Boyacá
Biblioteca Nacional de Colombia (BNC)
  Fondo Anselmo Pineda
  Fondo Manuel Ancízar
  Fondo José María Quijano Otero
Biblioteca de la Universidad Pedagógica y Tecnológica de Colombia (UPTC), Fondo Posada

Newspapers

El Boyacense, Tunja
El Constitucional de Boyacá, Tunja
Diario de Cundinamarca, Bogotá
La Fraternidad, Santa Rosa
Gaceta Oficial/Diario Oficial/Rejistro Oficial, Bogotá
El Loco, Bogotá
El Occidente, Chiquinquirá
El Patriota, Tunja
El Porvenir, Bogotá
El Rejenerador del Pueblo, Tunja, December 10, 1840, archived at BNC, Fondo Pineda, 888
El Tiempo, Bogotá
El Tornillo, Tunja

Memorias

PUBLISHED BY THE NATIONAL GOVERNMENT

Esposición del Director Jeneral de la Renta del Tabaco, 1844–45
Memoria de Fomento, 1881, 1884
Memoria de Guerra y Hacienda, 1819

Memoria de Hacienda, 1823, 1826–27, 1831, 1833–35, 1837–42, 1844–45, 1848, 1850–51, 1853–54, 1856–61, 1881–85, 1888, 1890, 1892, 1894, 1896, 1898, 1904, 1919, 1928
Memoria de Hacienda y Fomento, 1864–68, 1871–80
Memoria del Ministerio de Interior, 1841
Memoria de Gobierno, 1898

PUBLISHED BY GOVERNMENT OF BOYACÁ

Memoria de Presidente de Boyacá, 1869–70, 1872
Memoria de Gobierno de Boyacá, 1869, 1874, 1877–78, 1880, 1896, 1903

Other Published Primary Sources

Alcance el último número del Patriota (pamphlet). 1837. Archived at BNC, Fondo Pineda, Vol. 888, item 52.

Ancízar, Manuel. *Peregrinación de Alpha.* 2 volumes. Introduction by José María Samper. Indexes and summaries by C. Adames Santos. Biblioteca Banco Popular, Vol. 7. Bogotá: Fondo de promoción de la cultura del Banco Popular, 1984.

Auto del Tribunal de Boyacá declarando con lugar la formación de causa a los empleados de Salinas de Chita. 1838. Archived at BNC, Fondo Pineda, 466.

Camacho Roldán, Salvador. *Escritos varios.* 3 vols. 2d ed. Bogotá: Editorial Incunables, 1983.

———. "Negociación con los acredores extranjeros para la dación, en pago de la deuda exterior, de las salinas de Zipaquriá, Nemocón y Tausa." In Camacho Roldán, *Escritos varios,* 3:91–116.

———. "Presupuestos de rentas y gastos del estado de Cundinamarca en el año 1873 y 1874." In Camacho Roldán, *Escritos varios,* 3:3–30.

Carillo Batalla, Tomás Enrique. *Hemerografía económica del siglo xix.* 2 vols. Caracas: Biblioteca de Academia de Nacional de la Historia, 2002.

———. *El pensamiento económico de Francisco Aranda.* Prologue by D. F. Maza Zavala. 4 vols. Caracas: Academia Nacional de Ciencias Económicas, 1992.

———. *El pensamiento económico de Santos Michelena.* Prologue by Luis Pastoria. 4 vols. Caracas: Academia Nacional de Ciencias Económicas, 1993.

Caro, Miguel Antonio. *Mensaje que el vice-presidente de la república, encargado del poder ejecutivo, dirije al Congreso Nacional* (Bogota: Zalamea Hermanos, 1894). Archived at UPTC, Fondo Posada, Vol. 3640, item 4.

Carrasquilla, Tomás. *Immigración u colonización: informe que presenta Tomás Carrasquilla al Ministro de Obras Públicas* (1906). UPTC, Fondo Posada, 3739/10.

Codazzi, Agustin. *Geografía física y política de la Confederacíon Granadina.* 6 vols. Edited by Camilo A. Domínguez Ossa et al. (Bogotá: Universidad Nacional de Colombia/ Universidad Pedagógica y Tecnológica de Colombia, 2003).

Corredor, Jacinto. *Informe del administrador principal de salinas de Boyacá.* Archived at AGN, SR, Salinas de Boyacá, leg. 4, fols. 146–162, December 1871.

Cuervo, Angel. *Cómo se evapora un ejercito.* Bogotá: Editorial Incunables, 1984.

Cuervo, Rufino. "Noticia biográfica." In Angel Cuervo, *Cómo se evapora un ejercito,* v–xxxvi. Bogotá: Editorial Incunables, 1984.

de Paula Borda, Francisco. *Conversaciones con mis hijos.* 3 vols. Introduction by Juan Lozano y Lozano. Edited by Jose M. de Mier. Bogotá: Banco Popular, 1974.

"Documentos." *Caribarare* 4, no. 4 (Feb. 1992): 8–12.

Donadíu y Puignau, Delfín, ed. *Novísmo diccionario enciclopédico de la lengua Castellana.* Barcelona: J. Espasa, 1890.

Gaitán, Jorge. "Informe del administrador de las salinas de Chita y Muneque." In *Memoria de Hacienda de 1890* (Felipe Paúl), "Documentos sobre salinas," 165–69.

Giraldo Jaramillo, Gabriel, ed. *Relaciones de mando de los Virreyes de la Nueva Granada:*

memorias económicas. Archivo de la Economía Nacional, Vol. 13. Bogotá: Banco de la República, 1954.
Humboldt, F. E. Alexander von. *Memoria raciocinado de las salina de Zipaquirá*. Bogotá: Banco de la República, 1952.
López Domínguez, Luis Horacio, ed. *Administraciones de Santander*. 6 vols. Bogotá: Fundación para la Conmemoración del Bicentenario del Natalicio y el Sesquicentenario de la Muerte General Francisco de Paula Santander, 1990.
———, ed. *De Boyacá a Cucuta: memoria administrativa, 1819–1821*. Bogotá: Biblioteca de la Presidencia de la República, 1990.
Marroquín, José Manuel. *Escritos históricos*. Biblioteca Banco Popular, Vol. 109. Bogotá: Banco Popular, 1982.
Mollien, Gaspard-Théodore. *Viaje por la República de Colombia en 1832*. Introduction by Carlos José Reyes. Biblioteca V Centenario Colcultura: Viajeros por Colombia. Bogotá: Instituto Colombiano de Cultura, 1992.
Moreno y Escandón, Francisco Antonio. *Indios y mestizos de la Nueva Granada a finales del siglo XVIII*. Edited by Jorge Melo. Biblioteca Banco Popular, edited by Germán Colmenares and Alonso Valencia, Vol. 124. Bogotá: Fondo de Promoción de la Cultura del Banco Popular, 1985.
"Nota oficial que el JENERAL PATRIA DIRIJIO al presidente J. Y. Marques." September 20, 1840. Archived at BNC, Fondo Pineda, 888, 64.
Novoa Zerda, Benjamín. *Los montes de la Salina de Chita y los intereses de la nación*. Bogotá: La Luz, 1889.
Oficina de Estadística Nacional de Colombia. *Anuario de estadística de Colombia*. 1875, 1876, and 1883. Bogotá.
Ortiz, Nicolas. *Estadística de las salinas de la Republica*. Bogotá, 1889.
Parra, Aquileo. *Memorias de Aquileo Parra*. Bogotá, 1912.
Pérez, Enrique. *Vida de Felipe Pérez*. Bogotá: La Luz, 1911.
Perez, R. "Informe del Administrador de Salinas de Zipaquirá." In *Memoria de Hacienda de 1864*.
Pombo, Manuel Antonio, and José Joaquín Guerra, eds. *Constituciones de Colombia: recopiladas y precedidas de una breve reseña histórica*. 4 vols. Bogotá: Ministerio de Educación Nacional, 1951.
Pombo, Rafael. *Renta de Salinas*. Bogotá, 1863. Archived at BNC, Fondo Quijano.
Restrepo Echavarría, Emiliano. *Una excursión al territorio de San Martín en Diciembre de 1869*. Archivo de la economía nacional, Vol. 14. Bogotá: Banco de la República, 1955.
Rivera, Pacifico E., compiler. "Informes de las administraciones de Salinas." In *Memoria de Hacienda de 1859*.
Rodríguez Freile, Juan. *El Carnero (segun el manuscrito de yerbabuena)*. Edited by Mario Germán Romero. Bogotá: Instituto Caro y Cuervo, 1984.
Soto, Foción. *Memorias sobre el movimiento de resistencia a la dictadura de Rafael Núñez, 1884–85*. 2 vols. Bogotá: Editorial Incunables, 1986.
Triunfo de las armas del gobierno contra los rebeldes de las provincias del Norte. Archived at BNC, Fonda Pineda 888, January 1841.
"Tunja en 1610: descripción de la ciudad de Tunja." *Reportorio Boyacense* 4, no. 40 (1916): 193.
Uribe Uribe, Rafael. *La Regeneración conservadora de Núñez y Caro*. Edited and with an introduction by Otto Morales Benítez. Bogotá: Instituto Para el Desarrollo de la Democracia Luis Carlos Galan, 1995.
Ustio, S. A. L. (pseud.). *Las minas de la sal (artículos publicado en El Relator y El Diario de Cundinamarca)*. Bogotá: Imprenta de Llera, 1893. Archived at BNC, Miscelánea, Vol. 850, item 3.
Varon, Estevan. *Alegato de conclusión defendiendo al juez del distrito de la salinas*. May 1860. Archived at BNC, Fondo Quijano, Vol. 201, item 4.

Vergara Velaso, Francisco Javier. *Nueva geografía de Colombia: escrita por regiones naturales*. Bogotá, 1901.
Vindicación de los ex-empleados de Chita. Pamphlet. 1836. Archived at BNC, Fondo Pineda, 466[a].

Secondary Sources

Adelman, Jeremy. "Unfinished States: Historical Perspectives on the Andes." In *State and Society in Conflict: Comparative Perspectives on Andean Crises,* edited by Paul W. Drake and Eric Hershberg, 41–73. Pittsburgh, Pa.: University of Pittsburgh Press, 2006.
Adelman, Jeremy. *Republic of Capital: Buenos Aires and the Legal Transformation of the Atlantic World*. Stanford, Calif.: Stanford University Press, 1999.
Adshead, S. A. M. *Salt and Civilization*. New York: St. Martin's, 1992.
Afehortúa, Adolfo León, and Humberto Vélez R. *Estado y fuerzas armadas en Colombia: 1886–1953*. Cali: Pontificia Universidad Javeriana, 1994.
Aguilera Peña, Mario. *Los comuneros: guerra social y lucha anticolonial*. Bogotá: Universidad Nacional de Colombia, 1985.
Aguilera Peña, Mario, and Renan Vega Cantor. *Ideal democrático y revuelta popular: Bosquejo histórico de la mentalidad política popular en Colombia 1781–1948*. Bogotá: Ismac, 1991.
Amaya Roldán, Martín. *Historia de Chita*. Tunja: Imprenta Oficial, 1930.
Anderson, Benedict. *Imagined Communities: Reflections on the Origin and Spread of Nationalism*. Rev ed. New York: Verso, 2006.
Andrews, Anthony P. *Maya Salt Production and Trade*. Tucson: University of Arizona Press, 1983.
Andrien, Kenneth J., and Lyman L. Johnson, eds. *The Political Economy of Spanish America in the Age of Revolution, 1750–1850*. Albuquerque: University of New Mexico Press, 1994.
Antei, Fiorgio. *Mal de América: las obras y las días de Agustín Codazzi, 1793–1859*. Bogotá: Museo Nacional, 1993.
Appelbaum, Nancy P. *Muddied Waters: Race, Region, and Local History in Colombia, 1846–8*. Durham, N.C.: Duke University Press, 2003.
Arango, Mariano. *Café e industria 1850–1930*. Bogotá: Carlos Valencia Editores, 1979.
Arboleda, Gustavo. *Historia contemporánea de Colombia*. 12 vols. Bogotá: Banco Central, 1990.
Arbulu, Javier Tantalean. *Política económico-financiera y la formación del estado: siglo xix*. Lima: Centro de Estudios Para el Desarrollo y la Participación, 1983.
Arcila Farias, Eduardo. *Historia de un monopolio: es estanco del tabaco en Venezuela (1779–1833)*. Caracas: Facultad de Humanidades y Educación, Instituto de Estudio Hispanoamericanos, Universidad Central, 1977.
Archila Neira, Mauricio. *Cultura e identidad obrera: Colombia 1910–1945*. Bogotá: CINEP, 1991.
Ardant, Gabriel. "Financial Policy and Economic Infrastructure of Modern States and Nations." In *The Formation of National States in Western Europe,* edited by Charles Tilly, 164–242. Princeton, N.J.: Princeton University Press, 1975.
Arenas Paz, Belisario. *Guía geográfica de Colombia (diccionario breve de geografía física, económica, política, histórica y matemática)*. 3d ed. Bogotá: Editorial Santafe, 1940.
Arrom, Silvia M., and Servando Ortoll, eds. *Riots in the Cities: Popular Politics and the Urban Poor in Latin America, 1765–1910*. Wilmington, Del.: Scholarly Resources, 1996.
Avellaneda, José Ignacio. *The Conquerors of the New Kingdom of Granada*. Albuquerque: University of New Mexico Press, 1995.
Báez Osorio, Myriam. *La educación radical en Boyacá: fundamento social y político*. Tunja: Academica Boyacense de Historia, 1996.
Bakewell, Peter. *Miners of the Red Mountain: Indian Labor in Potosí, 1545–1650*. Albuquerque: University of New Mexico Press, 1985.

Bejarano, José Antonio, and Orlando Pulido. *El tabaco en una economía regional: Ambalema siglos XVIII y XIX*. Bogotá: Universidad Nacional de Colombia, 1986.
Bergquist, Charles. *Coffee and Conflict in Colombia*. Durham, N.C.: Duke University Press, 1986.
Bonilla de Pica, Alba Luz. "Espacio poblamiento en el resguardo de Chita en la segunda mitad del siglo xvii." *Historia Critica* 26 (July-Dec. 2003): 77–92.
Bonney, Richard, ed. *The Rise of the Fiscal State in Europe, 1200–1815*. Oxford: Oxford University Press, 1999.
Botero Herrera, Fernando. *Estado, nación y provincia de Antioquia: guerras civiles e invención de la región 1829–1863*. Medellín: Hombre Nuevo Editores, 2003.
Brading, D. A., and Harry E. Cross. "Colonial Silver Mining: Mexico and Peru." *Hispanic American Historical Review* 52, no. 4 (1972): 545–79.
Braudel, Fernand. *The Structures of Everyday Life*. Vol. 1 of *Civilization and Capitalism 15th-18th Century*. Translated by Siân Reynolds. New York: Harper and Row, 1981.
Braun, Rudolph. "Taxation, Sociopolitical Structure, and State-Building: Great Britain and Brandenburg-Prussia." In *The Formation of National States in Western Europe*, edited by Charles Tilly, 243–327. Princeton, N.J.: Princeton University Press, 1975.
Brew, Roger. *El desarrollo de Antioquia desde la independencia hasta 1920*. Bogotá: Banco de la República, 1977.
Brungardt, Maurice Philip. "Tithe Production and Patterns of Economic Change in Central Colombia, 1764–1883." PhD diss., University of Texas, 1974.
Buitrago, Francisco Leal. *Estado y política en Colombia*. Bogotá: CEREC/Siglo XIX, 1984.
Burns, E. Bradford. *The Poverty of Progress: Latin America in the Nineteenth Century*. Berkeley: University of California Press, 1983.
Bushnell, David. *The Making of Modern Colombia: A Nation in Spite of Itself*. Berkeley: University of California Press, 1993.
———. *The Santander Regime in Gran Colombia*. Westport, Conn.: Greenwood, 1970.
———. "Voter Participation in the Colombian Election of 1856." *Hispanic American Historical Review* 51, no. 2 (May 1971): 237–49.
Caballero Calderon, Eduardo. *El Cristo de Espaldas*. New York: Macmillan, 1967.
———. *Siervo sin tierra*. Madrid: Ediciones Rodas, 1973.
Calderón, Climaco. *Elementos de la hacienda pública*. Bogotá: La Luz, 1911.
Camargo Pérez, Gabriel. *Pueblos y jornadas boyacenses*. Tunja: Academia Boyacense de Historia y Gobernación de Boyacá, 1994.
Cardale de Schrimpf, Marianne. *Las salinas de Zipaquirá: su explotación indígena*. Bogotá: Banco de la República, 1981.
Cárdenas Támara, Felipe, Antoine M. Cleef, and Abdón Cortés Lombana, eds. *El páramo: un ecosistema de alta montaña*. Available at http://www.banrepcultural.org/blaavirtual/faunayflora/paramo/indice.htm.
Carrillo Batalla, Tomás Enrique. *Hemerografía económica del siglo xix*. 2 vols. Caracas: Biblioteca de Academia de Nacional de la Historia, 2002.
———. *El pensamiento económico de Francisco Aranda*. Prologue by D. F. Maza Zavala. Historia de Pensamiento Económico Venezolano, Vol. 3. Caracas: Academia Nacional de Ciencias Económicas, 1993.
———. *El pensamiento económico de Santos Michelena*. Prologue by Luis Pastoria. Historia de Pensamiento Económico Venezolano, Vol. 4. Caracas: Academia Nacional de Ciencias Económicas, 1993.
Castañeda, Julio Mondragon, *Las ideas de paz y de constitucionalidad de las adalides boyacenses en el radicalismo colombiano, 1850–1885: con el texto de la Constitución Política para el Estado de Boyacá (Septiembre 3, 1869)*. Tunja: Publicaciones del Magister en Historia, UPTC, 1991.
Castellanos, Jorge. *La abolición de la esclavitud en Popayán, 1832–52*. Cali: Universidad de Valle, 1980.

Castillo Muñoz, Pablo. "Ferrería de Samacá: Samacá–Boyacá." Monumentos Nacionales de Colombia. Online exhibition available at Biblioteca Virtual, Biblioteca Luis Ángel Arango, http://www.lablaa.org/blaavirtual/exhibiciones/monu/samaca.htm. Accessed June 28, 2010.

Cataño, Gonzalo. "Los radicales y la educación." *Revista Credencial Historia* 66 (June 1995). Available at http://www.lablaa.org/blaavirtual/revistas/credencial/junio1995/junio2.htm.

Centeno, Miguel. *Blood and Debt: War and the Nation-State in Latin America*. University Park: Pennsylvania State University Press, 2002.

Centeno, Miguel, and Fernando López-Alvez, eds. *The Other Mirror: Grand Theory through the Lens of Latin America*. Princeton, N.J.: Princeton University Press, 2001.

Choudhury, Sadananda. *Economic History of Colonialism: A Study of British Salt Policy in Orissa*. Delhi: Inter-India, 1979.

Clark, Meri Linnea. "Education for a Moral Republic: Schools, Reforms, and Conflict in Colombia, 1780–1845." PhD diss., Princeton University, 2003.

Clavijo Ocampo, Hernán. "Monopolio fiscal y guerras civiles en el Tolima, 1865–1899." *Boletín Cultural y Bibliográfico* 30, no. 32 (1993). Available at Biblioteca Virtual, Biblioteca Luis Ángel Arango, http://www.banrep.gov.co/blaavirtual/publicacionesbanrep/boletin/boleti5/bol32/monopo1.htm. Accessed February 18, 2011.

Coatsworth, John H., and Alan Taylor, eds. *Latin America and the World Economy since 1800*. Cambridge, Mass.: Harvard University, David Rockefeller Center for Latin American Studies, 1998.

Colburn, Forrest D., ed. *Everyday Forms of Peasant Resistance*. New York: M. E. Sharpe, 1989.

Colmenares, Germán. *Partidos politicos y clases sociales*. Bogotá: Ediciones Universidad de los Andes, 1968.

———. *La Provincia de Tunja en el Nuevo Reino de Granada: ensayo de historia social (1539–1800)*. Bogotá: Universidad de los Andes, 1970.

Contraloría General de la República de Colombia. *Geografía económica de Colombia*. Vol. 3, *Boyacá*. Bogotá: Imprenta Nacional, 1936.

Cook, David. *Born to Die: Disease and New World Conquests, 1492–1650*. Cambridge: Cambridge University Press, 1998.

Cooney, Jerry W., and Thomas L. Whigham. "Paraguayan Commerce with the Outside World, 1770–1850." In *The Political Economy of Spanish America,* edited by Kenneth Andrien and Lyman L. Johnson, 215–42. Albuquerque: University of New Mexico Press, 1994.

Correa, Ramón C. *Monografías de los pueblos de Boyacá*. 2 vols. Biblioteca de la Academia Boyacense de Historia, Serie Obras Fundamentales, vols. 6–7. Tunja: Publicaciones de la Academia Boyacense de Historia, 1987–89.

Coronil, Fernando. *The Magical State: Nature, Money, and Modernity in Venezuela*. Chicago: University of Chicago Press, 1997.

Corrigan, Philip. "State Formation." In *Everyday Forms of State Formation: Revolution and Negotiation of Rule in Modern Mexico,* edited by Gilbert M. Joseph and Daniel Nugent, xvii–xix. Durham, N.C.: Duke University Press, 1994.

Corrigan, Philip, and Derek Sayer. *The Great Arch: English State Formation as Cultural Revolution*. Oxford: Basil Blackwell, 1985.

Corsi Otálora, Luis. *Boyacá: Atlántida andina*. Tunja: Academia Boyacense de Historia, 2005.

Craib, Raymond B. *Cartographic Mexico: A History of State Fixations and Fugitive Landscapes*. Durham, N.C.: Duke University Press, 2004.

Curry, Glenn Thomas. "The Disappearance of the Resguardos Indigenas of Cundinamarca, Colombia, 1800–1863." PhD diss., Vanderbilt University, 1981.

Diacon, Todd A. *Stringing Together a Nation: Cândido Mariano da Silva Rondon and the Construction of Modern Brazil, 1906–1930*. Durham, N.C.: Duke University Press, 2004.

Dalence, José María. *Bosquejo estadístico de Bolivia*. La Paz: Universidad Boliviana/Universidad Mayor de San Andrés, 1975.

Deans-Smith, Susan. *Bureaucrats, Planters, and Workers: The Making of the Tobacco Monopoly in Bourbon Mexico*. Austin: University of Texas Press, 1992.

———. "State Enterprise, Work and Workers in Mexico: The Case of the Tobacco Monopoly, 1765–1850." In *The Political Economy of Spanish America,* edited by Kenneth J. Andrien and Lyman L. Johnson, 63–94. Albuquerque: University of New Mexico Press, 1994.

Deas, Malcolm. *"Del poder y la gramática" y otros ensayos sobre historia, política y literatura colombianas*. Bogotá: Tercer Mundo, 1993.

———. "The Fiscal Problems of Nineteenth-Century Colombia." *Journal of Latin American Studies* 14, no. 2 (Nov. 1982): 287–328.

———. "La presencia de la política nacional en la vida provinciana, pueblerina y rural de Colombia en el primer siglo de la república." In *"Del poder y la gramática" y otros ensayos sobre historia, política y literatura colombianas,* 175–206. Bogotá: Tercer Mundo, 1993.

Del Castillo, Lina. "The Science of Nation Building: A History of Geographic Sciences in Colombia, 1821–1921." PhD diss., University of Miami, 2007.

Del Castillo Mathieu, Nicolás, et al. *Historia económica y social del Caribe colombiano*. Bogotá: Ediciones Uninort, Centro de Estudios Regionales, 1994.

Del Rio M., Alejandro. "Deslizamientos de Jerico (Departamento de Boyacá)." *Compilación de los estudios geológicos oficiales en Colombia* 7 (1946).

Delpar, Helen. "Aspects of Liberal Factionalism in Colombia, 1875–1885." *Hispanic American Historical Review* 51, no. 2 (May 1971): 250–74.

———. *Red against Blue: The Liberal Party in Colombian Politics, 1863–1899*. Tuscaloosa: University of Alabama Press, 1981.

Diaz de Zuluaga, Zamira. *Guerra y economía en las haciendas: Popayán, 1780–1830*. Vol. 2 of *Sociedad y economía en el Valle del Cauca*. 5 vols. Bogotá: Banco Popular, 1983.

Díaz Díaz, Fernando. *La desamortización de bienes eclesiásticos en Boyacá*. Tunja: Ediciones la Rana y el Aguila, 1977.

Didsbury, Brian. "Cheshire Saltworkers." In *Miners, Quarryman and Saltworkers,* edited by Raphael Samuel, 137–203. London: Routledge and Kegan Paul, 1977.

Dobado González, Rafael. "El monopolio estatal del mercurio en Nueva España durante el siglo XVIII." *Hispanic American Historical Review* 82, no. 4 (Nov. 2002): 685–718.

Donghi, Tulio Halperin. *Guerra y finanzas en los origenes del estado Argentino (1791–1850)*. Buenos Aires: Editorial de Belgrano, 1982.

Dore, Elizabeth. "One Step Forward, Two Steps Back: Gender and the State in the Long Nineteenth Century." In *Hidden Histories of Gender and the State in Latin America,* edited by Elizabeth Dore and Maxine Molyneux, 3–32. Durham, N.C.: Duke University Press, 2000.

Dore, Elizabeth, and Maxine Molyneux, eds. *Hidden Histories of Gender and the State in Latin America*. Durham, N.C.: Duke University Press, 2000.

Drake, Paul W., and Eric Hershberg, eds. *State and Society in Conflict: Comparative Perspectives on Andean Crises*. Pittsburgh, Pa.: University of Pittsburgh Press, 2006.

Ducy, Michael T. "Liberal Theory and Peasant Practice: Land and Power in Northern Veracruz, Mexico, 1836–1900." In *Liberals, the Church, and Nineteenth-Century Spanish America,* edited by Robert H. Jackson, 65–94. Albuquerque: University of New Mexico Press, 1997.

Dueñas Vargas, Guillermo. "Algunas hipótesis para el estudio de la resistencia campesina en la región central de Colombia, siglo XIX." *Anuario Colombiano de Historia Social y de la Cultura* 2 (1992): 90–106.

Duque Gómez, Luis. *Tribus indigenas y sitios arqueloicos, la cordillera Oriental los Chibchas, el grupos de los Muiscas: Los Laches, Los Guanes, los Chitareros.* Historia Extensa en Colombia, edited by Luis Martínez Delgado, Vol. 2. Bogotá: Editorial Lerner, 1967.
Eakin, Marshall C. *British Enterprise in Brazil: The St. John d'el Rey Mining Company and the Morro Velho Gold Mine.* Durham, N.C.: Duke University Press, 1989.
Earle, Rebecca. *Rumours of War: Civil Conflict in Nineteenth-Century Latin America.* London: ILAS, 2002.
———. "*Sobre Héroes y Tumbas*: National Symbols in Nineteenth-Century Spanish America." *Hispanic American Historical Review* 85, no. 3 (Aug. 2005): 375–416.
Ehrensaft, Philip. "The Political Economy of Informal Empire in Pre-colonial Nigeria, 1807–1884." *Canadian Journal of African Studies* 6, no. 3 (1972): 451–90.
Ellis, Myriam. *O monopólio do sal no estado de Brasil (1631–1801): contribução estudo do monopólio comercial português no Brasil durante o periodo colonial.* Boletins da Faculdade de Filosofia, Ciencieas e Letras, Historia da Civilizacão Brasileira, Vol. 14. São Paulo: Universidad do São Paulo, 1955.
Enciso Díaz, Humberto. *Historia del Municipio de Chita.* Tunja: n.p., 1992.
Escobar, Carmen. *La revolución liberal y la protesta del artesanado.* Bogotá: Fundación Universitaria Autónoma de Colombia, 1990.
Escobar, Ines Pinto. *Progreso, industrialización y utopía en Boyacá.* Tunja: Publicaciones de Magister en Historia, UPTC, 1997.
Escorcia, José. *Desarrollo político, social y económica 1800–1854.* Sociedad y Economía en el Valle del Cauca, Vol. 3. Bogotá: Biblioteca Banco Popular/Universidad del Valle, 1983.
España, Gonzalo. *La Guerra civil de 1885: Núñez y la derrota del radicalismo.* Bogotá: El Áncora Editores, 1985.
Evans, Peter B., Dietrich Rueschemeyer, and Theda Skocpol, eds. *Bringing the State Back In.* Cambridge: Cambridge University Press, 1985.
Ewald, Ursula. *The Mexican Salt Industry 1560–1980.* Stuttgart: Gustav Fischer Verlag, 1985.
Fajardo, Darío. *Campesino y capitalismo en Colombia.* Bogotá: Centro de Investigación y Educación Popular, 1981.
Fals Borda, Orlando. *Historia doble de la Costa.* 4 vols. Bogotá: Carlos Valencia Editores, 1979–86.
———. *El hombre y la tierra en Boyacá: bases socio-históricas para una reforma agraria.* Bogotá: Ediciones Documentos Colombianos, 1957.
———. *El Presidente Nieto.* Vol. 3 of *Historia doble de la Costa.* Bogotá: Carlos Valencia Editores, 1981.
———. "El reordenamiento territorial: itinerario de una idea." In *Territorios, regiones, sociedades,* edited by Renán Silva, 229–42. Bogotá: CEREC, 1994.
———. *Taller sobre sistemas de ordenamiento territorial, autonomía y descentralización, Bogotá, 1993.* Bogotá: Comisión de Ordenamiento Territorial/Instituto Geográfica Agustín Codazzi, 1994.
Fisher, John R., Allan J. Kuethe, and Anthony McFarlane, eds. *Reform and Insurrection in Bourbon New Granada and Peru.* Baton Rouge: Louisiana State University Press, 1990.
Florescano, Enrique. *Precios del maíz y crisis agrícolas en México (1708–1810): ensayo sobre el movimiento de los precios y sus consecuencias económicas y sociales.* Mexico, D.F.: Colegio de México, 1969.
Flores G., Lenin, and Adolfo Atehortúa C. *Estudios sobre la Regeneración.* Cali: Imprenta Departamental del Valle, 1987.
Friede, Juan. *La Batalla de Boyacá—7 de Agosto de 1819—a través de los Archivos Españoles.* Bogotá: Banco de la República, 1969.
———. "La conquista del territorio y el poblamiento." In *Manual de historia de Colombia,*

edited by Jaime Jaramillo Uribe, 4th ed., 3 vols., 1:119–224. Bogotá: Tercer Mundo, 1992.

———. "Demographic Changes in the Mining Community of Muzo after the Plague of 1629." *Hispanic American Historical Review* 47, no. 3 (Aug. 1967): 338–43.

Fuentes Combita, José de Jesús. "Las salinas de Zipaquirá y Nemocón en la historia social y cultural de Colombia." Master's thesis, Universidad Externado de Colombia, Bogotá, 1990.

Gómez F., Augusto. *Indios, colonos, y conflictos: una historia regional de los llanos orientales 1780–1970.* Bogotá: Siglo Veintiuno, 1992.

González, Fernán E. "A Propósito de 'Las palabras de la guerra': los comienzos conflictivos de la construcción del estado nación y las guerras civiles de la primera mitad del siglo XIX." *Estudios Políticos* 24 (July–Dec. 2004): 37–70.

———. *Iglesia y estado en Colombia durante el siglo XIX.* Bogotá: Centro de Investigación y Educación Popular, 1985.

———. *Para leer la política: ensayos de historia política colombiana.* 2 vols. Bogotá: CINEP, 1997.

González, Francisco L. G. "El páramo en paisaje deshumanizado: el caso de las provincias del Norte y Gutiérrez (Boyacá—Colombia)," 2.1.4, "Período Muisca (1.200–500 años A.P.)." In *El páramo: un ecosistema de alta montaña,* edited by Felipe Cárdenas Támara, Antoine M. Cleef, and Abdón Cortés Lombana (Bogotá: Instituto de Estudio Ambientales para el Desarrollo, 1996). Available at http://www.banrepcultural.org/blaavirtual/faunayflora/paramo/para214.htm.

González, Margarita. "Las rentas del estado." In *Era republicana,* edited by Jaime Jaramillo Uribe, 185–96. Vol. 2 of *Nueva historia de Colombia,* edited by Alvaro Tirado Mejía. Bogotá: Planeta, 1989.

———. *El resguardo en el Nuevo Reino de Granada.* 3d. ed. Prologue by Bernardo Tovar. Bogotá: El Áncora Editores, 1992.

Gootenberg, Paul. *Imagining Development: Economic Ideas in Peru's "Fictitious Prosperity" of Guano, 1840–1880.* Berkeley: University of California Press, 1993.

Graham, Lance. *The Political Economy of Smuggling: Regional Informal Economies in Early Bourbon New Granada.* Boulder, Colo.: Harper Collins, 1997.

Guerra Vilaboy, Sergio. *La república artesana en Colombia.* La Habana, Colombia: Editorial de Ciencias Sociales, 1980.

Guerrero Baron, Javier. *Los años del olvido: Boyacá y los orígenes de la Violencia.* Bogotá: Tercer Mundo, 1991.

Guha, Ranajit. *Elementary Aspects of Peasant Insurgency in Colonial India.* Foreword by James Scott. Oxford: Oxford University Press, 1983.

Guillén Martinez, Fernando. *La Regeneración: primer frente nacional.* Bogotá: Carlos Valencia, 1986.

Guss, David M. *The Festive State: Race, Ethnicity, and Nationalism as Cultural Performance.* Berkeley: University of California Press, 2000.

Gutiérrez Cely, Eugenio. "Las luchas populares in Bogotá en el siglo XIX: el Motín del Pan." *Revista Universidad Distrital Francisco José de Caldas* 3–4 (1988): 33–48.

Guzmán Ramirez, Nohora Beatriz. "Economía extractiva: la explotación de la sal mina en el territorio de San Martín, Colombia, siglo XIX." Paper presented at the Segundo Congreso Latinoamericano de Historia y Económica. Available at http://www.economia.unam.mx/cladhe/index.php. Accessed June 2010.

Haber, Stephen, ed. *How Latin America Fell Behind: Essays on the Economic Histories of Brazil and Mexico, 1800–1914.* Stanford, Calif.: Stanford University Press, 1997.

Helg, Aline. *Liberty and Equality in Caribbean Colombia, 1770–1835.* Chapel Hill: University of North Carolina Press, 2004.

Henderson, James D. *When Colombia Bled.* Tuscaloosa: University of Alabama Press, 1985.

Hernández Vargas, Carlos. "La Salina: una riqueza en el olvido." *Caribarare* 4, no. 4 (Feb. 1992): 5–6.

Holden, Robert. *Armies without Nations: Public Violence and State Formation in Central America, 1821–1960*. New York: Oxford University Press, 2004.

Horna, Hernán. "Transportation Modernization and Entrepreneurship in Nineteenth Century Colombia." *Journal of Latin American Studies* 14, no. 1 (May 1982): 33–54.

Huck, Eugene. "Economic Experimentation in a Newly Independent Nation: Colombia under Francisco de Paula Santander 1821–1840." *The Americas* 29, no. 1 (July 1972): 17–29.

Hunefeldt, Christine. "Reinventing Old Power: State, Tribute and Indians in Puno." *Fronteras* 4, no. 4 (1999): 11–53.

Hunt, Stacy. "Languages of Stateness: A Study of Space and El Pueblo in the Colombian State." *Latin American Research Review* 41, no. 3 (2006): 88–121.

Husson, Patrick. "Changement social et insurrections indiennes: 'la révolte du sel' à Huanta, Pérou, en 1896." *Cahiers des Amériques Latines* 23 (1981): 102–40.

Instituto de Estudios Colombianos. *Historia económica de Colombia: un debate en marcha*. Bogotá: Biblioteca Banco Popular, 1979.

Jacobson, Nils, and Cristóbal Aljovín de Losada. "The Long and Short of It: A Pragmatic Perspective on Political Cultures, Especially for the Modern History of the Andes." In *Political Cultures in the Andes 1750–1950*, edited by Jacobsen and Aljovín de Losada, 1–24. Durham, N.C.: Duke University Press, 2005.

———. "State- and Nation-Building Projects and Their Limitations." In *Political Cultures in the Andes 1750–1950*, edited by Jacobsen and Aljovín de Losada, 69–73. Durham, N.C.: Duke University Press, 2005.

Jackson, Robert H., ed. *Liberals, the Church, and Indian Peasants: Corporate Lands and the Challenge of Reform in Nineteenth-Century Spanish America*. Albuquerque: University of New Mexico Press, 1997.

Jaramillo Uribe, Jaime. "La economía colonial del virreinato (1740–1810)." In *Historia económica de Colombia*, edited by José Antonio Ocampo, 49–86. Bogotá: Siglo Veintiuno, 1987.

———. *Era republicana*. Vol. 2 of *Nueva historia de Colombia*, edited by Alvaro Tirado Mejía, 8 vols. Bogotá: Planeta, 1989.

———, ed. *Manual de historia de Colombia*. 3 vols. Bogotá: Tercer Mundo, 1992.

———. *El pensamiento colombiano en el siglo XIX*. 3d. ed. Bogotá: Editorial Temis, 1982.

Jeyaraj, K. V. *A History of the Salt Monopoly in Madras Presidency (1805–1878)*. Madurai, India: Ennes, 1984.

Jiménez, Michael. "Class, Gender, and Peasant Resistance in Central Colombia, 1900–1930." In *Everyday Forms of Peasant Resistance*, edited by Forrest D. Colburn, 122–50. New York: M. E. Sharpe, 1989.

———. "Traveling Far in Grandfather's Car: The Life-Cycle of Coffee Estates in Central Colombia; The Case of Viotá, Cundinamarca, 1900–30." *Hispanic American Historical Review* 69, no. 2 (May 1989): 185–219.

Jimeno, Myriam. "Región, nación y diversidad cultural en Colombia." In *Territorios, regiones, sociedades*, edited by Renán Silva, 65–78. Bogotá: CEREC, 1994.

Johnson, David. *Santander: siglo XIX—cambios socioeconómicos*. Bogotá: Carlos Valencia Editores, 1984.

Joseph, Gilbert M., and Daniel Nugent, eds. *Everyday Forms of State Formation: Revolution and Negotiation of Rule in Modern Mexico*. Durham, N.C.: Duke University Press, 1994.

Kalmanovitz, Salomón. *Economía y nación: una breve historia de Colombia*. Bogotá: Siglo Veintiuno, 1986.

———. "El régimen agrario durante el siglo XIX." In *Manual de historia de Colombia*, edited by Jaime Jaramillo Uribe, 3 vols., 2:211–326. Bogotá: Tercer Mundo, 1992.

Keremitsis, Dawn. "Latin American Women Workers in Transition: Sexual Division of the Labor Force in Mexico and Colombia in the Textile Industry." *The Americas* 154 (Apr. 1984): 491–504.
Kittleson, Roger A. *The Practice of Politics in Postcolonial Brazil: Porto Allegre, 1845–1895*. Pittsburgh, Pa.: University of Pittsburgh Press, 2005.
Klein, Herbert S. "Structure and Profitability of Royal Finance in the Viceroyalty of the Río de la Plata in 1790." *Hispanic American Historical Review* 53, no. 3 (May 1973): 440–69.
Klein, Herbert S., and Jacques A. Barbier. "Recent Trends in the Study of Spanish American Colonial Public Finance." *Latin American Research Review* 23, no. 1 (1988): 25–62.
Kline, Harvey F. "Colombia: Lawlessness, Drug Trafficking, and Carving Up the State." In *State Failure and State Weakness in a Time of Terror,* edited by Robert I. Rotberg, 161–82. Washington, D.C.: Brookings Institution Press, 2003.
Knight, Franklin W. *Slave Society in Cuba during the Nineteenth Century*. Madison: University of Wisconsin Press, 1970.
Kuethe, Allan. *Military Reform and Society in New Granada, 1773–1808*. Gainesville: University Press of Florida, 1978.
———. "More on 'The Culmination of the Bourbon Reforms': A Perspective from New Granada." *Hispanic American Historical Review* 58 (Aug. 1978): 477–80.
Kurlansky, Mark. *Salt: A World History*. New York: Walker, 2002.
Larson, Brooke. *Trials of Nation Making: Liberalism, Race, and Ethnicity in the Andes, 1810–1910*. New York: Cambridge University Press, 2004.
Lasso, Marixa. *Myths of Harmony: Race and Republicanism during the Age of Revolution, Colombia 1795–1831*. Pittsburgh, Pa.: University of Pittsburgh Press, 2007.
LeGrand, Catherine. *Frontier Expansion and Peasant Protest in Colombia, 1850–1936*. Albuquerque: University of New Mexico Press, 1986.
Lemaitra, Eduardo. *El General Juan José Nieto y su epoca*. Bogotá: Carlos Valencia Editores, 1983.
López-Alvez, Fernando. *State Formation and Democracy in Latin America, 1810–1900*. Durham, N.C.: Duke University Press, 2000.
———. "The Transatlantic Bridge: Mirrors, Charles Tilly, and State Formation in the River Plate." In *The Other Mirror: Grand Theory through the Lens of Latin America,* edited by Miguel Angel Centeno and Fernando López-Alvez, 153–76. Princeton, N.J.: Princeton University Press, 2001.
López Garavito, Luis Fernando. *Historia de la Hacienda e el tesoro en Colombia, 1821–1900*. Bogotá: Banco de la República, 1992.
———. *Origen y fundamentos de la Hacienda y la economía colombianas*. Bogotá: Universidad Externado de Colombia, 2000.
———. *Pensamiento económico y fiscal colombiano*. Bogotá: Universidad Externado de Colombia, 1998.
Lovejoy, Paul E. *Salt of the Desert Sun: A History of Salt Production and Trade in the Central Sudan*. Cambridge: Cambridge University Press, 1986.
Loy, Jane Meyer. "Forgotten Comuneros: The 1781 Revolt in the Llanos of Casanare." *Hispanic American Historical Review* 61, no. 2 (1981): 235–57.
———. "Primary Education during the Colombian Federation: The School Reform of 1870." *Hispanic American Historical Review* 51, no. 2 (May 1971): 282–315.
Mallon, Florencia. *Peasant and Nation: The Making of Postcolonial Mexico and Peru*. Berkeley: University of California Press, 1995.
Man Bun, Kwan. *The Salt Merchants of Tianjin: State-Making and Civil Society in Late Imperial China*. Honolulu: University of Hawai'i Press, 2001.

Marroquín, José Manuel. *Escritos históricos*. Biblioteca Banco Popular, Vol. 109. Bogotá: Banco Popular, 1982.
Martínez Delgado, Luis, ed. *Historia extensa en Colombia*. Vol. 1. Bogotá: Editorial Lerner, 1967.
Martz, John D. *The Politics of Clientelism: Democracy and the State in Colombia*. New Brunswick, N.J.: Transition, 1997.
Marulanda, Elsy. *Colonización y conflicto: las lecciones del Sumapaz*. Bogotá: Tercer Mundo, 1991.
Marzahl, Peter. *Town in the Empire: Government, Politics, and Society in Seventeenth-Century Popayán*. Latin American Monographs no. 45. Austin: Institute of Latin American Studies, University of Texas, 1978.
Mayor Mora, Alberto. "Las fábricas de hierro en Colombia, en el siglo XIX." *Boletín Cultural y Bibliográfico* (May 2005). Available at Biblioteca Virtual, Biblioteca Luis Ángel Arango, http://www.lablaa.org/blaavirtual/revistas/credencial/julio1993/julio2.htm. Accessed June 28, 2010. Originally published in *Revista Credencial Historia* 43 (July 1993).
McDougall. E. Ann. "Salts of the Western Sahara: Myths, Mysteries, and Historical Significance." *International Journal of African Historical Studies* 23, no. 2 (1990): 231–57.
McFarlane, Anthony. *Colombia before Independence: Economy, Society and Politics under Bourbon Rule*. Cambridge: Cambridge University Press, 1993.
———. "The 'Rebellion of the Barrios': Urban Insurrection in Bourbon Quito." *Hispanic American Historical Review* 69, no. 2 (May 1989): 283–330.
McGreevy, William Paul. *An Economic History of Colombia 1845–1930*. Cambridge: Cambridge University Press, 1971.
Means, Robert Charles. *Underdevelopment and the Development of Law: Corporations and Corporation Law in Nineteenth-Century Colombia*. Chapel Hill: University of North Carolina Press, 1980.
Melo, Jorge Orlando. "La evolución económica de Colombia 1830–1900." In *Manual de historia de Colombia*, edited by Jaime Jaramillo Uribe, 4th ed., 3 vols., 2:135–210. Bogotá: Tercer Mundo, 1992.
———. "La idea del progreso en el siglo xix, ilusiones y desencantos, 1780–1930." *Revista de Estudios Colombianos* 36 (2010): 16–29.
———. *Las origines de los partidos políticos en Colombia*. Bogotá: Instituto Colombiano de Cultura, 1978.
———. "Las vicisitudes del modelo liberal (1850–1899). In *Historia económica de Colombia*, edited by José Antonio Ocampo, 119–72. Bogotá: Siglo Veintiuno, 1988.
Méndez, Cecilia. *The Plebeian Republic: The Huanta Rebellion and the Making of the Peruvian State, 1820–1850*. Durham, N.C.: Duke University Press, 2005.
Miller, Shawn W. "Fuelwood in Colonial Brazil: The Economic and Social Consequences of Fuel Depletion for the Bahian Recôncavo, 1549–1820." In *Forest and Conservation History* 38, no. 4 (Oct. 1994): 181–92.
Molina, Gerardo. *Las ideas liberales en Colombia, 1849–1914*. Bogotá: Tercer Mundo, 1982.
Mondragon Castañeda, Julio. *Las ideas de paz y de constitucionalidad de los adalides boyacenses en el radicalismo colombiano, 1850–1885: con el texto de la Constitución Política para el Estado de Boyacá (Septiembre 3 de 1869)*. Tunja: Publicaciones del Magister en Historia, UPTC, 1991.
Morales Benítez, Otto. *Origen, programas, y tesis del liberalismo*. Bogotá: Lerner, 1997.
Mörner, Magnus. *Region and State in Latin America's Past*. Baltimore, Md.: Johns Hopkins University Press, 1993.
Moyano Rodríguez, Néstor. *Vélez: la gobernación, 1832–1857*. Bogotá: Ediciones Carare, 2003.
Muecke, Ulrich. *Political Culture in Nineteenth-Century Peru: The Rise of the Partido Civil*. Translated by Katya Andrusz. Pittsburgh, Pa.: University of Pittsburgh Press, 1998.

Múnera, Alfonso. *Ensayos costeños: de la colonia a la república: 1770–1890*. Biblioteca de Autores Costeños, Vol. 2. Bogotá: Colcultura, 1994.
———. *El fracaso de la nación: región, clase y raza en el Caribe colombiano (1717–1821)*. Bogotá: Banco de la República/El Áncora Editores, 1998.
Murray, Pamela S. *Dreams of Development: Colombia's National School of Mines and Its Engineers, 1887–1970*. Tuscaloosa: University of Alabama Press, 1987.
Multhauf, Robert P. *Neptune's Gift: A History of Common Salt*. Baltimore, Md.: Johns Hopkins University Press, 1978.
Needell, Jeffrey D. *The Party of Order: The Conservatives, the State, and Slavery in the Brazilian Monarchy, 1831–1871*. Stanford, Calif.: Stanford University Press, 2006.
Negrete B., Victor. *Origen de los luchas agrarias en Córdoba*. Montería, Colombia: Fundación del Caribe, 1981.
Nieto Arteta, Luis Eduardo. *Economía y cultura en la historia de Colombia*. Bogotá: Banco de la República/El Áncora Editores, 1996 [1941].
Ocampo, José Antonio, ed. *Colombia y la economía mundial, 1830–1910*. Bogotá: Siglo Veintiuno, 1984.
———. *Historia económica de Colombia*. Bogotá: Siglo Veintiuno, 1987.
Ocampo López, Javier. *Colombia en sus ideas*. 3 vols. Bogotá: Fundación Universidad Central, 1999.
———. *Historia del pueblo boyacense: de los origines paleonindígenas y míticos a la culminación de la independencia*. Tunja: Instituto de Cultura y Bellas Artes de Boyacá, 1983.
———. *Los hombres y las ideas en Boyacá*. Tunja: Publicaciones de UPTC, 1989.
———. "José María del Castillo y Rada." www.lablaa.org.
———. "Santander y el civilismo colombiana." *Boletín de Historia y Antigüedades* 69, no. 738 (July–Sept. 1982): 641–63.
Ortiz Mesa, Luis Javier. *Fusiles y plegarias: guerra de guerrillas en Cundinamarca, Boyacá y Santander, 1876–1877*. Medellín: Universidad Nacional de Colombia, 2004.
———. "La sociedad colombiana en el siglo XIX." In *Mujeres y sociedad*, 169–203. Vol. 2 of *Las mujeres en la historia de Colombia*, edited by Magdala Velásquez Toro. Bogotá: Grupos Editorial Norma, 1995.
Ospina Vásquez, Luis. *Industria y protección en Colombia, 1810–1930*. 4th ed. Medellín: FAES, 1987.
Oszlak, Oscar. "The Historical Formation of the State in Latin America: Some Theoretical and Methodological Guidelines for Its Study." *Latin American Research Review* 16, no. 2 (1981): 3–32.
Otero Muñoz, Gustavo. *Wilches y su época*. Colección Memoria Regional. Bucaramanga, Colombia: Gobernación de Santander, 1990.
Pacheco G., Margarita R. *La Fiesta Liberal en Cali*. Cali, Colombia: Ediciones Universidad del Valle, 1992.
Palacios, Marco. *Between Legitimacy and Violence: A History of Colombia, 1875–2002*. Translated by Richard Stoller. Durham, N.C.: Duke University Press, 2006.
———. *El café en Colombia, 1850–1970: una historia económica, social y política*. 2d ed. Bogotá: El Áncora Editores, 1983.
———. *La delgada corteza de nuestra civilización*. Bogotá: Procultura, 1986.
———. "El estado liberal colombiano y la crisis de la civilización en el siglo XIX." *Boletín de Historia y Antigüedades* 753 (1986): 401–18.
Park, James William. *Rafael Núñez and the Politics of Colombian Regionalism, 1863–1886*. Baton Rouge: Louisiana State University Press, 1985.
Peña Díaz, Anuar Herna. *Sal, sudor y fisco: el proceso de institucionalización del monopolio de la sal en las salinas de Chámeza, Recetor y Pajarito, 1588–1950*. Bucaramanga, Colombia: División de Publicaciones UIS/Universidad Industrial de Santander, 2008.
Pérez Aguirre, Antonio. *Los Radicales y la Regeneración*. Bogotá: Editorial Cromos, 1941.

———. *25 años de historia colombiana 1853 a 1878: del centralismo a federación*. Bogotá: Editorial Sucre, 1959.

Pérez, Felipe, Santiago Pérez, Tomás Cuenca, Felipe Zapata, and Fidel Cano. *Periodistas liberales del siglo XIX*. Biblioteca Aldeana de Colombia, Vol. 63. Bogotá: Editorial Minerva, 1936.

Pérez Angel, Hector Publio. "Economía y demografía de la Salina: fuentes para un studio." *Caribarare* 4, no. 4 (Feb. 1992): 57–64.

Phelan, John Leddy. *The People and the King: The Comunero Revolution in Colombia, 1781*. Madison: University of Wisconsin Press, 1978.

Piccato, Pablo. *The Tyranny of Opinion: Honor in the Construction of the Mexican Public Sphere*. Durham, N.C.: Duke University Press, 2010.

Pinto, José Miguel. "División política del Departamento de Boyacá de 1821 a 1905." *Boletín de Historia y Antigüedades* 6, no. 69 (Feb. 1911): 578–81.

Pinto Escobar, Ines. *Progreso, industrialización y utopía en Boyacá: el caso de la Siderúrgica de Samacá*. Tunja: Publicaciones del Magister en Historia, UPTC, 1997.

Platt, Tristan. *Estado boliviano y ayllu andino: tierra y tribute en el norte de Potosí*. Lima: Instituto de Estudios Peruanos, 1982.

Plazas Olarte, Guillermo. *La guerra civil de los Mil Días*. Tunja: Publicaciones de la Academia Boyacense de Historia, 1985.

———. *Sugamuxi: crónicas y leyendes*. Bogotá: Fuerzas Militares, 1961.

Plazas Pérez, Martin Javier. "La Salina: el pueblo de la sal, una puerta de la Casa del Sol." *Caribarare* 4, no. 4 (Feb. 1992): 49–56.

Pombo, Manuel Antonio, José Joaquin Guerra, and Carlos Restrepo Piedrahita, eds. *Constituciones de Colombia*. 4th ed. 4 vols. Bogotá: Banco Popular, 1986.

Posada, Jaime, ed. *Manual de historia de Colombia*. 3d ed. 3 vols. Bogotá: Procultura, 1982.

Posada-Carbo, Eduardo. "Elections and Civil Wars in Nineteenth-Century Colombia: The 1875 Presidential Campaign." *Journal of Latin American Studies* 26, no. 3 (Oct. 1994): 621–49.

Putnam, Lara, Sarah C. Chambers, and Sueann Caulfield. *Honor, Status, and Law in Modern Latin America*. Durham, N.C.: Duke University Press, 2005.

Ramos Peñuela, Aristides. "Colonos, empresarios y vagos: experiencias fronterizas en el Carare en la primera mitad del siglo XIX." *Fronteras* 2, no. 2 (1998): 65–92.

Rausch, Jane. *Colombia, Territorial Rule and the Llanos Frontier*. Gainesville: University Press of Florida, 1999.

———. *The Llanos Frontier in Colombian History 1830–1930*. Albuquerque: University of New Mexico Press, 1993.

———. *A Tropical Plains Frontier: The Llanos of Colombia, 1531–1832*. Albuquerque: University of New Mexico Press, 1984.

Reinhardt, Nola. "The Consolidation of the Import-Export Economy in Nineteenth-Century Colombia: A Political-Economic Analysis." *Latin American Perspectives* 13, no. 1 (Winter 1986): 75–98.

Robaya Avendano, Juan Manuel. "La propiedad rural como elemento del crédito en la colonia: la provincia de Tunja en la segunda mitad del siglo XVIII." Master's thesis, Universidad Pedagógica y Tecnológica de Colombia, 1977.

Roberts, Richard. *Two Worlds of Cotton: Colonialism and the Regional Economy in the French Soudan, 1800–1946*. Stanford, Calif.: Stanford University Press, 1996.

Rock, David. *State Building and Political Movements in Argentina, 1860–1916*. Stanford, Calif.: Stanford University Press, 2002.

Rodriguez, Linda Alexander. *The Search for Public Policy: Regional Politics and Government Finances in Ecuador 1830–1940*. Berkeley: University of California Press, 1985.

Rodríguez R., Gustavo Humberto. *Boyacenses en la historia de Colombia, bocetos biográficos*. Bogotá: Jurídicas Wilches, 1994.

Rodríguez Salazar, Oscar, and Desci Arévalo Hernandez. "La historiografía económica colombiana del siglo XIX." In *La historia al final del milenio: ensayos de historiografía colombiana y latinoamericana,* edited by Bernardo Tovar Zambrano and Mauricio Archila, 2 vols, 1:187–250. Bogotá: Editorial Universidad Nacional, 1994.
Rojas, Christina. *Civilization and Violence: Regimes of Representation in Nineteenth-Century Colombia.* Minneapolis: University of Minnesota Press, 2001.
Rotberg, Robert I., ed. *State Failure and State Weakness in a Time of Terror.* Washington, D.C.: Brookings Institution Press, 2003.
Rueda Mendez, David. *Introducción a la historia de la esclavitud negra en la provincial de Tunja: siglo XVIII.* Tunja: UPTC, 1989.
Safford, Frank. "Acerca de las interpretaciones socioeconómicos de la política en la Colombia del siglo xix: variaciones sobre un tema." *Anuario Colombianos de Historia Social y de la Cultura* 13–14 (1985–86): 91–150.
———. "Commerce and Enterprise in Central Colombia." PhD diss., Columbia University, 1965.
———. *The Ideal of the Practical: Colombia's Struggle to Form a Technical Elite.* Austin: University of Texas Press, 1976.
———. "Race, Integration, and Progress: Elite Attitudes and the Indian in Colombia, 1750–1870." *Hispanic American Historical Review* 71, no. 1 (Feb. 1991), 1–33.
Safford, Frank, and Marcos Palacios. *Colombia: Fragmented Land, Divided Society.* New York: Oxford University Press, 2002.
Samuel, Raphael, ed. *Miners, Quarrymen and Saltworkers.* London: Routledge and Kegan Paul, 1977.
Sanders, James E. *Contentious Republicans: Popular Politics, Race, and Class in Nineteenth-Century Colombia.* Durham, N.C.: Duke University Press, 2004.
Sayer, Derek. "Everyday Forms of State Formation: Some Dissident Remarks on 'Hegemony.'" In *Everyday Forms of State Formation: Revolution and Negotiation of Rule in Modern Mexico,* edited by Gilbert M. Joseph and Daniel Nugent, 367–78. Durham, N.C.: Duke University Press, 1994.
Schwartz, Stuart B. *Sugar Plantations in the Formation of Brazilian Society: Bahia, 1550–1835.* Cambridge: Cambridge University Press, 1985.
Scott, James C. *Seeing Like a State: How Certain Schemes to Improve the Human Condition Have Failed.* New Haven, Conn.: Yale University Press, 1999.
Scott, James C. *Weapons of the Weak: Everyday Forms of Peasant Resistance.* New Haven, Conn.: Yale University Press, 1985.
Silva, Renán, ed. *Territorios, regiones, sociedades.* Bogotá: CEREC, 1994.
Silva Celis, Eliécer. "Contribución al conocimiento de la civilización de los Laches." *Boletín de Arqueología* 1, no. 5 (1945): 369–424.
Soifer, Hillel David. "Authority over Distance: Explaining Variation in State Infrastructural Power in Latin America." PhD diss., Harvard University, 2006.
———. "Authority over Distance: Institutions and Long-Run Variation in State Development in Latin America." Unpublished manuscript. Temple University, Department of Political Science, 2011.
———. "The Sources of Infrastructural Power: Evidence from Nineteenth-Century Chilean Education." *Latin American Research Review* 44, no. 2 (2009): 158–80.
Sowell, David. *The Early Colombian Labor Movement: Artisans and Politics in Bogotá 1832–1919.* Philadelphia: Temple University Press, 1992.
———. "The 1893 *Bogotazo*: Artisans and Public Violence in Late Nineteenth-Century Bogotá." In *Riots in the Cities: Popular Politics and the Urban Poor in Latin America, 1765–1910,* edited by Silvia M. Arrom and Servando Ortoll, 137–54. Wilmington, Del.: Scholarly Resources, 1996.
Stoller, Richard. "Liberalism and Conflict in Socorro." PhD diss., Duke University, 1991.

Suarez Casteñeda, Gilberto. *Fondo Salinas: Sección Colonial, catálogo e indice*. Bogotá: AGN, 1992.

Tenenbaum, Barbara A. *The Politics of Penury: Debts and Taxes in Mexico, 1821–1856*. Albuquerque: University of New Mexico Press, 1986.

Thompson, E. P. "The Moral Economy of the English Crowd in the 18th Century." *Past and Present* 50 (1971): 76–136.

Thurner, Mark. *From Two Republics to One Divided: Contradiction of Postcolonial Nationmaking in Andean Peru*. Durham, N.C.: Duke University Press, 1997.

Tilly, Charles. *Coercion, Capital, and European States, AD 990–1990*. Cambridge, Mass., Basil Blackwell, 1990.

———, ed. *The Formation of National States in Western Europe*. Princeton, N.J.: Princeton University Press, 1975.

———. "Reflections on the History of European State-Making." In *The Formation of National States in Western Europe*, edited by Tilly, 3–83. Princeton, N.J.: Princeton University Press, 1975.

Tirado Mejía, Alvarado. *Aspectos sociales de las guerras civiles en Colombia*. Bogotá: Colcultura, 1976.

———. "El estado y la política en el siglo XIX." In *Manual de historia de Colombia*, edited by Jaime Posada, 3 vols., 3d ed., 2:327–86. Bogotá: Procultura, 1982.

Tisnés Jiménez, Roberto María. *María Martínez de Nisser y la revolución de los supremos*. 2 vols. Biblioteca Banco Popular, Vol. 111. Bogotá: Banco Popular, 1983.

Topic, Steven, Carlos Marichal, and Zephyr Frank, eds. *From Silver to Cocaine: Latin American Commodity Chains and the Building of the World Economy, 1500–2000*. Durham, N.C.: Duke University Press, 2006.

Torres Almeida, Jesus C. *Manuel Murillo Toro: caudillo radical y reformador social*. Bogotá: Ediciones El Tiempo, 1984.

Tovar Pinzón, Hermes. "La lenta ruptura con el pasado colonial (1810–1850)." In *Historia económica de Colombia*, edited by José Antonio Ocampo, 87–118. Bogotá: Siglo Veintiuno, 1987.

———. *El movimiento campesino en Colombia: durante los siglos XIX y XX*. Bogotá: Ediciones Libres, 1975.

Tovar Pinzón, Hermes, Camilo Tovar Pinzón, and Jorge Tovar Mora. *Convocatoria al poder del número: censos y estadísticas de la Nueva Granada 1750–1830*. Bogotá: Archivo General de la Nación, 1995.

Tovar Zambrano, Bernardo, and Mauricio Archila, eds. *La historia al final del milenio: ensayos de historiografía colombiana y latinoamericana*. 2 vols. Bogotá: Editorial Universidad Nacional, 1994.

Universided del Valle, ed. *Sociedad y economía en el Valle del Cauca*. 5 vols. Bogotá: Biblioteca Popular/Universidad del Valle, 1983.

Uribe de Hincapié, María Teresa, and Liliana María López Lopera. *Las palabras de la guerra; metáforas, narraciones y lenguajes políticos: un estudio sobre las memorias de las guerras civiles en Colombia*. Medellín: Instituto de Estudio Políticos, Universidad de Antioquia, 2006.

Uribe-Uran, Victor. "The Birth of a Public Sphere in Latin America during the Age of Revolution." *Comparative Studies in Society and History* 42, no. 2 (Apr. 2000): 425–57.

———. *Honorable Lives: Lawyers, Families, and Politics in Colombia 1780–1850*. Pittsburgh, Pa.: University of Pittsburgh Press, 2000.

———. "Rebellion of the Young Mandarins: Lawyers, Political Parties and the State in Colombia, 1780–1850." PhD diss., University of Pittsburgh, 1993.

Van Ausdal, Shawn Kenneth. "The Logic of Livestock: An Historical Geography of Cattle Ranching in Colombia, 1850–1950." PhD diss., University of California at Berkeley, 2009.

Vanderwood, Paul J. *Disorder and Progress: Bandits, Police, and Mexican Development.* Lincoln: University of Nebraska Press, 1981.

———. *The Power of God against the Guns of Government: Religious Upheaval in Mexico at the Turn of the Nineteenth Century.* Stanford, Calif.: Stanford University Press, 1998.

Varela B., Edgar. *La cultura de la violencia en Colombia durante el siglo XIX.* Cali, Colombia: Imprenta Departamental del Valle, 1990.

Vargas Martínez, Gustavo. *Colombia 1854: Melo, los artesanos y el socialismo.* Bogotá: Oveja Negra, 1972.

Vasques, Luis Ospina. *Industria y protección en Colombia, 1830–1910.* 4th ed. Medellín: FAES, 1987.

Velásquez Toro, Magdala, et al., eds. *Las mujeres en la historia de Colombia.* 3 vols. Bogotá: Editorial Norma, 1995.

Velosa, Junco. *Del fraude y la violencia al clientelismo: Boyacá, 1930–1990.* Tunja: UPTC, 1992.

Wade, Peter. *Blackness and Race Mixture: The Dynamics of Racial Identity in Colombia.* Baltimore, Md.: Johns Hopkins University Press, 1993.

Weisman, Alan. *Gaviotas: A Village to Reinvent the World.* White River Junction, Vt.: Chelsea Green, 1998.

Whitehead, Lawrence. *Latin America: A New Interpretation.* New York: Palgrave Macmillan, 2006.

Williams, Raymond. *Keywords: A Vocabulary of Culture and Society.* New York: Oxford University Press, 1983.

Williams, Robert G. *States and Social Evolution: Coffee and the Rise of National Governments in Central America.* Chapel Hill: University of North Carolina Press, 1994.

Young, Crawford. *The African Colonial State in Comparative Perspective.* New Haven, Conn.: Yale University Press, 1994.

Zambrano P., Fabrio. "Historiografía sobre los movimiento sociales en Colombia, siglo XIX." In *La historia al final del milenio: ensayos de historiografía colombiana y latinoamericana.* 2 vols., edited by Bernardo Tovar Zambrano and Mauricio Archila, 1:147–82. Bogotá: Editorial Universidad Nacional, 1994.

Index

Alarcón, Trinidad, 32–33, 55
almacén, in la Salina, 4, 26, 63, 71, 77–90, 98, 106–8, 143
almacenes, in Boyacá, 68, 106, 108, 124, 127
Ancízar, Manuel, 13, 16, 34, 43, 72

Barón, Estavan, 83–84, 86, 116, 118
Bogotá, 4, 14–5; artisans of 101–2, 131; during the Comuneros Rebellion, 6, 20; during the 1871 rebellion in Boyacá, 120, 123; during the Melista Rebellion, 80, 102–3; and Northern Railroad, 126; press in, 28, 49, 66, 70–71, 80, 132; politics in, 101, 115–16, 121, 123–24, 139–40, 142; and the Regeneration, 131; War of 1860–1862, 116–17; War of 1876–1877, 128; War of 1895, 133; War of the Supremes, 96–97
Bolívar, Simon, 22–24, 96
Bourbon reforms, 18, 21
Boyacá, 9–17, 30, 141–43; and development, 51–52, 58, 95, 124–27; during the Melista Rebellion, 102–3; politics in, 81, 94, 105–6, 109–10, 113–19; during the rebellion of 1871, 119–23; salt administration in, 84; salt distribution in, 68, 71, 87, 92; during the War of the Supremes, 96–98; War of 1876–1877, 127–28; War of 1885, 129–30; wealth in, 32, 69. *See also* Tunja
brine, 3, 42, 44, 54, 109; boiling of, 44–45; gleaning, 38; in Nemocón, 48
cabildo of la Salina, 29, 31, 57, 84
caked salt, 39, 42, 46–50, 68, 72, 90–93, 125, 130
Caicedo, Ignacio, 3–5, 21–22, 24, 27, 42–43
Casanare, 3, 6, 12, 36, 75, 94, 109, 111, 113, 139, 144; buyers of salt from, 87; during the Melista Rebellion, 104; migration to, 31, 137; ranching in, 14–15, 24, 56, 71–72; salt shortages, 73; during the Thousand Days War, 129–30; during the War of the Supremes, 96–97, 98
Casanare River, 3–4, 42–43, 61, 73, 94
Chámeza, 6, 36–37, 48, 107, 117
Chita, 15–16, 21, 90, 94–95, 103–5, 109, 129; *resguardo,* 19–20; roads, 29; and the salt trade, 75, 78, 82
ceramics, 19, 24, 36, 45, 50–54, 58, 76–78, 83, 107–9
Codazzi, Agustín, 14–16, 48, 53, 72, 140
Comuneros Rebellion, 6, 20–21
Conservative Party, 96, 100–103, 113–16, 119–21, 126, 133–37, 143
constitution: of Angostura, 22; of 1833, 95; of 1853, 101–2; of 1858, 106; of 1886, 70, 115, 128; of Rionegro (1886), 66, 81, 116–17, 119, 125, 128, 130
contraband, 24–25, 30, 66, 69, 91–92, 118, 138, 143; in Chámeza 37; and guards, 27
Córdoba, 3, 42–44, 53, 72–73, 134
Corredor, Jacinto, 84, 86, 124–25
cosecheros, 21, 57–58
Cuadra, Medardo, 38–39, 80, 83–84, 86

Cuenca, Tomás, 50, 56, 66, 118

Davidson, Diego, 34, 38, 76
de Márquez, Jose Ignacio, 13–14, 96
Del Castillo y Rada, José María, 23, 27, 63, 70, 95
direct administration, 4, 25–28, 34, 37, 42, 72–73, 76, 78, 99
Dovieski, José, 106–12

El Cocuy, 28, 91, 97, 109–10
El Cocuy (canton), 12, 14–16, 32, 76, 79, 104, 111, 129, 139
Espinosas, the, 28–29, 47; Bernardo Espinosa, 49; Bonifacio Espinosa, 30–31, 57–58, 78; Honorato Espinosa, 54, 58, 81, 83–86

fábrica, 44–46, 54, 76–77, 82–83, 99, 108–9, 134–39
forests, 28, 35, 50, 54–56, 99, 131–32, 138–39. *See also* woods
fuel, 24, 50, 52, 54, 70, 86, 109, 131, 133, 136, 138; efficiency, 45–46
Finance Ministry, 4–6, 9, 23–28, 70–71, 95, 118, 130–34, 138–40; authority in La Salina, 40, 43, 76–77, 80–84, 91–92, 106, 112, 115, 118; critiques of, 69, 89, 124; contracts for, 24–25, 28–29, 47, 56, 60, 78, 111; control over salt making, 4, 24, 38, 53–55, 105; employees of, 26, 57, 98–99; open production, 49, 68; owning land, 26, 34, 39–40, 55–57, 61, 99; and progress, 50, 66; and the rebellion of 1857, 81, 106, 108–9; representative in London, 44; revenue, 64–65, 67, 75, 96, 119, 128; the state, 9, 57, 61, 71, 99, 111, 143

Galindo, Anibal, 30, 34, 70, 130
Garcia Rovira, 12, 72, 78, 86
gleaning, 37–39, 139
guards, 27–28, 82–6, 92, 99, 108, 117, 130, 134–36, 138; requests for, 4, 21–22

independence, of Colombia, 5–6, 22

kettle salt, 47–50, 68, 92, 125

Lache, the, 17, 19, 42, 144
La Uvita, 89–91, 130, 135
Liberal Party, 56, 81, 96–97, 101–3, 105–6, 125–30, 133–37; regional politics 114–24, 131–32, 134–35
liberalism, 5, 67, 70, 78–79, 100–101, 111–12, 130, 141–43; economic liberalism, 119, 133
Luque, Alejo, 133–36

Malo, Antonio, 24–25, 100
Melista Rebellion of 1854, 80–81, 102–3
monopoly, 76, 78–85, 88–91, 136; colonial, 18, 20–21. *See also* speculation; resale of salt
Moreno y Escandón, Francisco Antonio, 19, 21
Mosquera, Tomás Cipriano de, 59, 97, 116–19, 121
mules, 3, 34, 73, 80, 90, 103–4, 125, 129–30
Muneque, 48, 53, 72, 75, 138
Murillo Toro, Manuel, 64, 97, 100, 122–23, 125

Nemocón, 36, 48, 59
Niño, Rafael, 55, 105, 110, 116–17, 123–24
Northern Railroad, 126–28, 141
Núñez, Rafael, 66, 128–30, 133

Obando, José María, 96, 102
open production, 47–50, 68, 89, 128, 134
ovens, 3–4, 43–46, 52–54, 58–60, 63, 85, 99, 104, 108, 137–38; and contraband, 92
Oszlak, Oscar, 7–9, 92

Pamplona, 12, 14, 18, 72, 86, 97, 104, 114, 125
Paipa, 103, 122
Parra, Aquileo, 69, 120–21, 125–26

ranching, 14–15, 24, 41–42, 71–72, 125, 139
rebellion of 1857 (La Salina), 81, 106–11
rebellion of 1871 (Boyacá), 119–24
reforms of 1850, 12, 48, 64–67, 72, 100–101
Regeneration, the, 70, 115, 128, 130–36, 141
resale of salt, 77, 79–80, 82–83, 88–89, 91, 132, 135. *See also* monopoly; speculation
resguardos (community land), 19, 36–37; of Chita, 19–21; of La Salina, 19, 32–34, 53, 55
Reyes y Niño Company, 29, 38, 55–56, 97
Rodrigoque, 19, 32, 53, 55, 58–59, 129
Roldán, Antonio, 52, 70

Roldán, Salvador Camacho, 68, 89, 124–25
royal treasury, 4, 18–19, 21, 36

Sácama, 40, 104, 108, 116, 129
Sáenz de Santamaria, Mariano, 29, 78
Samacá, 58–59, 127; ironworks 51–52
salt making, 3–4, 18–20, 30, 42–43, 46, 61–62, 141; causing poverty, 22, 35; and demography, 18–19, 32; government control of, 23–24, 55, 58; inferior salt, 45; and local conflict, 91–92; and the Melista Rebellion, 104; and open production, 47–49; and the rebellion of 1857, 108–10; and taxation, 5, 7; and the War of 1895, 134; women's participation, 37–38
salt monopoly (Colombia), 140–42; colonial, 3, 18; critiques of, 63–64, 66–71, 118–19, 132–33; legislation about, 64, 119; and the Northern Railroad, 124–26, 128; returns from, 5, 65, 75, 137–38; and the republican salt monopoly, 5–6, 22–25, 27, 48, 99–100, 106, 111
salt monopolies, 5–6, 41
Santander, 10–16, 20, 66, 69, 114–19, 122–30, 142–43
Santander, Francisco de Paula, 25, 95–96, 144
Santa Rosa, 27, 103, 107–8, 115, 122, 124, 128, 131
Santa Rosa (canton), 12, 14–15, 72, 78, 106
schools, 31, 98
Sesquilé, 48, 59, 77
sheds, 3, 45–6, 92
Soata, 78, 103, 106
Soata (canton), 12, 14, 72, 106, 120
Socorro, 12, 14, 20, 72, 97, 114, 125
Sogamoso, 27, 97, 102–3, 115–17, 124, 127–29, 143
Sogamoso (canton), 12, 14–15, 72, 78
Solano, Juan Nepomuceno, 28, 98
speculation, 71, 75–76, 80–81, 89, 98, 133. *See also* monopoly; resale of salt
springs, 3–4, 18–21, 27–28, 42–43, 45, 66, 90, 92, 108–11, 118, 135–37; in Córdoba, 42–43; and gleaning, 38–39. See also *vertientes*
state, the, 4–9, 33, 41, 61–63, 69–72, 84, 92–94, 99, 111, 139–44; authority of, 56, 109; building of, 21–22, 26, 37, 45, 55, 58; the colonial state, 18, 21; and guard units, 27–28; the institutional state, 25, 57; as land owner, 55; and the Regeneration, 70; and the salt monopoly, 63–64, 76, 81–82, 89, 91, 125, 132
steam power salt production, 47–50, 53

taxes, 5–6, 18, 20–21, 23, 63–67, 69–71, 95, 100–102, 132; special contributions, 14, 32, 35; in Tundama, 105
Tensa Valley, 12, 14–5
Terán, General Guillermo, 40, 90–91, 132
textiles, 13–5, 36–7, 150n33
Thousand Days War, 4, 129, 136–38
timber, 3, 24, 30, 34, 52, 54–55, 57–58, 78, 81, 131–32, 138; carrying of, 38, 50, 59, 111, 139; contractors, 91; and the rebellion of 1857, 107, 110; shortages of, 47, 86; theft of, 83; and the War of the Supremes, 99; in Zipaquirá, 36. *See also* fuel; forest; wood; woods
tobacco, 64–66, 95, 100–101, 125
Tundama, 12–13, 15, 48, 81; during the Melista Rebellion, 102–5; politics in, 114–16, 118, 127, 143; during the rebellion of 1857, 105–7, 109–10; during the War of 1885, 128, 131–32; during the War of the Supremes, 97
Tunebos, 7, 43, 176n49
Tunja, 13–14, 58; politics, 115, 117–18, 131, 142–43; province of, 12, 99, 109; during the rebellion of 1871, 120–23; during the War of 1885, 129; during the War of the Supremes, 97

Valderáma, Antonio, 106, 108, 110, 116, 124
Valderrama, Pedro Ignacio, 24, 32–33, 37, 55–56
Vargas, Vicente, 38–39, 57, 78, 82
Várgas Santos, Gabriel, 56, 128–29, 137
Velez, 12, 20, 96, 113
Venezuela, 10, 12, 22–23, 95, 98, 117, 131
vertientes, 3, 42, 82. *See also* springs
viceroy of New Granada, 10, 19, 21
viceroyalty of New Granada, 17–18
viceroyalty of Peru, 18, 20
Von Humboldt, Alexander, 4, 6, 21, 54

wages, 31, 35, 39, 57–9
War of the Supremes, 96–100
War of 1860–1862, 116–17

War of 1876–1877, 128
War of 1885, 128–30
War of 1895, 133–34
wall (for protecting springs), 4, 21–22, 43, 61, 77, 84–85, 137
wood, 3, 29, 33–34, 44–46, 54, 58, 72, 85, 109–11, 136
woods, 29, 33–34, 83. *See also* forests

Zipaquirá 20–1, 23, 26, 42; bundles of wood at, 54; families of, 80; gleaning, 38–39; guard unit, 27; market, 71; during the Melista Rebellion, 102–3; and monopoly (speculation), 79; open production, 48; plan to rent, 69, 125; poverty in, 36–37; profits at, 59, 75, 119, 138; and protected market, 24, 111; salt from, 6, 17, 68, 106, 108, 124–25; saltworks of, 4, 6, 23, 132; steam power, 50; theft at, 77; during the War of the Supremes, 97

Printed and bound by CPI Group (UK) Ltd, Croydon, CR0 4YY
16/09/2026
14963485-0001